The Global Antimicrobial Resistance Epidemic - Innovative Approaches and Cutting-Edge Solutions

Edited by Guillermo Tellez-Isaias

Published in London, United Kingdom

The Global Antimicrobial Resistance Epidemic - Innovative Approaches and Cutting-Edge Solutions
http://dx.doi.org/10.5772/intechopen.98171
Edited by Guillermo Tellez-Isaias

Contributors
Jesus A. Maguey-González, Sergio Gómez-Rosales, María de Lourdes Angeles, Guillermo Téllez-Isaías, Yashar Jalali, Juraj Payer, Monika Jalali, Kothandapani Sundar, Ramachandira Prabu, Gopal Jayalakshmi, Daniel A. Charlebois, Shamanth A. Shankarnarayan, Joshua D. Guthrie, Abigarl Ndudzo, Sakhile Ndlovu, Nesisa Nyathi, Angela Sibanda Makuvise, Istifanus Anekoson Joshua, Mathew Bobai, Clement Sokfa Woje, Robert Bragg, Samantha Mc Carlie, Gunther Staats, Bernadette Belter, Boudine Van Der Walt, Patrícia Poeta, Carla Miranda, Gilberto Igrejas, Rosa Capita, Carlos Alonso-Calleja, Magaly de Chial, Carlos Ramos, Julio Castillo, Brenda de Mayorga, Olga Chen, Ovidio Durán, Eghe Izevbizua, Jane M. Morrell, Pongpreecha Malaluang, Aleksandar Cojkic, Ingrid Hansson, Hussein Al-Dahmoshi, Noor Salman Al-Khafaji, Sahar A. Ali, Samuel Kojo Kwofie, Cynthia Ayefoumi Adinortey, Michael Wilson

First published in London, United Kingdom, 2022 by IntechOpen
IntechOpen is the global imprint of INTECHOPEN LIMITED, registered in England and Wales, registration number: 11086078, 5 Princes Gate Court, London, SW7 2QJ, United Kingdom

British Library Cataloguing-in-Publication Data
A catalogue record for this book is available from the British Library

Additional hard and PDF copies can be obtained from orders@intechopen.com

The Global Antimicrobial Resistance Epidemic - Innovative Approaches and Cutting-Edge Solutions
Edited by Guillermo Tellez-Isaias
p. cm.
Print ISBN 978-1-80356-041-0
Online ISBN 978-1-80356-042-7
eBook (PDF) ISBN 978-1-80356-043-4

For EU product safety concerns:
IN TECH d.o.o., Prolaz Marije Krucifikse Kozulić 3, 51000 Rijeka, Croatia,
info@intechopen.com or visit our website at intechopen.com.

Meet the editor

Guillermo Tellez-Isaias received his DVM and MS in Veterinary Sciences from the National Autonomous University of Mexico (UNAM), and his Ph.D. from Texas A&M University. He worked as a professor at UNAM for sixteen years, eight as head of the Avian Medicine Department, College of Veterinary Medicine. He was previously president of the National Poultry Science Association of Mexico and is a member of the Mexican Veterinary Academy and the Mexican National Research System. Currently, he works as a research professor at the Center of Excellence in Poultry Science, University of Arkansas. His research is focused on poultry gastrointestinal models to evaluate the beneficial effects of functional foods to enhance intestinal health and disease resistance.

Contents

Preface

Antibiotic resistance is a global issue and a public health emergency worldwide. The misuse of antibiotics in the human, animal, food, and agricultural sectors has exacerbated the current crisis. Globally, the rapid emergence of resistant bacteria jeopardizes the efficacy of antibiotics, which have revolutionized medicine and saved millions of lives. Bacterial infections have reemerged decades after the first patients were treated with antibiotics. Antibiotic resistance has been attributed to overuse and misuse of these medications as well as a lack of new drug development by the pharmaceutical industry because of diminished economic incentives and onerous regulatory requirements. The Centers for Disease Control and Prevention (CDC) has identified several bacteria as posing urgent, serious, and concerning threats, with many already imposing significant clinical and financial burdens on the United States' healthcare system, patients, and families. Similarly, the World Health Organization (WHO) has developed a five-point global action plan to combat antimicrobial resistance and coordinate the efforts of numerous international sectors. Coordinated efforts to implement new policies, reinvest in research, and pursue crisis management strategies are critical. We now face a long-term and possibly never-ending battle against multidrug-resistant bacteria. A more comprehensive approach to bacterial infection is required, which may include non-compound approaches (products other than conventional antibacterial agents) that target bacteria or any approach that targets the host such as antibodies, probiotics, phytobiotics, and vaccines, all of which are discussed in this book.

The editors express their sincere appreciation to all the authors who contributed to this book for their hard work and dedication, as well as to the IntechOpen editorial team for allowing us to complete this project.

Guillermo Tellez-Isaias
Department of Poultry Science,
University of Arkansas Agricultural Experiment Station,
Fayetteville, AR, USA

Chapter 1

Introductory Chapter: The Antibiotic Resistance Epidemic

Guillermo Tellez-Isaias

1. Introduction

Antibiotic resistance is a global problem that has triggered a global human and animal health crisis worldwide. Antibiotics have changed the face of medicine and saved millions of lives, but resistant bacteria are threatening their usefulness.

The pharmaceutical industry is experiencing a shortage of new drug development due to declining economic incentives and demanding regulatory requirements. Governing agencies worldwide have identified several "super-bugs," bacteria resistant to all antibiotics known by man.

A more comprehensive approach to bacterial infection is required, including alternatives to conventional antibacterial agents.

Several antibiotics were discovered from the 1950s to the 1970s to cure previously incurable diseases, including tuberculosis and syphilis [1]. Since then, no new antibiotic classes have been discovered, which is concerning given the bacterial resiliency [2] and the continuous abuse and misuse of antibiotics [3].

Alexander Fleming recognized the phenomena of resistance to antimicrobial agents, by the misuse of antibiotics, in 1945 when he stated, "The time may come when anyone in the shops can buy penicillin. Then there is the danger that the ignorant man may easily under-dose himself and by exposing his microbes to nonlethal quantities of the drug make them resistant" [4].

In the absence of novel and more potent drugs, we risk a future where minor injuries and illnesses can be fatal and essential operations like surgery and chemotherapy become unmanageable. Antimicrobial resistance poses a serious threat to our way of life and could lead to a global pandemic if we don't act to combat it. Longer hospital stays and higher medical costs are now the results of antimicrobial resistance [5].

Antibiotic resistance is eroding our ability to treat bacterial illnesses. Infections resistant to most, if not all, current medicines are becoming common. The nature of significant acute bacterial infections and the economic realities of this field makes developing novel antibacterial medication difficult.

Because of the induction, amplification, and transmission of aspects of antimicrobial resistance among microbes, adequate management of a novel antibacterial agent is required for both the patient and the community when a new antibacterial agent is introduced. Furthermore, most antibiotic treatment regimens are brief (sometimes lasting only a week or two), and antimicrobial management aims to reduce the use of broader spectrum agents whenever possible to retain their usefulness, lowering the need for newer agents to enter the market.

 IntechOpen

In contrast, in many other treatment areas, such as diabetes, hypertension, and hyperlipidemia, long-term daily usage by patients does not add to the agent's loss of efficacy, and there is no medical reason to delay administration. While antimicrobial stewardship is critical, it will almost certainly diminish the economic benefits for a medication developer. Financial pressures associated with the development of antimicrobial medicines are not new [6].

There aren't enough reasons to invest in developing new antimicrobials under the current intellectual property innovation system. Since 1980, pharmaceutical corporations have made investments in cancer and chronic disease treatments for three reasons:

1. There is a better possibility of financial success than ever before.

2. They have realized that this is a lost war. The time and money invested in developing a new antimicrobial drug are not justified because profit will not be as attractive as other cancer or chronic disease treatments. Pharmaceutical corporations know that the improper use and abuse of the new drug will lead to the development of resistance by bacteria.

3. Those corporations are very well aware that bacteria are equipped with more advanced and sophisticated intelligent mechanisms to mute resistant clones than we humans have to make new drugs.

It is as simple as that. Where is the profit in a lost war?

Like eukaryotes, prokaryotes, including bacteria and other microorganisms, have membranes surrounding a droplet of cytoplasm. Prokaryotes acquire nutrients, communicate, excrete, and even process information in a "neurological" way [7, 8]. They can detect the presence of nutrients, toxins, and predators and adopt powerful escape techniques to preserve their viability [9]. From the evolutionary standpoint, bacteria have millions of years in advance against eukaryotes.

2. Novel alternatives to combat super-bugs

In 1908 Eli Metchnikoff obtained the Nobel Prize and was regarded as the originator of innate immunity, offering the breakthrough idea of consuming live bacterial cultures (yogurt) to improve the health and longevity of people more than a century ago [6, 10]. This principle is more appropriate than ever, as many antibiotic-resistant microorganisms threaten animal and human health [11].

In some countries around the world, pressures from society have resulted in establishing guidelines on antibiotic use in the feedstock industries. In other countries, like in the USA, it has become a commercial strategy for poultry companies and fast-food restaurants to label their products with no antibiotics ever (NAE). A similar commercial approach is observed in Brazil, which has dominated the chicken meat export market for over a decade, simply listening to the demands of countries in Europe, Asia and the middle east. Food animal production systems such as the poultry industry have been using alternative antibiotics to enhance disease resistance and productivity. According to new studies, nutritional treatments for stress, disease, and chronic inflammation may be more effective than antibiotics in some cases [12–14].

Disease resistance improvement in non-antibiotic raised animals has been shown to benefit animal products' health, welfare, production, and food safety. Alternative feed additives are being researched and developed in response to rising customer demand to remove growth-promoting antibiotics. Some of those alternatives include nutraceuticals such as probiotics [15, 16]; prebiotics [17]; organic acids [18, 19]; Phytochemicals [20–22]; enzymes [23, 24]; Vaccines and immunoglobulins [25, 26]. Innovative methodologies and cutting-edge technologies, like quorum sensing and quantum mechanics approaches, have the potential to shift the balance and help to minimize the epidemic problem in the future.

Conflict of interest

The author declares no conflict of interest.

Author details

Guillermo Tellez-Isaias
Department of Poultry Science, University of Arkansas, Fayetteville, AR, USA

*Address all correspondence to: gtellez@uark.edu

References

[1] Aminov RI. A brief history of the antibiotic era: Lessons learned and challenges for the future. Frontiers in Microbiology. 2010;**1**:134

[2] Carvalho G, Forestier C, Mathias J-D. Antibiotic resilience: A necessary concept to complement antibiotic resistance? Proceedings of the Royal Society B. 2019;**286**(1916):20192408

[3] Ventola CL. The antibiotic resistance crisis: Part 1: Causes and threats. Pharmacy and Therapeutics. 2015;**40**:277

[4] Fleming A. Penicillin: Nobel Lecture, December 11 1945. pp. 83-93

[5] Slama TG. Gram-negative antibiotic resistance: There is a price to pay. Critical Care. 2008;**12**(4):1-7

[6] Kaufmann SH. Immunology's foundation: The 100-year anniversary of the Nobel Prize to Paul Ehrlich and Elie Metchnikoff. Nature Immunology. 2008;**9**:705-712

[7] Desvaux M, Hébraud M, Talon R, Henderson IR. Secretion and subcellular localizations of bacterial proteins: A semantic awareness issue. Trends in Microbiology. 2009;**17**:139-145. DOI: 10.1016/j.tim.2009.01.004

[8] Lazcano A, Peretó J. Prokaryotic symbiotic consortia and the origin of nucleated cells: A critical review of Lynn Margulis hypothesis. Biosystems. 2021;**204**:104408. DOI: 10.1016/j. biosystems.2021.10440

[9] Liversidge A. Bacterial consciousness: Why spirochetes think as we do. Omni. 1993;**16**:10-11

[10] Gordon S. Elie Metchnikoff: Father of natural immunity. European Journal of Immunology. 2008;**38**:3257-3264

[11] Pollack A. To Fight 'Superbugs,' Drug Makers Call for Incentives to Develop Antibiotics. Vol. 20. New York Times; 2016

[12] Björkman I, Röing M, Sternberg Lewerin S, Stålsby Lundborg C, Eriksen J. Animal production with restrictive use of antibiotics to contain antimicrobial resistance in Sweden—A Qualitative Study. Frontiers in Veterinary Science. 2021;**7**:1197

[13] Wang B, Ye X, Zhou Y, Zhao P, Mao Y. Glycyrrhizin attenuates salmonella typhimurium-induced tissue injury, inflammatory response, and intestinal dysbiosis in C57BL/6 mice. Frontiers in Veterinary Science. 2021;**8**:648698. DOI: 10.3389/ fvets.2021.648698

[14] Chalvon-Demersay T, Luise D, Floc'h L, Tesseraud S, Lambert W, Bosi P, et al. Functional amino acids in pigs and chickens: Implication for gut health. Frontiers in Veterinary Science. 2021;**8**:496

[15] Adhikari B, Hernandez-Patlan D, Solis-Cruz B, Kwon YM, Arreguin MA, Latorre JD, et al. Evaluation of the antimicrobial and anti-inflammatory properties of bacillus-dfm (norum™) in broiler chickens infected with salmonella enteritidis. Frontiers in Veterinary Science. 2019;**6**:282

[16] Hernandez-Patlan D, Solis-Cruz B, Latorre JD, Merino-Guzman R, Morales Rodriguez M, Ausland C, et al. Whole genome sequence and interaction analysis in the production of six enzymes from the three Bacillus strains present in a commercial direct fed microbial (Norum™) using a Bliss independence test. Frontiers in Veterinary Science; **9**:784387. DOI: 10.3389/fvets.2022.784387

[17] Praxedes-Campagnoni I, Vecchi B, Gumina E, Hernandez-Velasco X, Hall JW, Layton S. Assessment of novel water applied prebiotic to evaluate gut barrier failure and performance in two commercial trials in Brazil. A pilot study with an economic perspective. Frontiers in Veterinary Science. 2021;**8**:652730. DOI: 10.3389/fvets.2021.652730

[18] Mani-López E, Garcia H, López-Malo A. Organic acids as antimicrobials to control Salmonella in meat and poultry products. Food Research International. 2012;**45**:713-721

[19] Gomez-Osorio L-M, Yepes-Medina V, Ballou A, Parini M, Angel R. Short and medium chain fatty acids and their derivatives as a natural strategy in the control of necrotic enteritis and microbial homeostasis in broiler chickens. Frontiers in Veterinary Science. 2021;**8**:773372. DOI: 10.3389/fvets.2021.773372

[20] Asakura H, Nakayama T, Yamamoto S, Izawa K, Kawase J, Torii Y, et al. Long-term grow-out affects campylobacter Jejuni colonization fitness in coincidence with altered microbiota and lipid composition in the cecum of laying hens. Frontiers in Veterinary Science. 2021;**8**:657

[21] Anderson RC, Levent G, Petrujki'c BT, Harvey RB, Hume ME, He H, et al. Antagonistic effects of lipids against the anti-escherichia coli and anti-salmonella activity of thymol and thymol-beta-d-glucopyranoside in porcine gut and fecal cultures in vitro. Frontiers in Veterinary Science. 2021;**8**:773372. DOI: 10.3389/fvets.2021.773372

[22] Park I, Goo D, Nam H, Wickramasuriya SS, Lee K, Zimmerman NP, et al. Effects of dietary maltol on innate immunity, gut health, and growth performance of broiler chickens challenged with Eimeria maxima. Frontiers in Veterinary Science. 2021;**8**:508

[23] Nusairat B, Wang J-J. The effect of a modified GH11 xylanase on live performance, gut health, and clostridium perfringens excretion of broilers fed corn-soy diets. Frontiers in Veterinary Science. 2021;**8**:523

[24] Li H, Yin J, He X, Li Z, Tan B, Jiang Q, et al. Enzyme-treated soybean meal replacing extruded full-fat soybean affects nitrogen digestibility, cecal fermentation characteristics and bacterial community of newly weaned piglets. Frontiers in Veterinary Science. 2021;**8**:639039. DOI: 10.3389/fvets.2021.639039

[25] Juárez-Estrada MA, Tellez-Isaias G, Sánchez-Godoy FD, Alonso-Morales RA. Immunotherapy with egg yolk eimeria sp.-specific immunoglobulins in SPF leghorn chicks elicits successful protection against Eimeria Tenella infection. Frontiers in Veterinary Science. 2021;**8**:758379. DOI: 10.3389/fvets.2021.758379

[26] Jabif MF, Gumina E, Hall JW, Hernandez-Velasco X, Layton S. Evaluation of a novel mucosal administered subunit vaccine on colostrum IgA and serum IgG in sows and control of enterotoxigenic escherichia coli in neonatal and weanling piglets: Proof of concept. Frontiers in Veterinary Science. 2021;**8**:89

Chapter 2

Honey as a Natural Product Worthy of Re-Consideration in Treating MRSA Wound Infections

Cynthia Ayefoumi Adinortey, Michael Wilson and Samuel Kojo Kwofie

Abstract

The use of antibiotics to treat bacterial infections have largely been successful. However, the misuse and overuse of these precious drugs have led to the development of bacterial resistance and this seems to have jeopardized their effectiveness. Many antibiotics that hitherto were seen as "miraculous drugs", have witnessed a low efficacy and this has threatened the life of humanity as never before. The rapid emergence of antibiotic resistance in bacteria is the major cause of this sad development. One such superbug is methicillin-resistant *Staphylococcus aureus* (MRSA). MRSA is a general problem in most healthcare centers with a reported astronomical incidence of invasive MRSA infections causing death. Honey, a natural product, popular for its antibacterial activity is increasingly being used owing to its reported antibiotic potential against 'stubborn' bacteria. This review discusses the fact that though honey is an ancient remedy, it is still relevant and its application in modern medicine for the treatment of chronically infected wounds caused by MRSA should be re-visited. Furthermore, the *in vitro* antibacterial and antibiofilm activities of medical-grade honey on *S. aureus* infections and challenges encountered by Researchers in developing honey, into an acceptable medical, therapeutic antibacterial agent for wound care have also been highlighted.

Keywords: MRSA, medical-grade honey, *Staphylococcus aureus*, wound treatment, antibiotic resistance

1. Introduction

Staphylococcus aureus (S. aureus) is the most commonly isolated pathogen in complicated skin and soft tissue infections (cSSTIs) worldwide. This Gram-positive bacterium is reported to be responsible for a high rate of morbidity and mortality in humans and has become a major threat to clinical practice [1]. There appears to be a significant growing trend of cSSTIs both in the community and healthcare settings with a stagey increase of the economic burden among people with these tissue infections [2]. A number of factors have contributed to the realization of *S. aureus* as a pathogen. One of the factors that have enhanced the virulence potential has been the evolution of antibiotic resistance. Resistance to beta-lactam antibiotics among *S. aureus* is an increasingly important problem in patients admitted to hospitals. *S. aureus* bacteria that are not susceptible to β-lactam antibiotics - methicillin are referred as methicillin-resistant *Staphylococcus aureus* (MRSA), whereas those that are sensitive are termed methicillin-sensitive *Staphylococcus aureus* (MSSA) [3].

MRSA is a major cause of severe healthcare-associated (HA) infections. Although during the last decade, the incidence of HA invasive infections have dropped, accounts of community-associated MRSA (CA-MRSA) infections have soared among the general populace [3–7]. Globally, MRSA remains one of the most important multidrug-resistant bacteria reported to account for most cSSTIs such as surgical wounds [8] and is also responsible for the high morbidity and mortality cases due to HA infections. The use of orthodox antibiotics has been the mainstay of treatment for these infections. Though the world's encounter with antibiotics has been beneficial, several reports have suggested that many antibiotics that used to be seen as "miraculous drugs" have witnessed a low efficacy and this has threatened the life of humanity as never before. The rapid emergence of "superbugs" such as MRSA bacteria that are highly resistant to many classes of antibiotics is the major cause of this sad development. Infections due to MRSA, in comparison with MSSA, are associated with a greater risk of treatment failure, increased patient mortality, and higher costs. Owing to irrational and excessive antibiotic use, the bacterium- *S. aureus* has become a multiple drug-resistant (MDR) pathogen most threatening to human health. With the rise in the prevalence of resistance to orthodox antibiotics by MRSA, honey a natural product known for its antibacterial activity is increasingly being endorsed, due to the reported bactericidal and bacteriostatic capacity on 'stubborn' bacteria. Moreover, several antibiotics used for the treatment of *S. aureus* infections leave in their wake adverse effects thus the need for better alternatives.

Honey is a sugary thick fluid that has been in use for almost 5500 years back. In the Sumerian tablet, the earliest inscribed information on honey for nutrition and medicinal purposes was found in 2100–2000 B.C. This chapter discusses the fact that though honey is an ancient remedy, its relevance and application in complicated and chronic infected wounds in modern medicine is still relevant and should be re-visited. This piece provides data on the *in vitro* antibacterial and antibiofilm activities of medical-grade honey (MGH) namely manuka, medihoney, and wound dressings such as Revamil® produced from it with a focus on its impact on both MSSA and MRSA. This narrative was coupled with information on bioactive compounds responsible for this activity. The attention on MGH is due to the fact that honey intended for the management of wounds ought to undergo irradiation care in order to remove treatment inference from honey microbiota which is the case for all MGH [9]. The additive and synergistic upshot of the MGH admixture with other antibiotics are also captured in this chapter. Evidence-based information on the cell and molecular mechanisms of action of MGH on *S. aureus* has been presented with prepositions of possible targets. Some potential challenges in the effort to develop honey into an acceptable medical therapeutic antibacterial agent for wounds have also been highlighted.

2. Antibacterial activities of MGH on both MSSA and MRSA

Honey according to Codex Alimentarius, 2001 is "the natural sweet substance produced by honey bees from the nectar of plants or secretions of living parts of plants or excretions of plant-sucking insects on the living parts of plants, which the bees collect, transform by combining with specific substances of their own, deposit, dehydrate, supply and leave in the honeycomb to ripen and mature". Honey is classified based on many criteria. Based on its nutritional and medicinal applications, two main kinds are known, namely medical and non-medical grades. Fresh honey which has not been exposed to irradiation are referred to as non-medical grade whereas those with radiation exposure are usually called medical grade honey (MGH). Fresh honey can contain bacterial spores, mainly those of bacillus species, and spores of the notorious pathogen Clostridium botulinum, which can cause wound botulism or gangrene. The likely presence of these microorganisms calls for the sterilization of honey through gamma-irradiation to destroy these bacterial spores and other microorganisms. Honey that has passed through this process is referred to as medical-grade honey (MGH) and approved for use in clinical settings. Manuka and medihoney (Comvita, NZ), one of the few honeys that appear to have FDA approval for clinical applications.

The limited knowledge of anti-bacterial compounds in honey and the variability of anti-bacterial activity of other types of honey are however major obstacles to the applicability in clinical use. This section presents information on the antibacterial effects of MGH mentioned earlier and the contribution of individual components in its antimicrobial action. Studies have shown that MGH comprises primarily fructose, glucose, sucrose, water, organic acids, flavonoids, phenolic acids, as well as minor components such as peptides [bee defensin-1 and 2, hemenopectin, apidaecin], enzymes [diastase, invertase, glucose oxidase] amino acids and vitamins [10, 11]. It is important to note that the bioactive components of honey can vary due to the different botanic and geographic origins [12].

In the past four decades, there have been various studies regarding honey and its components and how it drives the anti-bacteria character. Several studies have shown that no "stubborn" or resistant bacteria can be isolated after subjecting isolates of MRSA obtained from wound to various concentrations of MGH or MGH-based dressings [13–16]. This is reported to be as a result of the fact that these types of honey contain various antibacterial constituents and conditions such as low pH, high sugar content, methylglyoxyl (MGO), hydrogen peroxide, antimicrobial peptides (bee defensin-1) and other active substances [10, 17, 18]. It is of interest to note that MGO has been identified as one of the principal antimicrobic ingredient of most MGHs [19, 20]. Cooper and colleagues in 2010, demonstrated that some honeys also contain an antimicrobial peptide (bee defensin-1), which contributes substantially to bactericidal activity [14].

The antimicrobial qualities are critical in dermatologic applications, owing to the presence of active constituents, like MGO and antimicrobial peptides. MGH-based dressings are appropriate for the dressing of wounds and burns and have also been included in treatment therapy against diseases such as diaper dermatitis, dandruff, pityriasis, and psoriasis. It also exerts emollient, humectant, soothing, and hair conditioning effects, keeps the skin juvenile and retards wrinkle formation, regulates pH, and prevents pathogen infections [21]. Studies have also shown that Community-associated methicillin-resistant *Staphylococcus aureus* (CA-MRSA) which is now described worldwide, as a clinically significant pathogen, predominantly linked to skin and soft tissue infections succumbs to the effects of honey [22]. Additionally, according to George and Cutting, when honey is standardized, it can have a major antibacterial impact on multi-resistant bacteria that are regularly found in wounds [23]. It is clear from studies published so far that the application of honey and honey-based dressings can promote the healing of infected wounds that do not respond to the conventional therapy. These shreds of evidence buttress recommendations on the use of honey in the management of chronic wounds infected with *S. aureus* as shown on **Table 1**.

Non-MRSA Clinical isolate[1]; MRSA[2]; Non–MRSA Lab strain[3]; Resistant to all antibiotics[4]; Isolates sensitive to all antibiotics[5].

MIC- Minimum Inhibitory Concentration; ZI- Zone of Inhibition; R- Resistant.

Type of Honey	MIC	*S. aureus*	Test method	Reference
Manuka	34 μg	04–277-3567[1]	Broth microdilution	[24]
Manuka	8.0 w/v %	NCTC8325[3]	Microdilution assay	[25]
	8.0 w/v %	RPAH18[1,2]		
	8.0 w/v %	MW2[1,2]		
Medihoney	4.2 w/v %	MRSA[2]	Agar dilution method	[15]
	4.1 w/v %	Non-MRSA Clinical isolate[1]		
	4.2 w/v %	Resistant to all antibiotics[4]		
	4.4 w/v %	sensitive to all antibiotics[5]		
Medihoney	8.0 w/v %	NCT8325[3]	Microdilution assay	[25]
	8.0 w/v %	RPAH18[1,2]		
	8.0 w/v %	MW2[1,2]		
Manuka	3.0 v/v	MRSA1	Broth dilution method	[26]
	3.0 v/v	MRSA2		
	3.0 v/v	MRSA3		
	3.0 v/v	MRSA4		
	3.0 v/v	MRSA5		
	3.0 v/v	MRSA6		
	3.0 v/v	MRSA7		
	3.0 v/v	MRSA8		
	3.0 v/v	MRSA9		
	3.0 v/v	MRSA10		
	3.0 v/v	MRSA11		
	3.0 v/v	MRSA12		
	3.0 v/v	MRSA13		
	3.0 v/v	MRSA14		
	3.0 v/v	MRSA15		
	3.0 v/v	MRSA16		
	2.7 v/v	MRSA17		
Manuka	12.5 v/v	MRSA ATCC 43300[3]	Broth dilution method	[26]
	12.5 v/v	MRSA 0791[2]		
	12.5 v/v	MRSA 28965[2]		
	12.5 v/v	MRSA 01322[2]		
	12.5 v/v	MRSA 0745[2]		

Type of Honey	MIC	*S. aureus*	Test method	Reference
Medihoney	4.0 v/v	BORSA mecA neg[1]	Broth dilution method	[23]
	4.0 v/v	Multiresistant mecA neg[1]		
	4.0 v/v	Multiresistant[2]		
	4.0 v/v	Non-multiresistant[2]		
Manuka	60000 mg/ml	EMRA-15 NCTC 13142[2]	E-test strip	[27]
	60000 mg/ml		Broth dilution	
	60000 mg/ml		Checker board	
	60000 mg/ml		Time kill curve	

Table 1.
Antibacterial action of manuka and medihoney against various strains of S. aureus.

3. Combinatorial effect of MGH and antibiotics on *S. aureus*

One of the ways to curb the upsurge in antimicrobial resistance is to introduce novel approaches to combat these pathogens. Merging antimicrobial drugs with other agents such as honey that counteract and obstruct the antibiotic resistant mechanisms expressed by these pathogens is a novel strategy. The natural product, honey, is gaining acceptance as an alternative antimicrobial agent. Medical grade honey offers a favorable alternative for topical use, as a single or a multi-component agent in combination with other antibiotics [28]. One of the approaches to fight antimicrobial resistance is combination drug therapy [29]. This upturns efficacy and enhances the value of existing antimicrobials in the dearth of new antibiotics discovery and development. There are instances where, combining antimicrobial agents have made drug action synergistic. This has the benefits of lessening both the treatment costs and the risk of possible side effects owing to the reduced concentrations of both agents used [29]. This phenomenon is especially vital for chronic wounds where antibiotic therapy is usually applied for long-term. Moreover, combining antibiotics and other drug agents is reported to exhibit different, modes of action and as such reduces the risk of resistance arising during treatment.

Honey is a natural product substance applied in different forms such as gels for topical management of infected chronic wounds [30]. MGH-based dressings have been licensed by some national health authorities and are available to health professionals in many countries. Honey has a multifaceted substance [20] with well-known, broad-spectrum antibacterial activity against a various microorganism, including those that are commonly associated with chronic wounds such as *S. aureus* [31]. These bacteria appear unable to develop resistance to MGH, even when small concentrations are applied [14, 15]. This is contrary to orthodox antibiotics, where resistance is readily induced with sub-inhibitory exposure [31]. This dearth of resistance is probably due to the multiple antibacterial properties of honey that overwhelm S. aureus bacterial stress responses [15]. A combination of approved antibiotics and MGH could result in a new range of antimicrobials with the potential to avert the emergence of resistant offering broad-spectrum coverage and subsequently improving the curative efficiency. According to Saeed a researcher, therapeutic efficacy studies of conventional treatment with antibiotics combined with Manuka honey in the treatment of diabetic foot ulcers yields a reassuring healing process [32].

Type of Honey	Antibiotics	*S. aureus*	FICI	Test method	Reference
Manuka	Rifampicin	RPAH18[2]		Checkerboard	[33]
		MW2[2]	0.45		
		IMVS67[2]		Microdilution assay	
		O4–277-3567[1]			
Manuka		NCTC8325[3]	0.445	Agar diffusion test	[25]
	Rifampicin	RPAH18[1,2]	0.405		
		MW2[1,2]	0.435		
		O4–277-3567[2]	0.445		
	Clindamycin	NCTC8325[3]	0.405		
		RPAH18[1,2]	2		
		MW2[1,2]	0.275		
		O4–277-3567[2]	0.405		
	Gentamycin	NCTC8325[3]	0.8782		
		RPAH18[1,2]	2		
		MW2[1,2]	1.07		
		O4–277-3567[2]	0.955		
	Oxacillin	NCTC8325[3]	0.405		
		RPAH18[1,2]	0.8782		
		MW2[1,2]	0.753		
		O4–277-3567[2]	0.407		
Medihoney	Rifampicin	NCTC8325[3]	0.445	Agar diffusion test	[25]
		RPAH18[1,2]	0.405		
		MW2[1,2]	0.435		
		O4–277-3567[2]	0.445		
	Clindamycin	NCTC8325[3]	0.405		
		RPAH18[1,2]	2		
		MW2[1,2]	0.405		
		O4–277-3567[2]	0.405		
	Gentamycin	NCTC8325[3]	1.197		
		RPAH18[1,2]	2		
		MW2[1,2]	1.195		
		O4–277-3567[2]	0.955		
Medihoney	Oxacillin	NCTC8325[3]	0.405		
		RPAH18[1,2]	0.8782		
		MW2[1,2]	0.753		
		O4–277-3567[2]	0.407		
Manuka	Oxacillin	EMRSA-15 NCTC 13142[2]	0.001	Etest strip	[27]
				Broth dilution	
				Checker board	
				Time kill curve	

Table 2.
Combinatorial effects of manuka and medihoney and antibiotics against S. aureus strains.

In vitro studies combining MGH with antibiotics recorded a synergistic effect with tetracycline, oxacillin, imipenem and mupirocin against the growth of MRSA strain as shown on **Table 2** [27, 33]. Interestingly, small concentrations of MGH in combination with oxacillin restored the MRSA strain to oxacillin susceptibility. Convincing synergistic action between manuka honey and rifampicin against *S. aureus* strains, has also been reported, and the inclusion of honey was seen to have averted the emergence of rifampicin resistance *in vitro* [33]. This is of clinical consequence as rifampicin penetrates into cells tissues and sores and is normally used to treat superficial S. aureus infections, but rapidly induces resistance and thus has to be used together with another agent.

The fractional inhibitory concentration index (FICI) range of 0.5 to 4.0 is usually used to define additivity results in most combination studies. The fractional inhibitory concentration index (FICI) is calculated as the sum of the minimum inhibitory concentration (MIC) of each compound divided by the MIC of one compound used alone. Synergy and antagonism are defined by FICI ≤ 0.5 and FICI > 4 respectively, while FICI > 0.5 or $4 \leq$ is considered indifferent.

It is clear that MGH combining with gentamicin or clindamycin recorded the highest FICI compared with other antibiotics used as shown on **Table 2**.

Non-MRSA Clinical isolate[1]; MRSA[2]; Non–MRSA Lab strain[3].

4. Mechanisms of action of honey on *Staphylococcus aureus* strains

Understanding how MGH impacts the action of orthodox antibiotics and their mode on action may broaden our knowledge of how honey affects these pathogens. It is important to state that much evidence would be displayed using data on manuka honey because much research has been undertaken on it since sterile preparations of it is available commercially. According to a publication by Majtan, in 2014 there is no evidence of damage to host cells, when MGH is either consumed orally or used as a wound dressing. Honey appears to stimulate healing and reduce scarring when applied to wounds [34].

The antibacterial property of honey has been linked to some mechanisms. The high osmolarity of honey as a result of its high sugar content inhibits microbial growth [35]; the sugar molecules within the honey hold onto the water molecules, thereby denying the bacteria enough water to support their growth. However, this effect gets lessened as the honey becomes more diluted by wound exudates.

There is another mechanism that lies in the fact that its antimicrobial properties are retained, even when the honey is diluted by wound exudate. This is partly due to the presence of hydrogen peroxide, which is slowly released as a result of the action of glucose oxidase present in the honey. The latter gets diluted by the exudate and in the process becomes activated [36].

S. aureus is one of the Gram-positive strains that is susceptible to honey-mediated inhibition. The antimicrobial action of MGH, is made possible through the combined action bee defensin-1 (antimicrobial peptide), MGH (phytochemical), and hydrogen peroxide. Moreover, the high sugar contents of medical grade honey could also be helpful to inhibit and eliminating bacteria biofilm through osmosis as reported in a study using eucalyptus honey [37], observed enlarged cells containing septa when S. aureus was exposed to manuka honey. An indication that cell division was interrupted as a mechanism of action. In *S. aureus* strains, the primary mechanism of action has been reported to involve the interruption of the cell cycle, whereby bacteria fail to divide leading to an accumulation of arrested cells with fully formed septum [38]. The cleavage of the septum is normally controlled by autolysins, which digest peptidoglycan to produce two daughter cells [39]. In addition to this, it has been demonstrated that treatment with MGH leads to down regulation of the universal stress protein, UspA, in MRSA, reducing the ability of bacteria to survive conditions of cellular and metabolic stress [40].

Biofilms are generally populations of cells typically covered in a self-produced extracellular matrix and usually clinging on surfaces such as teeth, implanted devices and wounds. Generally, these microbes in biofilms are safeguarded from antimicrobial agents which can lead to stubborn infections. MGH is reported to disrupt cellular aggregates and averts the formation of biofilms formed by many pathogens, including *S. aureus* [18]. Honey has been reported to disrupt biofilms and kill resident cells, though a higher concentration is required than for planktonic cells [41]. Honey acts as a bactericidal negotiator, penetrates biofilms, recovers aggressive infection, and eradicates colonies. Honey has exhibited bactericidal outcome against biofilms of pathogenic reference strains such as MRSA. This has clinical implications for using honey as dressing on wounds containing biofilms of *S. aureus*.

5. Proof of efficacy from animal and case studies, and hurdles so far encountered

MGH has been used on animals with accidental or surgical wounds, such as rhinos, Horses, etc., with positive outcomes [42, 43]. Case reports using MGH for non-healing wounds and ulcers have recorded substantial improvement where conventional antibiotics had failed [44–48].

In the intervention studies, the treatment of non-healing wounds with MGH-based dressings resulted in a higher number of completely healed wounds and a faster rate of wound size reduction. Also, it was observed that wound odor was neutralized with the provision of topical protection, and a reduction in wound pain intensity.

There is evidence from case studies of MGH-based dressing used to treat wounds that have failed with treatment with orthodox antibiotics. Though MGH has shown potent antibacterial actions in the past three decades and has high potential for use to treat chronic wounds caused by *S. aureus*. MGH-based dressing creates a moist and anti-inflammatory wound environment, neutralizes wound odor although promoting almost all facets of the wound healing processes such as angiogenesis and re-epithelialization. The efficacy of MGH has constantly been confirmed in other cases and in literature as well [49].

Meanwhile there are challenges that cannot be overlooked. There are countless reasons for this, including technical challenges in performing a double-blind placebo-controlled trial on a distinctive substance like honey. The issue of ethical considerations and lack of interest by clinical practitioners are other worries that affect patronage. Additionally, the use of honey ointment has been portrayed as messy as it leaves behind a sticky residue [50]. There are also apprehensions with the application of honey to vertical wounds in ambulatory patients. The newer honey-dressings have resolved such issues and there appears to be good ease of use, retention and removal and patient comfort with MGH- impregnated tulle.

6. Conclusion

There has been a collective effort to find or develop novel agents with antimicrobial activity in order to increase the collection of drugs against methicillin-resistant *S. aureus* infections and biofilm-forming resistant strains which are responsible for chronic wounds. Data collated indisputably demonstrate that honey possesses bacteriostatic, bactericidal and antibiofilm effects on various strains of *staphylococcus aureus*. Based on evidence gathered so far, MGH and MGH-based dressings seem to be an effective medication that could be considered a suitable therapy for wounds. Clinical application has proven it to be specifically beneficial in the treatment of wounds that are nonresponsive to conventional therapies and wounds infected with antibiotic-resistant bacteria such MRSA. Though several challenges are experienced in the effort to developing honey into an acceptable therapeutic agent, *staphylococcus aureus* strains appear unable to develop resistance to MGH such as Manuka honey and Medihoney, when exposed to various concentrations. MGH therefore offers a promising alternative for topical use in wound dressings, both as a single multicomponent agent as well as in combination with antibiotics. This evidence-based data has supported the fact that though honey is an ancient remedy, so far it is the most effective and efficacious drug agent for stubborn bacteria like MRSA. It is believed that this information would contribute to greater credibility of MGH-based dressings and their consideration in the management of non-healing wounds caused by MRSA.

Conflict of interest

The authors declare no conflict of interest.

Thanks

We are grateful to Professor Michael D. Wilson for continuous support.

Author details

Cynthia Ayefoumi Adinortey[1], Michael Wilson[2,3] and Samuel Kojo Kwofie[4,5*]

1 Department of Molecular Biology and Biotechnology, School of Biological Sciences, College of Agriculture and Natural Sciences, University of Cape Coast, Cape Coast, Ghana

2 Department of Parasitology, Noguchi Memorial Institute for Medical Research (NMIMR), College of Health Sciences (CHS), University of Ghana, Accra, Ghana

3 Department of Medicine, Loyola University Medical Center, Maywood, IL, USA

4 Department of Biomedical Engineering, School of Engineering Sciences, College of Basic and Applied Sciences, University of Ghana, Accra, Ghana

5 West African Centre for Cell Biology of Infectious Pathogens, Department of Biochemistry, Cell and Molecular Biology, College of Basic and Applied Sciences, University of Ghana, Accra, Ghana

*Address all correspondence to: skkwofie@ug.edu.gh

References

[1] Olaniyi R, Pozzi C, Grimaldi L, Bagnoli F. *Staphylococcus aureus*-Associated Skin and Soft Tissue Infections: Anatomical Localization, Epidemiology, Therapy and Potential Prophylaxis. Current Topics in Microbiology and Immunology. 2016;**409**:1-29

[2] Frei CR, Makos BR, Daniels KR, Oramasionwu CU. Emergence of community-acquired methicillin-resistant Staphylococcus aureus skin and soft tissue infections as a common cause of hospitalization in United States children. Journal of Pediatric Surgery. 2010;**45**(10):1967-1974. DOI: 10.1016/j.jpedsurg.2010.05.009

[3] Jin Y, Zhou W, Yin Z, Zhang S, Chen Y, Shen P, et al. The genetic feature and virulence determinant of highly virulent community- associated MRSA ST338-SCCmec Vb in China. Emerging Microbes & Infections. 2021;**2021**(10):1052-1064

[4] David MZ, Daum RS. Community-Associated Methicillin-Resistant Staphylococcus aureus: Epidemiology and Clinical Consequences of an Emerging Epidemic. Clinical Microbiology Reviews. 2010;**23**(3):616-687

[5] Otto M. Community-associated MRSA: What makes them special? International Journal of Medical Microbiology. 2013;**1-7**:324-330. DOI: 10.1016/j.ijmm.2013.02.007

[6] Mediavilla R, Chen L, Mathema B, Kreiswirth BN. Global epidemiology of community-associated methicillin resistant Staphylococcus aureus (CA-MRSA). Current Opinion in Microbiology. 2012;**2**(15):588-595

[7] Sergelidis D, Angelidis AS. Methicillin-resistant Staphylococcus aureus: a controversial food-borne pathogen. Letters in Applied Microbiology. 2017;**64**:409-418

[8] Bassetti M, Carnelutti A, Righi E. The role of methicillin-resistant Staphylococcus aureus in skin and soft tissue infections. Current Opinion in Infectious Diseases. 2017;**30**(2):150-157

[9] Carnwath R, Graham EM, Reynolds K, Pollock PJ. The antimicrobial activity of honey against common equine wound bacterial isolates. Veterinary Journal. 2013;**1-5**:110-114. DOI: 10.1016/j.tvjl.2013.07.003

[10] Alvarez-suarez JM, Gasparrini M, Forbes-hernánandez TY, Mazzoni L, Giampieri F. The Composition and Biological Activity of Honey: A Focus on Manuka Honey. Food. 2014;**3**:420-432

[11] Kwakman PHS, Zaat SAJ. Antibacterial components of honey. IUBMB Life. 2012;**64**:48-55

[12] Ajlouni S, Sujirapinyokul P. Hydroxymethylfurfuraldehyde and amylase contents in Australian honey. Food Chemistry. 2010;**119**(3):1000-1005. DOI: 10.1016/j.foodchem.2009.07.057

[13] Cabassi CS, Bertocchi M, Spadini C, Denti L, Flisi S, Schiano E, et al. honey on bacterial isolates from infected skin lesions of non-traditional companion animals. Veterinaria Italiana. 2021;**57**(2):119-126

[14] Cooper RA, Jenkins L, Henriques AFM. Absence of bacterial resistance to medical-grade manuka honey. European Journal of Clinical Microbiology & Infectious Diseases. 2010;**29**:1237-1241

[15] Blair SE, Cokcetin NN, Harry EJ, Carter DA. The unusual antibacterial

activity of medical-grade Leptospermum honey: Antibacterial spectrum, resistance and transcriptome analysis. European Journal of Clinical Microbiology & Infectious Diseases. 2009;**28**:1199-1208

[16] Kwakman PHS, Van Den AJPC, Gu A, Aslami H, Binnekade JM, De BL, et al. Medical-Grade Honey Kills Antibiotic-Resistant Bacteria In Vitro and Eradicates Skin Colonization. Clinical Infectious Diseases. 2008;**46**:1677-1682

[17] Johnston M, McBride M, Dahiya D, Owusu-Apenten R, Singh NP. Antibacterial activity of Manuka honey and its components: An overview. AIMS Microbiology. 2018;**4**(4):655-664

[18] Carter DA, Blair SE, Cokcetin NN, Bouzo D, Brooks P, Schothauer R, et al. Therapeutic Manuka Honey: No Longer So Alternative. Frontiers in Microbiology. 2016;**7**(April):1-11

[19] Adams CJ, Boult CH, Deadman BJ, Farr JM, Grainger MNC, Manley-harris M, et al. Isolation by HPLC and characterisation of the bioactive fraction of New Zealand manuka (Leptospermum scoparium) honey. Carbohydrate Research. 2008;**343**:651-659

[20] Mavric E, Wittmann S, Barth G, Henle T. Identification and quantification of methylglyoxal as the dominant antibacterial constituent of Manuka (Leptospermum scoparium) honeys from New Zealand. Molecular Nutrition & Food Research. 2008;**52**:483-489

[21] Burlando B, Cornara L. Honey in dermatology and skin care: a review. Journal of Cosmetic Dermatology. 2013;**12**:306-313

[22] Maeda Y, Loughrey A, Earle JAP, Millar BC, Rao JR, Kearns A, et al. Antibacterial activity of honey against *Staphylococcus aureus* (CA-MRSA). Complementary Therapies in Clinical Practice. 2008;**14**:77-82

[23] George NM, Cutting KF. Antibacterial Honey (Medihoney™): in-vitro Activity Against Clinical Isolates of MRSA, VRE, and Other Multiresistant Gram-negative Organisms Including Pseudomonas aeruginosa. Wounds. 2007;**19**(9):231-236

[24] Allen KL, Molan PC. The sensitivity of mastitis - causing bacteria to the antibacterial activity of honey. New Zealand Journal of Agricultural Reseasrch. 1997;**40**(4):537-540

[25] Liu M, Lu J, Müller P, Turnbull L, Burke CM, Schlothauer RC, et al. Antibiotic-specific differences in the response of Staphylococcus aureus to treatment with antimicrobials combined with manuka honey. Frontiers in Microbiology. 2015;**5**(779):1-9

[26] Cooper RA, Molan PC, Harding KG. The sensitivity to honey of Gram-positive cocci of clinical significance isolated from wounds. Journal of Applied Microbiology. 2002;**93**:857-863

[27] Jenkins RE, Cooper R. Synergy between oxacillin and manuka honey sensitizes methicillin-resistant Staphylococcus aureus to oxacillin. The Journal of Antimicrobial Chemotherapy. 2012;**67**(March):1405-1407

[28] Mcloone P, Tabys D, Fyfe L. Honey Combination Therapies for Skin and Wound Infections: A Systematic Review of the Literature. Clinical, Cosmetic and Investigational Dermatology. 2020;**13**:875-888

[29] Fischbach MA. Combination therapies for combating antimicrobial resistance. Current Opinion in Microbiology. 2011;**14**(5):519-523

[30] Cooper RA, Molan PC, Harding KG. Antibacterial activity of honey against strains of Staphylococcus aureus from infected wounds. Journal of the Royal Society of Medicine. 1999;**92**(June):283-285

[31] Packer JM, Irish J, Herbert BR, Hill C, Padula M, Blair SE, et al. International Journal of Antimicrobial Agents Specific non-peroxide antibacterial effect of manuka honey on the Staphylococcus aureus proteome. International Journal of Antimicrobial Agents. 2012;**40**:43-50

[32] Saeed M. Al, Therapeutic Efficacy of Conventional Treatment Combined with Manuka Honey in the Treatment of Patients with Diabetic Foot Ulcers: A Randomized Controlled Study. The Egyptian Journal of Hospital Medicine. 2013;**53**(January 2011):1064-1071

[33] Müller P, Alber DG, Turnbull L, Schlothauer RC, Carter DA, Whitchurch CB, et al. Synergism between Medihoney and Rifampicin against Methicillin-Resistant Staphylococcus aureus (MRSA). PLoS One. 2013;**8**(2):1-9

[34] Majtan J. Honey: An immunomodulator in wound healing. Wound Repair and Regeneration. 2014;**22**:187-192

[35] Molan PC. Re-introducing Honey in the Management of Wounds and Ulcers–Theory and Practice. Ostomy/Wound Management. 2002;**48**(11):28-40

[36] Molan PC, Betts JA. Clinical usage of honey as a wound dressing: an update. Journal of Wound Care. 2004;**13**(9):353-356

[37] Proaño A, Coello D, Villacrés-Granda I, Ballesteros I, Debut A, Vizuete K, et al. The osmotic action of sugar combined with hydrogen peroxide and bee-derived antibacterial peptide Defensin-1 is crucial for the antibiofilm activity of eucalyptus honey. LWT. 2021;**136**(Part 2):1-37

[38] Henriques AF, Jenkins RE, Burton NF. The intracellular effects of manuka honey on Staphylococcus aureus. European Journal of Clinical Microbiology & Infectious Diseases. 2010;**29**:45-50

[39] Jenkins R, Burton N, Cooper R. Manuka honey inhibits cell division in methicillin-resistant *Staphylococcus aureus*. The Journal of Antimicrobial Chemotherapy. 2011;**66**(September):2536-2542

[40] Jenkins R, Burton N, Cooper R. Effect of manuka honey on the expression of universal stress protein A in meticillin-resistant Staphylococcus aureus. International Journal of Antimicrobial Agents. 2011;**37**(4):373-376. DOI: 10.1016/j.ijantimicag.2010.11.036

[41] Lu J, Turnbull L, Burke CM, Liu M, Carter DA, Schlothauer RC, et al. Manuka-type honeys can eradicate biofilms produced by Staphylococcus aureus strains with di ff erent biofilm-forming abilities. Peer Journal. 2014;**2**(e326):1-25

[42] Bischofberger AS, Dart CM, Horadagoda N, Perkins NR, Jeffcott LB, Little CB, et al. Effect of Manuka honey gel on the transforming growth factor β1 and β3 concentrations, bacterial counts and histomorphology of contaminated full-thickness skin wounds in equine distal limbs. Australian Veterinary Journal. 2016;**94**(1):27-34

[43] Dart AJ, Bischofberger AS, Dart CM, Jeffcott LB. Review Article A review of research into second intention equine wound healing using manuka

honey: Current recommendations and future applications. Equine Veterinary Education. 2015;**27**(12):658-664

[44] Smith SE, Roukis TS. Bone a nd Wound Healing Augmentation with Platelet - Rich Plasma. Clinics in Podiatric Medicine and Surgery. 2009;**26**:559-588

[45] Jeffery S, Henry N, Radotra I. Properties and use of a honey dressing and gel in wound management. The British Journal of Nursing. 2019;**28**(6):31-35

[46] Vyhlídalová D, Kozáková R, Zeleníková R. Management of non-healing wounds with honey dressings: a literature review. Central European Journal of Nursing and Midwifery. 2018;**9**(3):880-888

[47] Jull A, Cullum N, Dumville J, Westby M, Deshpande S, Walker N. Honey as a topical treatment for wounds (Review). Cochrane Database of Systematic Reviews. 2015;**3**:1-98

[48] Holland LC, Norris JM. Medical grade honey in the management of chronic venous leg ulcers. International Journal of Surgery. [Internet]. August 2015;**20**:17-20. DOI: 10.1016/j.ijsu.2015.05.048

[49] Marais HJ, Glyphis ZG, Cremers NAJ. Medical grade honey: Hope for wounded white rhinos. Vetenary and Animal Science. 2021;**13**(June):1-8. DOI: 10.1016/j.vas.2021.100196

[50] Alcaraz A, Kelly J. Treatment of an infected venous leg ulcer with honey dressings. The British Journal of Nursing. 2002;**11**(13):859-866

Chapter 3

Unlocking the Potential of Ghost Probiotics in Combating Antimicrobial Resistance

Abigarl Ndudzo, Sakhile Ndlovu, Nesisa Nyathi and Angela Sibanda Makuvise

Abstract

Antimicrobial resistance is a global concern that requires immediate attention. Major causes of development of antimicrobial resistance in microbial cells are overuse of antimicrobials along the food chain especially in livestock, in preventing infections as well as misuse of antimicrobials by patients. Probiotics could be a viable alternative to antibiotics in the fight against antimicrobial resistance. Probiotic strains can act as a complement to antimicrobial therapy, improving antimicrobial function and enhancing immunity. However, there are safety concerns regarding the extensive use of live microbial cells especially in immunocompromised individuals; these include microbial translocation, inhibition of other beneficial microorganisms and development of antimicrobial resistance, among other concerns. Inevitably, ghost probiotics have become the favored alternative as they eliminate the safety and shelf-life problems associated with use of probiotics. Ghost probiotics are non-viable microbial cells (intact or broken) or metabolic products from microorganisms, which when administered in adequate amounts have biologic activity in the host and confer health benefits. Ghost probiotics exert biological effects similar to probiotics. However, the major drawback of using ghost probiotics is that the mechanism of action of these is currently unknown, hence more research is required and regulatory instruments are needed to assure the safety of consumers.

Keywords: ghost probiotics, antimicrobial, resistance, potential, combat

1. Introduction

1.1 Antimicrobial resistance

Misuse and abuse of antimicrobials are key contributors to the introduction of selective pressures in our natural environments, resulting in the rapid increase of antimicrobial resistant microbial strains. Random antimicrobial use has impelled microorganisms to adapt and survive by acquiring antimicrobial resistance genes that lead to antimicrobial resistant strains [1]. Antimicrobials are drugs or medicines, including antibacterials, antivirals, antifungals and antiparasitics, used to prevent and

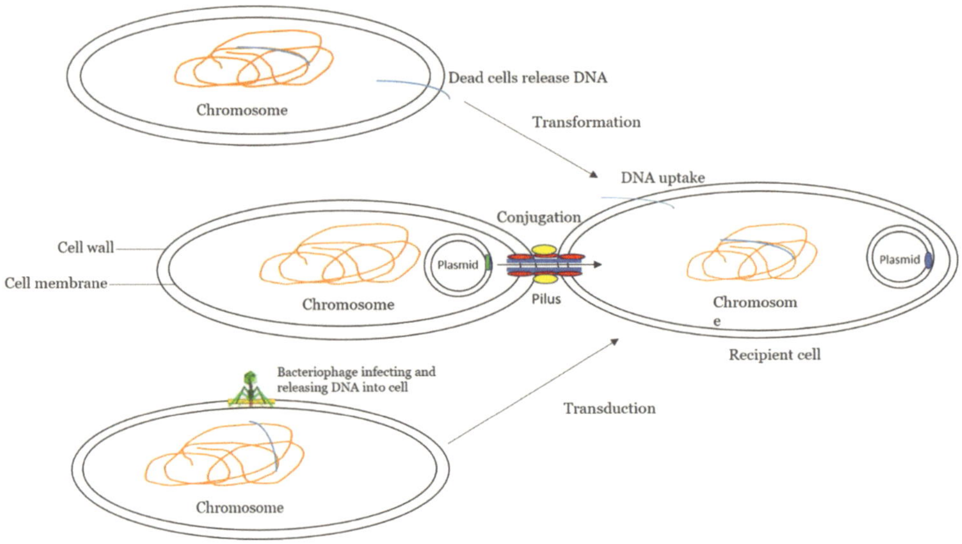

Figure 1.
Mechanisms of horizontal gene transfer in bacteria. Acquired antimicrobial resistance genes can pass between related and unrelated species by transformation, transduction and conjugation.

treat infections in humans, plants and animals [2]. Antimicrobial resistance occurs when a microbial strain is no longer susceptible to antimicrobials that would normally inhibit their growth and allows them to withstand the drugs [3]. Chromosomal or plasmid DNA encoding antimicrobial resistance is implicated in the rapid spread of multiple resistance through horizontal gene transfer [4], as shown in **Figure 1**.

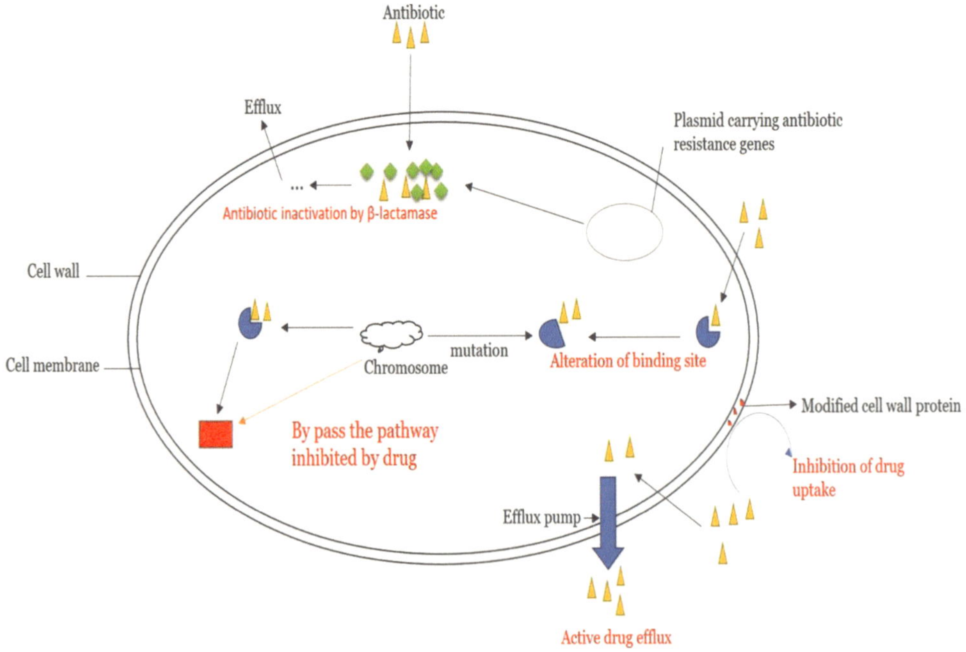

Figure 2.
Mechanisms of antibiotic resistance.

Mobile genetic elements including plasmids and transposons are instrumental in horizontal gene transfer [5]. Chromosomal resistance is caused by mutations in the developing spontaneous bacterial chromosome [6] while extra chromosomal resistance depends on the extra chromosomal genetic material that can be transferred in ways such as plasmids, transposons and integro [7]. Different types of resistance occur including natural resistance, acquired resistance, cross resistance and multi drug resistance [8, 9].

1.2 Mechanisms of antimicrobial resistance

The major problem with antimicrobial resistance is the selection and stabilization of mechanisms directed by foreign genes taken up by susceptible and resistant strains [2]. Microorganisms can evade the effects of the antimicrobial agents throughdecreased influx (limiting uptake of a drug), alterationof drug target site, drug inactivation using enzymes and active drug efflux (efflux pump) [10, 11] as detailed in **Figure 2**.

As a result of antimicrobial resistance, antibiotics and other antimicrobial medicines have become ineffective and infections are becoming increasingly difficult or impossible to treat increasing the risk of disease spread, severe illness and death. Alternative strategies are being employed in order to combat antimicrobial resistance. Such strategies include the use of probiotics.

2. Probiotics

Probiotics are live microorganisms, which when administered in correct proportions confer a health benefit on a host [12]. These microbes are a combination of bacteria, fungi, viruses and protozoa [13], and the commonly used probiotics are Lactobacillus and Bifidobacterium [14]. Beneficial probiotics are found in several locations on the body such as the gut, mouth, urinary tract, skin, lungs [8, 15]. Probiotic microorganisms can be isolated from plants, food products, environment, human and animal sources. Probiotics can be administered as supplements in a variety of forms, including in foods, drinks, capsules or pills, powders and liquids [16].

2.1 Characteristics of probiotics

Microorganisms must possess a number of characteristics in order to be classified as probiotics, that is the microbes should be easily isolated from humans, have the ability to live in the gut after consumption, have a proven benefit and must be safe to consume [17, 18]. Some of the characteristics are shown in **Figure 3**.

2.2 Mechanisms of action of probiotics

Probiotics exert their biological effect using different mechanisms of action as shown in **Figure 4**, these include; inhibition of the growth of pathogenic bacteria (competitive exclusion), reduction of bacterial and/or toxin translocation, modulation of the intestinal immune system, production of specific substance such as bacteriocins, modifications of the structure and function of intestinal epithelium, competitive adhesion to epithelial receptors, vitamin absorption and provision of other nutrients [19, 20].

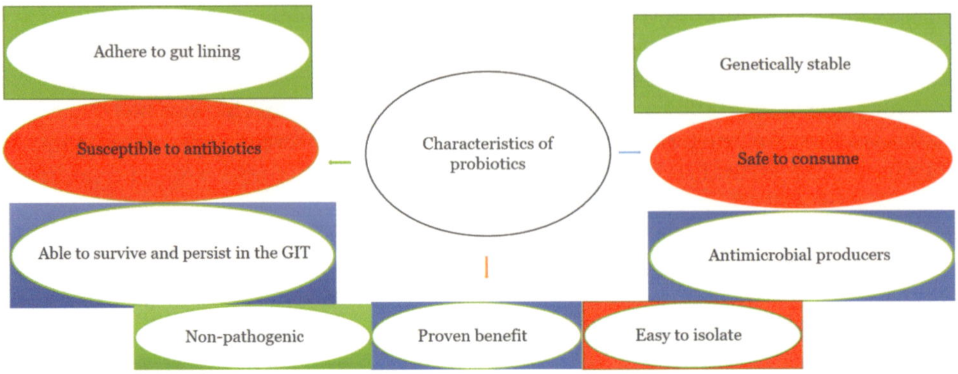

Figure 3.
Characteristics of probiotics.

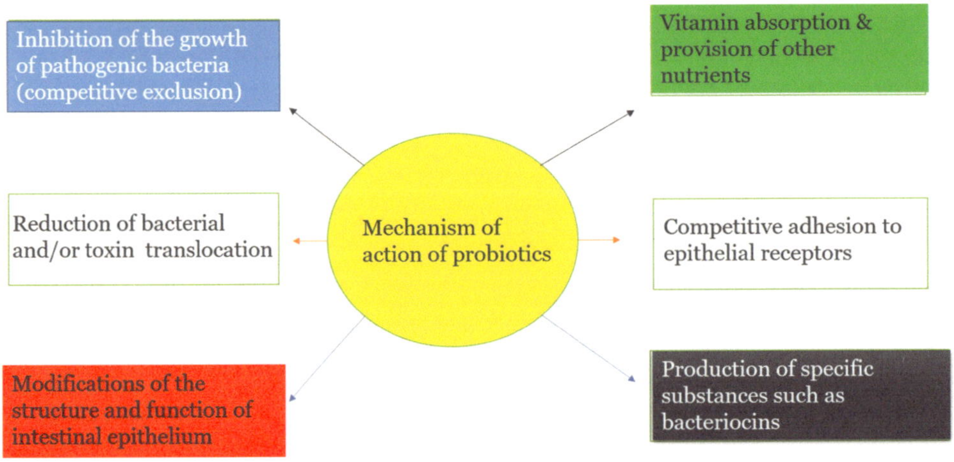

Figure 4.
Mechanisms of action of probiotics.

2.3 Probiotic health effects

Probiotics have a number of health benefits, including lowering the risk of some infectious diseases and reducing the need for antimicrobials to treat secondary infections. For example, the use of probiotics with antimicrobials reduces the incidence, duration and severity of antimicrobial-associated diarrhea, thereby reducing the evolution of resistance [21–23] and unlike antimicrobials which kill untargeted microbials, probiotics help to keep the gut microbiota in check. Some of the health benefits are shown in **Table 1** below.

Probiotics aren't perfect, there are many safety challenges associated with the use of probiotics such as ability to acquire antimicrobial resistance and virulence genes [24], chances of microbial translocation from gut to the blood stream [25], high risk of allergic reactions [26] and that biological effects of probiotics are strain specific, therefore proper strain identification is required [27] for a specific condition. As a result, ghost probiotics have become the preferred alternative to probiotics in order to solve the majority of these safety issues [14].

Probiotic health effects	
Metabolic effects	**Microbiota & Immunomodulation effects**
• Reduce risk of colon cancer. • Hydrolyse lactose to improve lactose tolerance. • Lower levels of mutagenic and/or toxigenic reactions in the gut. • Lower serum cholesterol & Supply vitamins like folate to colon epithelium.	• Control of irritable bowel syndrome. • Colonization resistance which leads to suppression of endogenous & exogenous pathogens e.g., Antibiotic associated diarrhea & travelers' diarrhea respectively. • Strengthened innate immunity& • Balanced immune response – to alleviate food allergy symptoms in infants & control inflammatory bowel diseases.

Table 1.
Probiotics beneficial health effects.

3. Ghost probiotics

Health benefits observed for physiologically active probiotics are not associated with their viability only [28]. Probiotic products containing dead cells can produce effective biological responses. This proves that probiotics merely have an expiry date and can be used beyond that. This phenomenon is known as the probiotic paradox, that is, both live and dead cells produce the same biological response [29]. Ghost probiotics (inactivated probiotics, non-viable probiotics, paraprobiotics) are inactive microbial cells or cell fractions that, when administered in adequate amounts, confer a health benefit to the consumer, [14, 30, 31]. They consist of molecules present on the cell surfaces such as peptidoglycan, teichoic acid, cell wall polysaccharides and cell surface-associated proteins [32]. These trigger the human immune system, stimulating a positive immune response and anti-inflammatory effects in animals and humans [33].

3.1 Technologies used in the production of ghost probiotics

The methods used in producing ghost probiotics are similar to the techniques used for bacterial inactivation such as thermal processing, irradiation, UV rays, high pressure and ultrasound [14, 34] as shown in **Table 2** below. Thermal treatment is the most common technique for the production of ghost probiotics in laboratories [14]. The cell membranes are damaged, leading to leakage of nutrients and ions, ribosome aggregation and DNA breakage. Ohmic heating has been proposed for ghost probiotics production. It involves an electric current passing through the sample, leading to fast and uniform heating [35]. Therefore, bacterial inactivation can be caused by thermal and non-thermal damage (electroporation). Inactivation methods have an impact on the beneficial effects. This means that ghost probiotics obtained with different technologies could exhibit different functional features [36].

3.2 Characteristics of ghost probiotics

They are quite safe, they are well-tolerated and associated with reduced risk for adverse effects in vulnerable individuals [37]. They have no risk for transferring antibiotic-resistant genes to pathogenic or commensal bacteria [38]. Their effectivity is independent of the cell viability, which ensures longer stability and improved shelf-life [39]. They present an easy industrial large-scale production [36]. They provide a wide range of health-promoting effects, some of which can be reinforced in

Methods of cell inactivation	Activities that lead to cell inactivation
Thermal/ heat treatment	• Cell membrane damage • Leakage of nutrients and ions • Protein denaturation • Ribosome aggregation • DNA breakage
High pressure treatment	• Membrane rupture • Alteration of ribosomes • Protein denaturation & coagulation • Reduction of intracellular pH • Inactivation of enzymes • Loss of solutes
Ultra Violet (UV) Irradiation	• Formation of DNA photoproducts
Ionizing radiations	• Damage of nucleic acids caused by oxidative radicals
High intensity ultrasound	• Cell wall shearing • Production of free radicals • DNA damage • Membrane breakdown & cell lysis

Table 2.
Methods used in the production of ghost probiotics.

comparison with the effect of intact viable microbial cells [40]. Another very interesting feature of ghost probiotics, is that, due to their nature, it appears feasible that they could be used with concurrent administration with antibiotic and antifungal agents.

Ghost probiotics are categorized into peptidoglycan, teichoic acid, cell wall polysaccharides, cell surface-associated proteins and proteinaceous filaments. These are the ones that mediate beneficial effects to the host [41]. Some bacterial cell walls such as *Lactobacilli* contain a thick layer of peptidoglycan. This enhances the sensitivity to autolysis, hydrophobicity of the cell envelope and resistance to lysozymes [42]. This part in bacteria can improve innate and systemic adaptive immune responses as well as suppress interleukins which are associated with autoimmune and inflammatory bowel diseases.

Teichoic acids (TAs) are the second main constituent of the cell wall of the microbes. They possess immunomodulatory characteristics and exert anti-inflammatory effects on the intestinal epithelial cells of humans [43]. Cell-wall polysaccharides are common in Gram-positive bacteria surfaces for example exopolysaccharides (EPS). These have the ability to facilitate the interaction of the bacteria with pathogens, have immunoregulatory effects and act as a protective layer [43]. Cell surface proteins are one of the most important components of the outermost cell envelope structure. S-layer proteins, pili proteins, moonlight proteins are part of the surface proteins. These play a role in the host biological processes [44].

3.3 Possible uses of ghost probiotics in combating antimicrobial resistance

To counteract the phenomenon of antimicrobial resistance, there is a need to reduce the frequency in which they are administered. Ghost probiotics are used as a possible solution in fighting against antimicrobial resistance [45]. Due to the risks and concerns of administering probiotics to livestock, scientists are now opting to use ghost probiotics.

Cows tend to suffer from inflammation of the udder (mastitis). The main pathogens that stimulate the infection are *Staphylococcus aureus, Streptococcus uberis,* isolated strains of *Escherichia coli* and *Streptococcus dysgalactiae* [46]. There has been excessive use of intramammary antimicrobials to treat mastitis hence, increasing bacterial resistance. This has reduced their treatment efficacy resulting in a growing interest in replacement therapies without antibiotics such as ghost probiotics [47]. Different variants of ghost probiotics have shown antibacterial activity against *S. aureus* strains associated with bovine mastitis, increased response of the immune system and a decrease in the number of somatic cells in milk [48].

Increased animal production keeps animals crowded, facilitating the transmission of various diseases. The use of ghost probiotics on farms can naturally bring about a balance of gut microbes, reduce the growth of pathogens and reduce the use of antibiotics for disease prevention [49]. Thus, reducing the occurrence of resistance effects among pathogenic bacteria as the major spread of antimicrobial resistance is through food chains [50]. In a study carried out using *Lactobacillus,* the cure rate was doubled by the administration of inactivated lactobacilli. For example, the dose of amoxicillin required to kill uropathogenic *E. coli* was halved [51]. Hence, the use of ghost probiotics can reduce excessive use of antimicrobials. This can help decrease the chances of antibiotic resistance developing.

A variety of ghost probiotics from the *Bifidobacterium* species have been shown to act as antitumor agents through inhibiting the proliferation of tumor cells. This is still waiting for more human trials for approval. Some have been shown to be cardioprotective, antiulcer, antioxidants and reduce cholesterol in the body [52]. Ghost probiotics containing *Bifidobacterium breve* and *S.thermophilus* on preterm infants causes clinical tolerance to tumors necrosis, lowered digestive and respiratory infections, and consumption of these results in lower abdominal distention [53]. Ghost probiotics can treat acute diarrhea, which is one of the most common causes of death in infants and children in developing countries [54]. Ghost probiotic *L. paracei* can treat atopic dermatitis and other skin infections [55].

There is an innate immune response of macrophages to non-viable *Lactobacillus casei* cells. This results in an increase in the expression of pro-inflammatory cytokines and an enhanced transcription of toll-like receptors [56]. Ghost probiotics from *B. breve* and *Streptococcus thermophilus* cause prolonged dendritic cell survival and maturation hence can be used to enhance immune regulatory function, and improve epithelial barrier function [53]. *Lactobacillus paracasei* ghost probiotic, can act via the inhibition of immune cell inflammation and protect the host from pathobionts, enteric pathogens and protect a patient against colitis [57].

The capability of ghost probiotics to safeguard the host's health against serious infections induced by pathogens is fulfilled through various mechanisms such as inhibition of pathogenic adhesion, invasion, biofilm formation, and improvement of immune responses in thegut environment. Additionally, some ghost probiotics derived from *Lactobacillus* can provide effective protection against infections induced by certain viruses. *Lactobacillus* species possess antibacterial and anti-viral properties [58], inhibition of Gastric Corona, HIV, and Rotavirus in vitro along with a noticeable diminution in viral load in vivo [6, 30]. According to [59], ghost probiotics have also been shown to fortify endogenous beneficial microorganisms within the gut of the host. This activity by ghost probiotics is suggested to be better than the supplementation with unfamiliar microbial strains as probiotic, [60].

3.4 Advantages of ghost probiotics

To address the safety concerns surrounding probiotics, the attention has switched to the use of non-viable microbial cells, commonly known as ghost probiotics. Ghost probiotic and probiotic cells exhibit similar immunological responses by means of using the same or different mechanisms of action [14, 31, 32]. This has been demonstrated by an experiment done on the human epithelial colorectal adenocarcinoma Caco-2 cell line, both viable and UV-inactivated *Lactobacillus rhamnosus* GG cells were equally effective in decreasing thepro-inflammatory cytokine, interleukin-8 (IL-8), upon flagellin induction using different mechanisms of action [61]. Additionally, in a different study, RAW 264.7 macrophages were exposed to heat-killed *Lactobacillus acidophilus*, *Lactobacillus bulgaricus*, *Lactobacillus casei*, *Lactobacillus gasseri*, *Lactobacillus helveticus*, *Lactobacillus reuteri*, *Bifidobacterium* species, and *S. thermophilus* as well as the cell envelope components and cytoplasmic extracts of these bacteria. Whole inactivated cells, the cell envelope components, and cytoplasmic fractions from probiotic bacteria stimulated macrophages to producetumor necrosis factor-α (TNF-α), interleukin (IL-6), and nitric oxide [62]. This experiment demonstrated that bioactive compounds responsible for eliciting immune responses can be found anywhere in bacterial cells, and that it does not need an intact cell to elicit immune responses.

Furthermore, various inactivation methods for making ghost probiotics, such as high-pressure treatment and high-intensity ultrasound, have been reported to cause membrane rupture and cell lysis, respectively [14, 63]. Inactivation of microbial cells by cell lysis can produce additional beneficial effects, the contact between the released molecule and the host cells is improved [36, 40], increasing chances of MAMP-PRR interactions which are important for eliciting immunological responses, making ghost probiotics attractive than probiotics.

Misuse of antimicrobials in agriculture and medicine has resulted in development of antimicrobial-resistant microbes in animals. And also, interaction of microbes in gut might result in acquisition of antibiotic resistance and virulence genes in strains that previously lacked these through horizontal gene transfer. Alarmingly, antibiotic-resistant *Lactobacillus* and Bifidobacterium probiotics have been discovered in dairy products such as milk culture, yogurts, and cheese [27]. As a result of this, the focus has shifted towards using ghost probiotics that cannot acquire and transfer antibiotic resistantand virulence genes. Furthermore, ghost probiotics have a great advantage over probiotics in that they can be administered concurrently with antimicrobial agents without any bioactivity loss and/or development of antimicrobial resistance [64, 65] and thus making it attractive to immunocompromised and multi-diseased individuals.

Industrial processing and storage of probiotic products present viability and stability challenges, probiotic cultures should remain viable and sufficient numbers must reach the target site after thermal processing, storage, and gastrointestinal transit. To avoid these technological challenges, ghost probiotics are used. The dead inactivated cells, ghost probiotics do not require refrigeration to maintain the cultures in a stable and viable state. This reduces the cost of storing and transporting ghost probiotics, allowing them to be used by the poor in impoverished locations such as rural areas where refrigeration machines and facilities are lacking. [14], making them a cheaper and accessible option than probiotics.

Remarkably, ghost probiotics can remain stable in extreme environmental conditions, like water activity (A_w), temperature and pH which are considered stressful to probiotics and they have a longer shelf life. In addition, they can be supplemented into foods, other than dairy products like fruit juices and other cereal products [14], thus provide beneficial effects to lactose intolerant individuals. The heat-inactivated *L. gasseri* CP2305 strain is an example of ghost probiotics found in non-dairy goods. A sports drink treated with heat-inactivated *L. gasseri* CP2305 helped young athletes recover from exhaustion, anxiety, and negative moods [66]. Ghost probiotics can withstand the thermal processes during production and thus they can be added before thermal processing without any functionality loss [14]. Because ghost probiotics do not interact directly with food matrices, they have no effect on the organoleptic and/or sensory qualities of the food [34], hence there is no detrimental modification such as that observed in yogurt, which results in high acidification [35].

3.5 Disadvantages of ghost probiotics

Ghost probiotics, being dead and inactive cells, are unable to create metabolites such as bacteriocins, lactic acid, vitamins, and enzymes that are essential for probiotic health effects [14]. Additionally, the chemical mechanism of action of ghost probiotics is unknown; nevertheless, cell wall polysaccharides, peptidoglycans, surface proteins, and teichoic acids are known to activate immunological responses. Unlike postbiotic components which exist in purified form, ghost probiotics mechanism of action is unclear and is difficult to point out which molecule does what due to complex bacterial architecture [36]. Some methods of microbial inactivation such as thermal treatment affect the physiological activity of the resulting dead cells and the stability of their beneficial effects during shelf life, resulting in altered and non-identical biological responses [14, 36]. For instance, heat treatment at 121°C for 15 minutes of multispecies of lactic acid bacteriaie., *L. acidophilus, Lactobacillus plantarum, Lactobacillus fermentum,* and *Enterococcus faecium* demonstrated reduced adherent capacity to Caco-2 cells by more than 50% while heat-treatment at 100°C for 30 minutes did not alter the capacity of these strains to adhere to Caco-2 cells at all [67].

4. Current issues surrounding the adoption of ghost probiotics

Their side effects have not been fully understood. Studies have been done on how the microbiome of the gut reconstituted itself after antimicrobial treatment with and without ghost probiotic administration [68]. This means the impact ghost probiotics can have in the medical industry is questionable. There is an issue of, what is being studied is not exactly what would be administered to people [69]. For instance, when research is being carried out it involves a specific organism defined by genus, species and strain (these are pure and carefully dosed). But when buying off the shelf mixed with other products such as food products, people become skeptical about what they are getting dosage wise [70].

Research being carried out is claimed to be of low quality, small in size and often funded by companies with significant conflicts of interest [71].

The inactivation method of ghost probiotics functions can interrupt the bacterial cells and allow for an interface between intracellular bioactive compounds and the host cells on the administration of ghost probiotics. Delivery and formulation of ghost probiotics has been limited in the clinical field [31].

5. Regulation of ghost probiotics

In light of the safety and technological challenges associated with probiotics [14, 40], use of ghost probiotics will expand in near future. Therefore, there is an urgent need to clarify several points to support regulatory authoritiesin defining the requirements for the registration and approval of functional foods containing ghost probiotics and those that have health claims to protect consumers. There is currently an overlap of terminology in defining the biotics terms, ghost probiotics and post biotics which makes communication difficult among researchers, manufacturers, and customers [14]. Therefore, there is a need for internationally recognized clear-cut definitions to avoid confusion that currently exists in biotics, especially for probiotics, postbiotics and ghost probiotics [14]. As result of this mayhem, the ghost probiotics are currently marketed as probiotics [72]. Chiefly, ghost probiotics production, detection, and quantification methods need to be look into closely [14] and standardized [14], before regulations and/or requirements are laid out and implemented. The FDA should then layout the ghost probiotics specific requirements and specifications to iron out the mix-up.

Global commercialization of ghost probiotics is also one of the issues recognized from a regulatory view point because of the geographical differences, for example some traditional probiotics are classified differently across countries like Generally Regarded As Safe for USA and QPS for Europe and additionally some probiotics do not follow the same regulation globally. The regulatory process followed so as to launch a non-traditional probiotic is as complicated as one required for drugs [73].

The current regulations on probiotics are inadequate to protect the consumers and the prescribing doctors, there is abuse of the word probiotics and no specifics of microorganism are indicated in products [74]. Obviously, just like probiotics, ghost probiotics cannot be approved as drugs, even though they are sometimes used for the prevention, management or treatment of disease [75]. In the United States, and many regions of the world, probiotic products are marketed as dietary supplements (not drugs) and are therefore subject to different manufacturing and quality control standards than approved drugs are [75, 76], the same should apply to ghost probiotics. Exemption should be given to ghost probiotics with health claims, these should be treated aspharmaceutical products and regulated as such [75]. To assure safety to end-users, pharmacists should be aware of product quality when recommending these dietary supplements to risk populations like immunocompromised individuals [75] and infants and manufacturing quality control standards should be steeper especial for this vulnerable group [40].

Additionally, manufacturers should be in a position to provide evidence of quality criteriawhen required to and they should guide pharmacists on the safe use of specific products [75]. Manufacturers should have quality management systems in place, and third-party and/or regulatory organizations should verify compliance. Accordingly, the regulatory aspects that need to be considered for ghost probiotics are efficacy, safety, andquality control of manufacturing.

6. Future perspectives and conclusion

There is a need for large randomized placebo-controlled single strain trials with standard dosing, formulation and duration of treatment in various diseases to get consistent results. At this moment it is difficult to recommend any particular ghost probiotic for a particular disease as the preparation and dosing may not be available commercially. The interaction of the gut microbiota with its host and mutual regulation has become one of the important topics of biomedical research. Their relevance in human diseases require much more research. The popularity of ghost probiotics is fast increasing shortly they will be used in food, medicine, and agriculture. Additionally, the diet microbiota host interface can give rise to newer therapeutic approaches based on selective alteration of microbial metabolite production to support human health and prevent diseases. The metabolic profiling approach, suggests how mining the microbiota may lead to personalized treatment.

Conflict of interest

The authors declare no conflict of interest.

Author details

Abigarl Ndudzo[1*], Sakhile Ndlovu[1], Nesisa Nyathi[1] and Angela Sibanda Makuvise[2]

1 Faculty of Agricultural Sciences, Department of Crop and Soil Sciences, Lupane State University, Lupane, Zimbabwe

2 Faculty of Applied Sciences, Department of Biology and Biochemistry, National University of Science and Technology, Bulawayo, Zimbabwe

*Address all correspondence to: andudzo@lsu.ac.zw

References

[1] Imperial ICVJ, Ibana JA. Addressing the antibiotic resistance problem with probiotics: Reducing the Riskn of its double edged sword effect Frontier Microbiology 2016;**7**(1):1983. DOi: 10.3389/fmicb.2016.01983]

[2] Singh RH, Singh R, Prakash A. Antimicrobial Resistance: The Threat. Handbook of Marine Microbes: The Therapeutic Prospective and Applications. Naini, India: Immortal Publications. 2021. ISBN:978-93-5457-895-3.

[3] Qamar MU, Rasool MH, Jahan S, Shafique M, Aslam B. Antibiotic Resistance. London, UK: IntechOpen; 2019. DOI: 10.5772/intechopen.82303

[4] Malaluang P, Wilen E, Lidahl J, Hansson I, Morrell JM. Antimicrobial resistance in equine reproduction. Animals 2021;**11**:3035. DOI: 10.3390/ani11113035.

[5] Liyanapathiran VC, Thevanesam V. Combating antimicrobial resistance. Sri Lankan Journal of Infectious Diseases. 2016;**6**(2):72-82. DOI: 10.4030/sljid.v6i2.8119

[6] Hasan AT, AL-Harmoosh R. Mechanisms of antibiotics resistance in bacteria. Systematic Reviews in Pharmacy. 2020;**11**:817-823. DOI: 10.31838/srp.2020.6.118

[7] Girlich D, Bonnin RA, Dortet L, Naas T. Genetics of acquired antibiotic resistance genes in Proteus spp. Frontiers in Microbiology. 2020;**11**:256. DOI: 10.3389/fmicb.2020.00256

[8] Liu W, Chen M, Duo L, Wang J, Guo S, Sun H, et al. Characterization of potentially probiotic lactic acid bacteria and bifidobacteria isolated from human colostrum. Journal of Dairy Science. 2020;**103**(5):4013-4025. DOI: 10.3168/jds.2019-17602

[9] Haas L, Elewaut A, Gerard CL, Umkehrer C, Leiendecker L, Pedersen M, et al. Acquired resistance to anti-MAPK targeted therapy confers an immune-evasive tumor microenvironment and cross-resistance to immunotherapy in melanoma. Nature Cancer. 2021;**2**(7):693-708. DOI: 10.1038/s43018-021-00221-9

[10] Singh R, Smitha MS, Singh SP. The role of nanotechnology in combating multi-drug resistant bacteria. Journal of Nanoscience and Nanotechnology. 2018;**14**(7):4745-4756. DOI: 10.1166/jnn.2014.9527

[11] Wang X, Biswas S, Paudyal N, Pan H, Li X, Fang W, et al. Antibiotic resistance in salmonella typhimurium isolates recovered from the food chain through National Antimicrobial Resistance Monitoring System between 1996 and 2016. Frontiers in Microbiology. 2019;**10**:985. DOI: 10.3389/fmicb.2019.00985

[12] Popova T. Effect of probiotics in poultry for improving meat quality. Current Opinion in Food Science. 2017;**14**(2017):72-77. DOI: 10.1016/j.cofs.2017.01.008

[13] Anyogu A, Ayomide O, Christian A, Helen O, Esther A, Obadina A, et al. Microorganisms and food safety risks associated with indigenous fermented foods from Africa. Food Control. 2021;**129**:108-227. DOI: 10.1016/j.foodcont.2021.108227

[14] Siciliano RA, Reale A, Mazzeo MF, Morandi S, Silvetti T, Brasca M.

Paraprobiotics: A new perspective for functional foods and nutraceuticals. Nutrients. 2021;**13**(4):1-19. DOI: 10.3390/nu13041225

[15] Xu Y, Cui Y, Yue F, Liu L, Shan Y, Liu B, et al. Exopolysaccharides produced by lactic acid bacteria and Bifidobacteria: Structures, physiochemical functions and applications in the food industry. Food Hydrocolloids. 2019;**94**:475-499. DOI: 10.1016/j.foodhyd.2019.03.032

[16] Markowiak P, Śliżewska K. The role of probiotics, prebiotics and synbiotics in animal nutrition. Gut Pathogens. 2018;**10**(1):21. DOI: 10.1186/s13099-018-0250-0

[17] Pandey KR, Naik SR, Vakil BV. Probiotics, prebiotics and synbiotics- a review. Journal of Food Science and Technology. 2015;**52**(12):7577-7587. DOI: 10.1007/s13197-015-1921-1

[18] Al-Qysi L, Mohammad M, Al-iedani A, AbuKhader MM. Investigating the characteristics of probiotics marketed in the Middle East and pharmacists' perception of use in Muscat. Oman *Pharma Nutrition*. 2020;**13**:100202. DOI: 10.1016/j.phanu.2020.100202

[19] Tiwari G, Tiwari R, Pandey S, Pandey P. Promising future of probiotics for human health: Current scenario. Chronicles of Young Scientists. 2012;**3**(1):17. DOI: 10.4103/2229-5186.94308

[20] Panpetch W, Visitchanakun P, Saisorn W, Sawatpanich A, Chatthanathon P, Somboonna N, et al. Lactobacillus rhamnosus attenuates Thai chili extracts induced gut inflammation and dysbiosis despite capsaicin bactericidal effect against the probiotics, a possible toxicity of high dose capsaicin. PLoS One. 2021;**16**(12):1-23. DOI: 10.1371/journal.pone.0261189

[21] Abdelazez A, Heba A, Zhu Z, Sami R. Potential benefits of Lactobacillusplantarum as probiotic and its advantages in human health and industrial applications: A review. Advances in Environmental Biology. 2018;**12**(1):16-27. DOI: 10.22587/aeb.2018.12.1.4

[22] Mekonnen SA, Merenstein D, Fraser CM, Marco M. Molecular mechanisms of probiotic prevention of antibiotic-associated diarrhea. Current Opinion in Biotechnology. 2020;**61**:226-234. DOI: 10.1016/j.copbio.2020.01.005

[23] Maity C, Gupta AK. Therapeutic efficacy of probiotic Alkalihalobacillusclausii 088AE in antibiotic-associated diarrhea: A randomized controlled trial. Heliyon. 2021;7(9):993. DOI: 10.1016/j.heliyon.2021.e07993

[24] Daniali M, Nikfar S, Abdollahi M. Antibiotic resistance propagation through probiotics. Expert Opinion on Drug Metabolism\& Toxicology. 2020;**16**(12):1207-1215. DOI: 10.1080/17425255.2020.1825682

[25] Poindexter B, FETUS CON, NEWBORN, Cummings J, Hand I, Adams-Chapman I, et al. Use of probiotics in preterm infants. Pediatrics. 2021;**147**(6):485. DOI: 10.1542/peds.2021-051485

[26] Remes S, Kulmala P. The interplay between risk and preventive factors explains why some children develop allergies to certain foods and others show tolerance. ActaPaediatrica. 2018;**107**(10):1677-1683. DOI: 10.1111/apa.14391

[27] Wang X, Yang Y, Huycke MM. Risks associated with enterococci as probiotics. Food Research International. 2020;**129**:108788. DOI: 10.1016/j.foodres.2019.108788

[28] Nakamura S, Mitsunaga F. Anti-allergic effect of Para-probiotics from non-viable acetic acid bacteria in ovalbumin-sensitized mice. Food and Nutrition Sciences. 2018;**09**(12):1376-1385. DOI: 10.4236/fns.2018.912099

[29] Senkarcinova B, Graça Dias IA, Nespor J, Branyik T. Probiotic alcohol-free beer made with Saccharomyces cerevisiae var. boulardii. LWT. 2019;**100**:362-367. DOI: 10.1016/j.lwt.2018.10.082

[30] Aguilar-Toala JE, Garcia-Varela R, Garcia HS, Mata-Haro V, González-Córdova AF, Vallejo-Cordoba B, et al. Postbiotics: An evolving term within the functional foods field. Trends Food Science and Technology. 2018;**75**:105-114

[31] Akter S, Park JH, Jung HK. Potential health-promoting benefits of paraprobiotics, inactivated probiotic cells. Journal of Microbiology and Biotechnology. 2020;**30**(4):477-481. DOI: 10.4014/JMB.1911.11019

[32] Kumar, H., Schütz, F., Bhardwaj, K., Sharma, R., Nepovimova, E., Dhanjal, D. S., Verma, R., Kumar, D., Kuča, K and Cruz-Martins, N. Recent advances in the concept of paraprobiotics: Nutraceutical or functional properties for promoting children health. Critical Reviews in Food Science and Nutrition, 2021;**0**(0):1-16. DOI: 10.1080/10408398.2021.1996327

[33] Hasan A, Paray BA, Hussain A, Qadir FA, Attar F, Aziz FM, et al. A review on the cleavage priming of the spike protein on coronavirus by angiotensin-converting enzyme-2 and furin. Journal of Biomolecular Structure and Dynamics. 2021;**39**(8):3025-3033. DOI: 10.1080/07391102.2020.1754293

[34] Guimarães JT, Balthazar CF, Scudino H, Pimentel TC, Esmerino EA, Ashokkumar M, et al. High-intensity ultrasound: A novel technology for the development of probiotic and prebiotic dairy products. Ultrasonics Sonochemistry. 2019;**57**:12-21. DOI: 10.1016/j.ultsonch.2019.05.004

[35] Barros CP, de Toledo Guimarães J, Esmerino EA, Duarte MCKH, da Silva MC, Silva R, et al. Paraprobiotics and postbiotics: Concepts and potential applications in dairy products. Current Opinion in Food Science. 2020;**32**:1-8. DOI: 10.1016/j.cofs.2019.12.003

[36] Nataraj BH, Ali SA, Behare PV, Yadav H. Postbiotics-parabiotics: The new horizons in microbial biotherapy and functional foods. Microbial Cell Factories. 2020;**19**(1):1-22. DOI: 10.1186/s12934-020-01426-w

[37] Wang S, Ahmadi S, Nagpal R, Jain S, Mishra SP, Kavanagh K, et al. Lipoteichoic acid from the cell wall of a heat killed lactobacillus paracasei D3-5 ameliorates aging-related leaky gut, inflammation and improves physical and cognitive functions: From C. elegans to mice. Gero Science. 2020;**42**:333-352. DOI: 10.1007/s11357-019-00137-4

[38] Zolkiewicz J, Marzec A, Ruszczynski M, Feleszko W. Postbiotics - a step beyond pre-and probiotics. Nutrients. 2020;**12**:2189. DOI: 10.3390/nu12082189

[39] Salminen S, Collado MC, Endo A. The international scientific Association of Probiotics and Prebiotics (ISAPP) consensus statement on the definition and scope of postbiotics. National Resolution Gastroenterology & Hepatology. 2021;**18**(9):649-667. DOI: 10.1038/s41575-021-00440-6

[40] Pique N, Berlanga M, Miñana-Galbis D. Health benefits of

heat-killed-Tyndallized probiotics: An overview. *International Journal of Molecular Sci*ence. 2019;**20**(10):2534. DOI: 10.3390/ijms20102534-

[41] Irwin C, McCartney D, Desbrow B, Khalesi S. Effects of probiotics and paraprobiotics on subjective and objective sleep metrics: A systematic review and meta-analysis. European Journal of Clinical Nutrition. 2020;**74**(11):1536-1549. DOI: 10.1038/s41430-020-0656-x

[42] Kim H, Lim J-J, Shin HY, Suh HJ, Choi H-S. Lactobacillus plantarum K8-based paraprobiotics suppress lipid accumulation during adipogenesis by the regulation of JAK/STAT and AMPK signaling pathways. Journal of Functional Foods. 2021;**87**:104824. DOI: 10.1016/j.jff.2021.104824

[43] Molaee Parvarei M, Fazeli MR, Mortazavian AM, Sarem Nezhad S, Mortazavi SA, Golabchifar AA, et al. Comparative effects of probiotic and paraprobiotic addition on microbiological, biochemical and physical properties of yogurt. Food Research International. 2021;**140**:110030. DOI: 10.1016/j.foodres.2020.110030

[44] Kye Y-J, Lee S-Y, Kim H-R, Lee B-H, Park J, Park M-S, et al. Lactobacillus acidophilus PIN7 paraprobiotic supplementation ameliorates DSS-induced colitis through anti-inflammatory and immune regulatory effects. Journal of Applied Microbiology. 2021;**132**(4):3189-3200. DOI: 10.1111/jam.15406

[45] Cuevas-González P, Liceaga A, Aguilar-Toalá JE. Postbiotics and Paraprobiotics: From concepts to applications. Food Research International. 2020;**136**:109502. DOI: 10.1016/j.foodres.2020.109502

[46] Vailati-Riboni M, Coleman DN, Lopreiato V, Alharthi A, Bucktrout RE, Abdel-Hamied E, et al. Feeding a Saccharomyces cerevisiae fermentation product improves udder health and immune response to a streptococcus uberis mastitis challenge in mid-lactation dairy cows. Journal of Animal Science and Biotechnology. 2021;**12**(1):62. DOI: 10.1186/s40104-021-00560-8

[47] Sharun K, Dhama K, Tiwari R, Gugjoo MB. Mohd Iqbal Yatoo and Shailesh Kumar Patel. Advances in therapeutic and managemental approaches of bovine mastitis: A comprehensive review. Vetenary Quarterly. 2021;**41**(1):107-136. DOI: 10.1080/01652176.2021.1882713

[48] Delpech G, Ceci M, Lissarrague S, Allende LG, Baldaccini B, Sparo M. In vitro activity of the antimicrobial peptide AP7121 against the human methicillin-resistant biofilm producers Staphylococcus aureus and Staphylococcus epidermidis. Biofouling. 2020;**36**(3):266-275. DOI: 10.1080/08927014.2020.1756266

[49] Gabriella T. Postbiotics: a solution to protect tropical fruits towards postharvest adulteration. 2022;**10**(2):189-196. ISSN 2285-5718

[50] Ringø E, Van Doan H, Lee SH, Soltani M, Hoseinifar SH, Harikrishnan R, et al. Probiotics, lactic acid bacteria and bacilli: Interesting supplementation for aquaculture. Journal of Applied Microbiology. 2020;**129**(1):116-136. DOI: 10.1111/jam.14628

[51] Balzaretti S, Taverniti V, Rondini G, Marcolegio G, Minuzzo M, Remagni MC, et al. The vaginal isolate lactobacillus paracasei LPC-S01 (DSM 26760) is suitable for oral administration. Frontiers in Microbiology. 2015;**6**:952. DOI: 10.3389/fmicb.2015.00952

[52] Kitazawa H, Shahid M, Rajoka R, Teame T, Wang A, Xie M, et al. Paraprobiotics and Postbiotics of probiotic lactobacilli, their positive effects on the host and action mechanisms: A review Frontiers in Nutrition. 2020;7:44. DOI: 10.3389/fnut.2020.570344

[53] Misra S, Mohanty D, Mohapatra S. Food-based probiotics: Functional dietary ingredients. In: Brandelli A, editor. Probiotics. United States of America: Academic Press; 2022. pp. 257-275. DOI: 10.1016/B978-0-323-85170-1.00006-3

[54] Bhushan I, Sharma M, Mehta M, Badyal S, Sharma V, Sharma I. Singh, H and Sistla bioactive compounds and probiotics–a ray of hope in COVID-19 management. Food Science and Human Wellness. 2021;**10**(2):131-140. DOI: 10.1016/j.fshw.2021.02.001

[55] Bajpai VK, Chandra V, Kim N-H, Rai R, Kumar P, Kim K, et al. Ghost probiotics with a combined regimen: A novel therapeutic approach against the Zika virus, an emerging world threat. Critical Reviews in Biotechnology. 2018;**38**(3):438-454. DOI: 10.1080/07388551.2017.1368445

[56] Rocha-Ramírez LM, Hernández-Ochoa B, Gómez-Manzo S, Marcial-Quino J, Cárdenas-Rodríguez N, Centeno-Leija S, et al. Evaluation of immunomodulatory activities of the heat-killed probiotic strain Lactobacilluscasei IMAU60214 on macrophages In vitro. Microorganisms. 2020;**8**(1):79. DOI: 10.3390/microorganisms8010079

[57] Yeşilyurt N, Yılmaz B, Ağagündüz D, Capasso R. Involvement of probiotics and Postbiotics in the immune system modulation. Biologics. 2021;**1**(2):89-110. DOI: 10.3390/biologics1020006

[58] Sunmola AA, Ogbole OO, Faleye TO, Adetoye A, Adeniji JA, Ayeni FA. Antiviral potentials of lactobacillus plantarum, lactobacillus amylovorus, and enterococcus hirae against selected enterovirus. Folia Microbiologica. 2019;**64**(2):257-264. DOI: 10.1007/s12223-018-0648-6

[59] DoCarmo MS, dos Santos CI, Araujo MC, Giron JA, Fernandes ES, Monteiro-Neto V. Probiotics, mechanisms of action, and clinical perspectives for diarrhea management in children. Food & Function. 2018;**9**(10):5074-5095. DOI: 10.1039/c8fo00376a

[60] Homayouni Rad A, Maleki LA, Kafil HS, Zavoshti HF, Abbasi A. Postbiotics as novel health-promoting ingredients in functional foods. Health Promotion and Perspectives. 2020;**10**(1):3-4. DOI: 10.15171/hpp.2020.02

[61] Lopez M, Li N, Kataria J, Russell M, Neu J. Live and ultraviolet-inactivated lactobacillus rhamnosus GG decrease flagellin-induced interleukin-8 production in Caco-2 cells. The Journal of Nutrition. 2008;**138**(11):2264-2268. DOI: 10.3945/jn.108.093658

[62] Tejada-Simon MV, Pestka JJ. Proinflammatory cytokine and nitric oxide induction in murine macrophages by cell wall and cytoplasmic extracts of lactic acid bacteria. Journal of Food Protection. 1999;**62**(12):1435-1444. DOI: 10.4315/0362-028x-62.12.1435

[63] Pagnossa JP, Rocchetti G, Ribeiro AC, Piccoli RH, Lucini L. Ultrasound: Beneficial biotechnological aspects onmicroorganisms-mediated processes. Current Opinion in Food Science. 2020;**31**:24-30

[64] De Almada CN, Almada CN, Martinez RC, Sant'Ana AS.

Paraprobiotics: Evidences on their ability to modify biological responses, inactivation methods and perspectives on their application in foods. Trends in Food Science and Technology. 2016;**58**:96-114. DOI: 10.1016/j.tifs.2016.09.011

[65] Warda AK, Rea K, Fitzgerald P, Hueston C, Gonzalez-Tortuero E, Dinan TG, et al. Heat-killed lactobacilli alter both microbiota composition and behaviour. Behaviour Brain Research. 2019;**362**:213-223. DOI: 10.1016/j.bbr.2018.12

[66] Sawada D, Kuwano Y, Tanaka H, Hara S, Uchiyama Y, Sugawara T, et al. Daily intake of lactobacillus gasseri CP2305 relieves fatigue and stress-related symptoms in male university Ekiden runners: A double-blind, randomized, and placebo-controlled clinical trial. Journal of Functional Foods. 2019;**57**:465-476. DOI: 10.1016/j.jff.2019.04.022

[67] Chen Chih-Yuan, Tsen H.Y., Lin C.L., Lin, C.K., Chuang, L. T and Chen, C.S. "Enhancement of the immune response against salmonella infection of mice by heat-killed multispecies combinations of lactic acid bacteria." Journal of Medical Microbiology, Volume 62, Pt 11 2013: 1657-1664. doi: 10.1099/jmm.0.061010-0

[68] Enck P, Schemann M, Frieling T. Paraprobiotics for irritable bowel syndrome: All that glitters is not gold. The Lancet Gastroenterology & Hepatology. 2020;**5**(9):797. DOI: 10.1016/S2468-1253(20)30243-0

[69] Butera A, Gallo S, Pascadopoli M, Maiorani C, Milone A, Alovisi M, et al. Paraprobiotics in non-surgical periodontal therapy: Clinical and microbiological aspects in a 6-month follow-up domiciliary protocol for Oral hygiene. Microorganisms. 2022;**10**(2):337. DOI: 10.3390/microorganisms10020337

[70] Warzée J-P, Elli M, Fall A, Cattivelli, François. Supranational assessment of the quality of probiotics: Collaborative initiative between independent accredited testing laboratories. Microorganisms. 2021;**9**(7):456. DOI: 10.3390/microorganisms9071456

[71] Baradaran GS, Pourhamzeh M, Farmani M, Raftar SKA, Shahrokh S, Shpichka A, et al. Cross-talk between immune system and microbiota in COVID-19. Expert Review of Gastroenterology & Hepatology. 2021;**15**(11):1-14. DOI: 10.1080/17474124.2021.1991311

[72] Case I. The Regulations of Probiotics. Msc in Regulations Science. John Hopkins university; Iris. 2022

[73] Martín R, Langella P. Emerging health concepts in the probiotics field: Streamlining the definitions. Frontiers in Microbiology. 2019;**10**:1047. DOI: 10.3389/fmicb.2019.01047

[74] Hoffman FA, Heimbach JT, Sanders ME, Hibberd PL. Executive summary: Scientific and regulatory challenges of development of probiotics as foods and drugs. Clinical Infectious Diseases. 2008;**46**(Supplement_2):S53-S57. DOI: 10.1086/523342

[75] Sanders ME, Merenstein DJ, Ouwehand AC, Reid G, Salminen S, Cabana MD, et al. Probiotic use in at-risk populations. Journal of the American Pharmacists Association. 2016;**56**(6):680-686. DOI: 10.1016/j.japh.2016.07.001

[76] De Simone C. The unregulated probiotic market. Clinical Gastroenterology and Hepatology. 2019;**17**(5):809-817. DOI: 10.1016/j.cgh.2018.01.018

Chapter 4

Managing Antimicrobial Resistance beyond the Hospital Antimicrobial Stewardship: The Role of One Health

Istifanus Anekoson Joshua, Mathew Bobai and Clement Sokfa Woje

Abstract

Infections caused by micro-organisms affect the health of people and animals, causing morbidity and mortality, with Asia and Africa as the epicenters. Some of the infectious diseases are emerging and re-emerging in nature. Examples include viral hepatitis, Lassa fever, Ebola, yellow fever, tuberculosis, covid-19, measles, and malaria, among others. Antimicrobials have been playing an important role in the treatment of infections by these microbes. However, there has been a development of resistance to these antimicrobials as a result of many drivers. This write-up used secondary data to explore the management of antimicrobial resistance (AMR) beyond the hospital antimicrobial resistance steward using the one health concept. The findings showed AMR to be a transboundary, multifaceted ecosystem problem affecting both the developed and developing countries. It is also one of the top ten global public health threats facing mankind. Globally, AMR will cost over US$100 trillion in output loss by 2050, about 700,000 deaths a year, and 4,150,000 deaths in Africa by 2050. About 2.4 million people could die in high-income countries between 2015 and 2050 without a sustained effort to contain AMR. The drivers of AMR are beyond the hospital and hospital AMR stewardship. Therefore, the need for one health concept to manage it.

Keywords: antimicrobials, antimicrobial resistance, Hospital antimicrobial stewardship, infections, one health

1. Introduction

Infectious diseases are caused by microbes such as bacteria, viruses, fungi, or parasite, which often affect human and animal health. The mode of transmission can be direct, such as spread from person to person, or indirect contact via insect bites, food and water contaminations, among others [1].

Africa with the fastest growing population in the world, is now catching up with Asia as infectious diseases hotspot [2]. Infectious diseases have accounted for about one-quarter of deaths, and are globally responsible for at least ten million deaths annually, especially in tropical countries at the beginning of the 21st century. Some of the infectious diseases are emerging while some are re-emerging in nature. Examples of these diseases reported in Africa include meningococcal meningitis, hepatitis B, C, and E viruses, tuberculosis, Dengue fever, Lassa fever, yellow fever, Ebola virus, COVID-19, measles, HIV/AIDS, plague, avian influenza, chikungunya, syphilis and poliomyelitis, monkey pox, Marburg virus, Zika virus, rift valley fever, malaria, cholera, rickettsia, among others (**Table 1**).

Disease	Type	Countries
Covid-19	Emerging	Nigeria, among other African countries
Lassa fever	Emerging	Nigeria, Liberia, Benin Republic
Ebola virus	Emerging	Sierra Leone, Liberia, Guinea, Democratic Republic of Congo (DRC), Uganda, Senegal
HIV/AIDS	Emerging	All African Countries
Hepatitis B, C, E	Emerging	Nigeria, Kenya, Uganda, Rwanda, most African countries
Typhoid fever	Emerging	Nigeria, Ghana, Madagascar, Senegal, Ethiopia, Burkina Faso, Kenya, Guinea Bissau, South Africa, Sudan, and Tanzania
Dengue fever	Emerging	Burkina Faso, Cote d'ivoire, Egypt, Cape Verde
Monkey pox	Emerging	Nigeria, Cameroon, Central Africa Republic (CAR)
Chikungunya	Emerging	Congo, Sudan, Kenya, Senegal
Zika virus	Emerging	Cape Verde
Malaria	Emerging/ Re-emerging	Nigeria, Tanzania, Mozambique, DRC
Measles	Re-emerging	Nigeria, DRC, Zambia, Ethiopia
Cholera	Re-emerging	Nigeria, Guinea Bissau, Zimbabwe, Niger, Chad, Cameroon, DRC, Sierra Leone, South Sudan, Zambia, Kenya, Somalia and Mozambique
Tuberculosis	Re-emerging	Nigeria, South Africa
Meningococcal meningitis	Re-emerging	Nigeria, Liberia, Togo, Niger, Guinea, South Sudan, Benin Republic, Chad, Burkina Faso, Ghana, DRC, Uganda, Cote d'ivoire
Syphilis	Re-emerging	Nigeria, Morocco, Burkina Faso, Togo, Ghana, Tanzania, Ethiopia, Rwanda, Cameroon, Gabon, Mozambique, Swaziland, South Africa
Plague	Re-emerging	Madagascar
Trypanosomiasis	Emerging/ Re-emerging	Nigeria, DRC, South Sudan, Angola, CAR, Chad, Congo, Malawi, Guinea Bissau, Cameroon, Cote d'Ivoire, Equatorial Guinea, Tanzania, Uganda, Zambia, Zimbabwe, Burkina Faso, Ghana, Kenya
Anthrax	Re-emerging	Tanzania, Zimbabwe, Botswana, Uganda, Namibia, Chad

Table 1.
Some emerging and re-emerging infectious diseases reported in Africa [3–13].

The burden of infectious diseases in Africa is huge, and it has topped the list of diseases that frequently require consultation, hospitalization and also remain a major cause of morbidity and mortality. Antimicrobials play important roles in their treatment, emergence of resistance, persistence, and transmission. They have also saved hundreds of millions from infectious agents. However, antimicrobial resistant (AMR) organisms are increasing globally, threatening to render existing treatments ineffective. They prolong illness, increase case fatality, facilitate transmission, and increase treatment costs.

Antimicrobial resistance caused by bacteria and viruses are of greater public health significance. This is because they account for a large share of clinical infections observed. Their emergence has compromised the effectiveness of antimicrobials [14]. The use of antibiotics makes them serve as reservoirs of resistant genes with the propensity to spread via ecological niche through the human, animal, and environmental interactions [15, 16].

Some factors associated with antimicrobial resistance include microbial adaptation and change, human susceptibility to infection, poor environmental practices, human demographics and behavior, international travel and commerce, technology and industry, breakdown of public health measures, poverty and social inequality, war, and famine and lack of political will [3].

2. Antimicrobials and antimicrobial resistance

Antimicrobials are global public good that has improved health care, saved lives, and enhanced economic gains [17]; and they are the cornerstone on which the health system is standing on [18]. Antimicrobial resistance is the alteration of microbes when exposed to the antimicrobial making them not sensitive. These drugs become ineffective and infections persist in the body, increasing the risk of spread to others.

Antimicrobial resistance is the development of resistance in a microorganism to an antimicrobial agent to which it was previously sensitive [19]; and it is a multifaceted ecosystem problem that threatens the interdependent humans, animals, and environmental health [15, 20]. In view of this importance, the World Health Organization theme for 2011 was tagged "antimicrobial resistance: no action taken, no cure tomorrow".

3. Magnitude of the problem of antimicrobial resistance

The World Health Organization (WHO) has declared that antimicrobial resistance is one of the top ten global public health threats the world is battling with [4]. Antimicrobials such as antibacterial, antivirals, antifungals, and antiparasitics are used to prevent and treat infections in human, animals and plants [4].

United Nations General Assembly, World leaders of G7 and G20, and WHO declared AMR as a global health security challenge today. It is a transboundary problem that concerns every country irrespective of its level of income and development, where the organisms require no international passports [15, 20]. Antimicrobial resistance is a global crisis that risks reversing a century of progress in health [21]. Alarming levels of resistance have been reported in both developing and developed countries, with the result that common diseases are becoming untreatable, and lifesaving medical procedures more at risk to perform [21].

Antimicrobial resistance is also an ecosystem problem threatening the interrelated human-animal-environment health under the "One Health" framework. Resistant bacteria arising in one geographical area can spread via cross-reservoir transmission to other areas worldwide either by direct exposure or through the food chain and the environment [22, 23]. Sixty percent of pathogens harmful to humans are of animal origin; humans and animals share the same bacteria [17].

The economic burden of AMR is difficult to calculate due to insufficient data and the need to account for externalities, especially in Africa [24]. Globally, drug-resistant microbes account for at least 700,000 yearly deaths and 230,000 deaths from resistant mycobacteria are projected to increase to 10 million deaths globally by 2050 in no action is taken. Around 2.4 million people could die in high-income countries between 2015 and 2050 without a sustained effort to contain antimicrobial resistance [21]. Estimates of the impact of AMR on the US economy are exceedingly high, including $20 billion in direct health care costs with additional indirect costs as high as $25 billion, 2 million illnesses, and 23000 deaths per year [25].

The World Bank projected that 24 million people could fall into extreme poverty by 2030 because of AMR and most would come from low- and middle-income countries [15]. Globally, AMR will cost over US$100 trillion in lost output by 2050 [23] and about 4,150,000 deaths in Africa by 2050 [19, 23]. The problem of AMR is global but is particularly more serious in sub-Saharan Africa, second only to that of Asia.

The increase in AMR could lead to a reduction in options available to treat infectious diseases, support chemotherapy, and surgery, and this will have a significant impact on the Health System and economies [19]. Infections with resistant organisms have been associated with an increased hospital stay, increased morbidity and mortality, use of additional drugs, laboratory tests, and increased treatment cost [26, 27]. This has financial implications for the individuals, families, communities, and the health system (HS) [19]. This has increased poverty as it has been documented that millions of Africans fall into poverty due to high out-of-pocket health payments [28]. Antimicrobial resistance could lead to loss of productivity from the spread of diseases to other animals and death of the animals, thereby threatening the sustainability and security of food production and the livelihood of farmers. The proportion of antimicrobials resistance has at least doubled in chickens and in pigs in the past two decades [25].

Reports have identified significant gaps in surveillance, standard methodologies, and data sharing related to AMR; and Africa and South East Asia as regions without established AMR surveillance systems [29]. This results in a lack of quality data leading to treatment guidelines that are not adequate for the local situation. Consequently, the rise and spread of AMR threaten the effective control and treatment of various bacterial diseases world wide [15, 20]. In addition, the lack of consistency in the measurement and reporting of susceptibility data makes it difficult to compare findings among different countries and laboratories, sometimes even within one country [30].

Infections caused by antimicrobial resistance are now alarming globally, and the increasing rates of antimicrobial resistance are resulting in fewer treatment options [31]. The world's known antimicrobials are becoming increasingly ineffective as drug resistance spreads globally leading to more difficult to treat infections and deaths [4]. The problem is further compounded by the fact that very few new antibiotics have been developed within the last thirty years. We effectively do not have any new weapon in the fight against AMR. Therefore, new antimicrobials are urgently needed to treat especially carbapenem-resistant gram-negative bacterial infections as identified by the WHO priority pathogen list [4].

Without effective tools for the prevention and adequate treatment of drug-resistant infections, the maternal number of death due to drugs resistant infections will increase, and medical procedures such as surgery, including cesarean sections, hip replacements, cancer chemotherapy, and organ transplantation will become riskier [4].

Statistics indicated that malaria claims more than one million lives yearly, and African countries bear the brunt of malaria accounting for more than 90% of all cases occurring worldwide [32]. In Africa, malaria has devastating consequences on agricultural households. It is estimated that malaria cost Africa more than twelve billion United State dollar per year slowing its economic growth by 1.3% annually [33]. Tuberculosis is one of the top leading causes of mortality globally and the highest incidence rates are found in Africa and south-east Asia [34].

HIV/AIDS kills and disable adults in the productive part of their lives affecting businesses, investments, industries, agricultural sustainability, and African agricultural labor force in particular affected [35]. It is worth noting, that bacterial diarrhea, malaria, tuberculosis, and HIV infections, responsible for high mortality rates in sub-Saharan Africa, are also showing increased resistance to hitherto effective antimicrobials. *Candida auris* has shown increased resistance to antifungal drugs such as fluconazole, amphotericin B, Voriconazole, among others [4]. In Nigeria, there is a widespread antimicrobial resistance among enteric *Escherichia coli*, particularly to penicillins, aminoglycosides, cephaloporins, chloramphenicol, tetracycline, and cotrimoxazole [36].

4. Drivers of antimicrobial resistance transmission

Antimicrobial resistance is complex, multi-sectoral and a cross-boundary challenge being driven by clinical, biological, social-political, economical, and environmental drivers and exerts effect not only on humans, but also animals and the ecosystem. However, the key drivers of antimicrobial resistance include poverty, lack of access to clean water, sanitation, and hygiene for both human and animals; poor infections and diseases prevention and control in healthcare facilities and farms; changing population density; poor management of pharmaceutical and hospital wastes; antibiotic misuse and overuse; poor access to quality and affordable medicines, vaccines, and diagnostics; poor public knowledge about antimicrobials and its resistance; lack of enforcement of legislation; lack of surveillance systems; lack of food safety and control measures; poor environmental practices, poor documentation of AMR in animals, poor evidence-based data on the magnitude and economic burden of AMR in humans; poor rules and regulations to control counterfeit drugs in the market and unique transmission properties of antimicrobial resistant organism, chemical stressors in an environment, among others [37].

Bacteria usually adopt some mechanisms to resist antibiotic action against them. These mechanisms include the inactivation of the antibiotic through enzymatic degradation, or modification of the antibiotic targets, alteration of the permeability of the cell membrane, and the expression of efflux pumps to keep intracellular of antibiotic below inhibitory level [37].

Several unique properties of antimicrobial resistant bacteria enable their development and propagation in the environment. Autochthonous bacteria constitute environmental reservoirs of antibiotic resistance genes or "resistomes" that can subsequently be transferred to pathogens via horizontal gene transfer (HGT) [37, 38]. This

HGT can occur through conjugation, transduction or transformation. However, the key global concern is the development of resistance of last resort, such as the cephalosporins, carbapenems, and polymyximises [39]. Resistance to third-generation cephalosporins has increased worldwide to bacterial acquisition of the ability to produce extended-spectrum beta-lactamase enzymes (ESBL) that mediate resistance to most beta – lactams [40]. Bacteria and mobile genetic elements conferring resistance linger on animal skin and in feces and by various means can be transferred between bacteria, and these organisms can make their way to human beings [41]. Evidence of transmission from livestock to human beings ESBL and AmpC – B – Lactamase genes on plasmids and *Escherichia coli* clones, most likely through the food chain have been reported [41].

4.1 Environmental and related factors

In developing countries with scarce resources, poor sanitation, poor food safety measures, sales of antimicrobial over the counter, overcrowding, use of antimicrobials in animal and fish farming, and weak government regulations are some of the leading causes of antimicrobial resistance [42, 43]. There has been documentation of antibiotics being added directly to dairy products by vendors in order to increase shelf-life in Ethiopia [44]. Others showed high antimicrobial residues in eggs and meat in Nigeria [45], Ghana [46], Senegal [47], Kenya [48]' and Tanzania [49].

4.2 Changing of population density

Movement of people from rural to urban areas (urbanization) brings considerable negative and positive changes in their living and working conditions. In the urban areas, housing density increases, there is overcrowding, animals and humans may share dwelling places and drinking water, among others with resultant negative health consequences. One of the problems associated with rural-urban migration of people includes AMR infection transmission, which has been documented [50].

4.3 Use of antimicrobials in human and veterinary medicine

Antimicrobial are among the most commonly prescribed drugs in human and veterinary medicine but about 50% of these are considered unnecessary [51]. This is associated with misuse, overuse, and underuse especially in low, middle, and high-income countries (LMIC) [52, 53]. These consumptions could be a major driver of AMR. When antibiotics are used, either for medicinal purposes or for food animal production, they inevitably make their way into the environment [40].

Antibiotics have been in use in livestock, cattle, and aquaculture, among others to enhance production and growth for human consumption. A study showed that among different countries using veterinary antibiotics, Myanmar, Indonesia, Nigeria, Peru, and Vietnam have been projected to have the greatest increase by 2030 in that descending order [54].

Treatment of ailing fish with antibiotics used for human medicine and then dumping these treatments directly into the water or via fish food is one of the leading causes of bacterial resistance in the aquatic environment. Substantial evidence supports the link between antibiotic resistance in livestock and the emergence of bacterial resistance in humans [55, 56].

4.4 Counterfeit antibiotics

Counterfeit antibiotics are a type of substandard drug and the influx into the global pharmaceutical market is estimated at 5% [57]. The majority of these products originating from south-East-Asia and Africa, are destined mainly for emerging countries including South-East –Asia, sub-Saharan Africa, Europe, and North America [57]. Even though it is a worldwide problem, it is still not eradicated and it continues to exert a devastating negative impact mainly because of poverty, globalization, ease of international trade, the lack of regulations, and law enforcement, among others.

4.5 Non-prescription antibiotics

Globally, antibiotics are becoming more and more available over-the-counter or via unregulated supply chains [58, 59], which is a problem in both developing and developed countries [60]. This results from weak law enforcement or even the absence of policies and regulations [61]. In developing countries mainly Africa, the community is providing different unauthorized services like consulting, diagnosing, prescribing, and dispensing medications [62]. These illegal practices if no care is taken can increase selection pressure and consequently AMR.

5. Overview antimicrobial and hospital antimicrobial stewardship

Antimicrobial stewardship is the effort to measure and improve how antimicrobials are prescribed by clinicians and used by patients. Improving antimicrobials prescribing and use is critical to effectively treat infections, protect patients from harms caused by unnecessary antimicrobial use, and combat antimicrobial resistance. (www.cdc.gov/antibiotic-use/core-elements/index.html).

CDC's Core Elements of Antibiotic Stewardship offers providers and facilities a set of key principles to guide efforts to improve antibiotic use and, therefore, advance patient safety and improve outcomes. These frameworks complement existing guidelines and standards from key healthcare partner organizations, including the Infectious Diseases Society of America, Society for Healthcare Epidemiology of America, American Society of Health System Pharmacists, Society of Infectious Diseases Pharmacists, and The Joint Commission (CDC www.cdc.gov/antibiotic-use/core-elements/index.html).

It is the use of standard antibiotic regimens for the treatment of infections thus optimization of antibiotic use. This program has been implemented in some countries with impressive results [48], leading to a reduction in the use of antibiotics especially broad-spectrum antibiotics in addition to a decrease in healthcare costs and the improvement of patient outcomes and AMR containment [63, 64]. Similar programs in South Africa, a lower-middle-income country, in both the private and public hospital sectors, have shown reductions in inappropriate antibiotic use, among others [65].

The Core Elements of Hospital Antibiotic Stewardship Programmes [66] include:

- Hospital Leadership Commitment which dedicates necessary human, financial, and information technology resources.

- Accountability appoints a leader or co-leaders, such as a physician and pharmacist, responsible for program management and outcomes.

- Pharmacy Expertise (previously Drug Expertise) which appoints a pharmacist, ideally as the co-leader of the stewardship program, to help lead implementation efforts to improve antibiotic use.
- Action that implements interventions, such as prospective audit and feedback or preauthorization, to improve antibiotic use.
- Tracking which monitors antibiotic prescribing, impact of interventions, and other important outcomes, like *Clostridium difficile* infections and resistance patterns.
- Regularly reporting information on antibiotic use and resistance to prescribers, pharmacists, nurses, and hospital leadership.
- Education of prescribers, pharmacists, nurses, and patients about adverse reactions from antibiotics, antibiotic resistance, and optimal prescribing.

6. Deficiencies in the hospital antimicrobial stewardship program

Because the drivers of antimicrobial resistance lie in humans, animals, plants, food, and the environment (i.e., beyond the hospital), a sustained One Health response is essential to engage and unite all stakeholders around a shared vision and goals.

Human resources for health (HRH) are key in the hospital antimicrobial resistance containment. However, inadequate and inequity in the distribution of health workers is a huge problem, especially in Africa, and Nigeria [67]. The maldistribution of health workforces is central to the existing inequalities in health service coverage and the burden of disease for populations in need.

Weak health system: Although the battle of AMR is a global one, Africa is currently at a disadvantage in the fight because of weak healthcare systems and other factors that are slowing the continent's efforts in the fight. This will have serious negative human, social, economic, and developmental consequences in the region [15]. Africa is a continent bellied with challenges such as widespread poverty, armed conflicts, high level of illiteracy, poverty, and very weak medical and veterinary health institutions [68], that have made the continent poorly prepared to effectively fight this public health threat.

7. The role of One Health

One Health is an approach of multiple disciplines working locally, nationally, and globally to obtain better health for people, animals, and the environment. It has the potential to mitigate the negative externality of AMR [69].

Studies have shown that implementing one health, especially in low-income countries will save lots of money for the veterinary and medical health systems [44, 68]. This money can be used to enhance surveillance and improve capacities in medical and veterinary HS. Surveillance systems are the foundation for a better understanding of the epidemiology of AMR and the key for tackling this public health threat [46].

7.1 The benefits of One Health in tackling antimicrobial resistance

Tackling antimicrobial resistance from the "One Health" perspective is emaced by the WHO/FAO/OIE Tripartite, the Declaration from the 2016 high-level meeting on antimicrobial resistance at the United Nations General Assembly, and is supported by the World Bank [20, 70, 71]. This model engenders broad effectiveness and efficiency outcomes generating savings in operating costs. It is based on building veterinary/human public-health capacity and enhancing awareness in order to reach effective global governance. Capitalizing on these capacities or reducing the vulnerabilities, especially in low-income countries will prevent or mitigate the leading causes of antimicrobial resistance and infectious pandemic.

The adoption and implementation of laboratory-based surveillance and monitoring system in the African WHO regional office is poor. In LMICs, the challenges are enormous due to weak laboratory and communications infrastructure, lack of trained and qualified staff, and higher incidence of counterfeit antibiotics [72]. Current surveillance capabilities are variable across the world. Europe and the USA have the best surveillance coverage while Sub-Saharan Africa, South and Southeast Asia have the least developed [51]. Therefore, there is need for global public health awareness on the importance of rational antibiotic use and emergence of resistance.

8. Conclusion

The importance of antimicrobial resistance cannot be neglected in view of its consequences globally, regionally, nationally, and locally. It is a hazard that must be prevented and/or mitigated. Health Education of the general population and clinicians on wrong antibiotic choice, wrong dose, wrong dose interval, wrong route, wrong duration, and delayed administration could be helpful.

Multimodal strategies for the control of AMR, Research and Development, environmental control, market control, and manufacturing should be explored.

Establishment of laboratory for human and animal diseases research: Adequate funding is critical; however, the sources of funding can be from governmental and non-governmental entities.

Surveillance of antibiotic consumption in medical and veterinary medicine is fundamental; and a massive global public awareness is important to enhance knowledge about AMR in general and antibiotic uses and resistance in particular. Surveillance systems are the foundation for a better understanding of the epidemiology of AMR and the key for tackling this public health threat.

Medical prescriptions should be based on the local antibiogram. There is a need to explore alternatives to antimicrobials, such as phages and probiotics, among others.

Acknowledgements

I sincerely acknowledge my teachers, namely Professors JKP Kwaga, Junaidu Kabir, TO Aken'Ova, Clara Ladi Ejembi, Kabir Sabitu, Mohammed Bello, and Dr FJ Giwa of Ahmadu Bello University, Zaria, Nigeria.

Others are Late Professors Andrew Nok, James Adagadzu Kagbu and Stephen Nkom, Late Dr TT Gbem and Late Colonel (Dr) Chinedu John Camillus Igboanusi, Yahuza Suleiman, Bawa Egga, Eunice Azimheye Mamedu, and Esther Jonah.

I also acknowledge Mrs Wazi Istifanus, Ovye Istifanus, Ahogbresha Istifanus, Ashe-ulu Istifanus, and Abesla Istifanus for all their support.

Conflict of interest

The authors declare no conflict of interest.

Author details

Istifanus Anekoson Joshua[1*], Mathew Bobai[2] and Clement Sokfa Woje[1]

1 Department of Community Medicine, Faculty of Clinical Sciences, College of Medicine, Kaduna State University, Kaduna, Nigeria

2 Department of Microbiology, Faculty of Science, Kaduna State University, Kaduna, Nigeria

*Address all correspondence to: dristifanus@yahoo.com

References

[1] Gerald LM, John EB, Raphael D, Mandell D. Bennett Principle and Practice of Infectious Diseases. 7th Edition. Philadelphia, USA: Churchill Living Stone, Elsevier Limited. 2010

[2] Bornard B, Delia R, John M. Africa's Growing Risk of Diseases that Spread from Animals to People, 2020. IFFRI blogi - issue. Available at: https//www.iffri.org/blogi/africas-growing-risk-disease-spread-animals-people [Accessed on 26 January, 2022]

[3] Fenollar F, Mediannikov O. Emerging Infectious Diseases in Africa in the 21st century. Philadelphia, USA: Elsevier Ltd. 2018

[4] WHO. Antimicrobial Resistance. 2021. Available at https://www.who.int/news.room/fact.sheet/details/antimicrobialresistance [Accessed 20th January, 2022]

[5] Carey EM, Steele DA. The severe Typhoid fever in Africa program highlights the need for broad deployment of typhoid conjugate vaccine. Clinical Infections Diseases. 2019;**69**(6):S413-S416

[6] Marks F, Kalckreuth V, Aaby P, Adu-Sarkodie YJ, El-Tayeb MA, Ali M, et al. Incidence of Invasive *Salmonella* Disease in sub-Saharan Africa: A multicentre population-based surveillance study. The Lancet Global Health. 2017;**5**(3):e310-e323

[7] Mourembou G, Lekana-Douki JB, Mediannikov O, Nzondo SM, Kouna LC, Essone JC, et al. Possible role of Rickettsia felis in acute febrile illness among children in Gabon. Emerging Infectious Diseases. 2015;**21**:1808-1815

[8] Sands P, Mundaca-Shah C, Dzau VJ. The neglected dimension of global Security a framework for countering infectious-disease crises. The New England Journal of Medicine. 2016;**374**:1281-1287

[9] Federal Ministries of Agriculture Environment and Health (FMAEH). Antibacterial Use and Resistance in Nigeria: Situation Analysis and Recommendations. 2019. pp. 12-158

[10] Noah K, Jeffrey DK. An update on the global epidemiology of Syphilis. Current Epidemiology Records. 2019;**5**(1):24-28. DOI: 10.1007/s40471-018-0138-2

[11] WHO. Trypanosomiasis, Human African (sleepy sickness), 2022. www.who.int'l...factsheets,detail [Accessed 19 February, 2022]

[12] Eliberiki RM, Sol H, Robert F, Hezron EN, Robinson HM, Eystein S. Anthrax outbreaks in the humans-livestock and wildlife interface areas of Northern Tanzania: A Retrospective Record Review, 2006-2016. BMC Public Health. 2018;**18**(106):2018

[13] Richard M, Notion TG, Tapuwa M, Mufuta T. Investigation of an Anthrax outbreak in Makoni district, Zimbabwe. BMC Public Health. 2021;**21**(298):2021

[14] Presi P, Stärk KD, Stephan R, Breidenbach E, Frey J, Regula G. Risk scoring for setting priorities in monitoring of antimicrobial resistance in meat and meat products. International Journal of Food Microbiology. 2009;**130**:94-100

[15] World Bank Group. Operational Framework for Strengthening Human,

Animal, and Environmental Public Health Systems at Their Interface; International Bank for Reconstruction and Development/The World Bank: Washington, DC, USA, 2018

[16] Iskandar K, Molinier L, Hallit S, Sartelli M, Catena F, Coccolini F, et al. Drivers of antibiotic resistance transmission in low and middle-income countries from a "One Health" perspective—A review. Antibiotics. 2020;**9**:372

[17] Koch BJ, Hungate BA, Price LB. Food-animal production and the spread of antibiotic resistance: The role of ecology. Frontiers in Ecology and the Environment. 2017;**15**:309-318

[18] European Food Safety Authority; European Centre for Disease Prevention and Control. The European Union summary report on antimicrobial resistance in zoonotic and indicator bacteria from humans, animals and food in 2016. EFSA Journal. 2018;**16**:e05182

[19] Machado AR, Pinho DB, de Oliveira SA, Pereira OL. New occurrences of Botryosphaeriaceae causing black root rot of cassava in Brazil. Tropical in Plant Pathology. 2014;**39**:464-470

[20] White A, Hughes JM. Critical importance of a one health approach to antimicrobial resistance. EcoHealth. 2019;**16**:404-409

[21] No time to wait: Securing the future from drug-resistant infections; Report to the Secretary-General of United Nations, April, 2019. www.who.int/antimicrobial-re-sistance/interagency-coordination-group/IACG_final_summary_EN.pdf?ua=1 United Nations. Political Declaration of the High-level Meeting of the General Assembly on Antimicrobial Resistance, A/71/L.2. [Accessed 22 September, 2016]

[22] Holmes AH, Moore LS, Sundsford A, Steinbakk M, Regmi S, Karkey A, et al. Understanding the mechanisms and drivers of antimicrobial resistance. Lancet. 2016;**387**:176-187

[23] McEwen SA, Collignon PJ. Antimicrobial resistance: A One Health perspective. Microbiology Spectrum. 2018;**6**:521-547

[24] Wright GD. The antibiotic resistome: The nexus of chemical and genetic diversity. Nature Reviews. Microbiology. 2007;**5**:175

[25] Van Hoek AH, Mevius D, Guerra B, Mullany P, Roberts AP, Aarts HJ. Acquired antibiotic resistance genes: An overview. Frontiers in Microbiology. 2011;**2**:203

[26] Essack SY, Connolly C, Sturm AW. Antibiotic use and resistance in public-sector hospitals in KwaZulu-Natal. South African Medical Journal. 2005;**95**:865-870

[27] Agyepong N, Govinden U, Owusu-Ofori A, Essack SY. Multidrug-resistant gram-negative bacterial infections in a teaching hospital in Ghana. Antimicrobial Resistance and Infection Control. 2018;**7**:37

[28] Ramsamy Y, Mlisana KP, Allam M, Amoako DG, Abia AL, Ismail A, et al. Genomic analysis of carbapenemase-producing extensively drug-resistant Klebsiella pneumoniae isolates reveals the horizontal spread of p18-43_01 plasmid encoding blaNDM-1 in South Africa. Microorganisms. 2020;**8**:137

[29] World Health Organization (WHO). Antimicrobial Resistance Global Report on Surveillance. Geneva: WHO; 2014

[30] Cox G, Wright GD. Intrinsic antibiotic resistance: Mechanisms,

origins, challenges and solutions. International Journal of Medical Microbiology. 2013;**303**:287-292

[31] Schwarz S, Loeffler A, Kadlec K. Bacterial resistance to antimicrobial agents and its impact on veterinary and human medicine. Veterinary Dermatology. 2016; **28**(1):82-e19

[32] Ruxin J, Palizzi JE, Wilson PA, Tozan Y, Kruk M, Teklehaimanot A. Emerging consensus in HIV/AIDS, Malaria, Tuberculosis and access to essential medicines. The Lancet. 2005;**365**(9459):618-621

[33] Bartram J, Lewis K, Kenton R, Wright A. Focusing on improved water and sanitation for health. The Lancet. 2005;**365**(9461):810-812

[34] Laxminarayan R, Klein E, Dye C, Floyd K, Dareley S, Adeyi O. Economic Benefit of Tuberculosis Control. Washington, DC: The World Bank; 2007

[35] UNAIDS. The impact of AIDS on people and Societies 2006. Available at htt://data.UNAIDS.org [Accessed 20th January, 2022]

[36] Fagbamila IO, Barco L, Mancin M, Kwaga J, Ngulukun SS, Zavagnin P, et al. Samonella serovars and their distribution in Nigerian Commercial chicken layer farms. PLoS One. 2017;**12**(3):e0173097

[37] Vikesland P, Garner E, Gupta S, Kang S, Maile-Moskowitz A, Zhu N. Differential drivers of antimicrobial resistance across the world. Accounts of Chemical Research. 2019;**52**:1916-1924

[38] D'Costa VM, McGrann KM, Hughas DW, Wright GD. Sampling the antibiotic resistome. Science. 2006;**311**:374-377

[39] Johnson AP, Woodford N. Global spread of antibiotic resistance: The example of New Delhi Metallo-B-Lactamase (NDM) - mediated Carbapenem resistance. Journal of Medical Microbiology. 2013;**62**:499-513

[40] Aliison HH, Luke SPM, Arnfinin S, Martin S, Sadie R, Abhilasha K, Philippe JG. Laura JPP. Understanding the mechanisms and drivers of antimicrobial resistance. Lancet. 2016;**387**:176-187

[41] Pal C, Bentsson-Palme J, Kristansson E, Larsson DGJ. The structure and diversity of human, animal and environmental resistomes. Microbiome. 2016;**4**:54

[42] Rousham EK, Unicomb L, Islam MA. Human, animal and environmental contributors to antibiotic resistance in low-resource settings: Integrating behavioural, epidemiological and One Health approaches. Proceedings of the Royal Society B: Biological Sciences. 2018;**285**:20180332

[43] Nadimpalli M, Delarocque-Astagneau E, Love DC, Price LB, Huynh BT, Collard JM, et al. Combating global antibiotic resistance: Emerging one health concerns in lower-and middle-income countries. Clinical Infectious Diseases. 2018;**66**:963-969

[44] Carruth L, Roess AA, Terefe Y, Hosh FM, Salman MD. Antimicrobial resistance and food safety in Africa. The Lancet Infectious Diseases. 2017;**17**:575-576

[45] Kabir J, Umoh VJ, Audu-Okoh E, Umoh JU, Kwaga JKP. Veterinary drug use in poultry farms and determination of antimicrobial drug residues in commercial eggs and slaughtered chicken in Kaduna State, Nigeria. Food Control. 2004;**15**:99-105

[46] Donkor ES, Newman MJ, Tay SC, Dayie NT, Bannerman E, Olu-Taiwo M. residues contaminating meat and egg in Ghana. Food Control. 2011;**22**:869-873

[47] Abiola FA, Diop MM, Teko-Agbo A, Delepine B, Biaou FC, Roudaut B, et al. Résidus d'antibactériens dans le foie et le gésier de poulets de chair dans les régions de Dakar et de Thiès (Sénégal). Revista Médicina véterinaria. 2005;**156**:264-268

[48] Kangethe EK, Aboge GO, Arimi SM, Kanja LW, Omore AO, Mcdermott JJ. Investigation of risk of consuming marketed milk with antimicrobial residues in Kenya. Food Control. 2005;**16**:349-355

[49] Kurwijila LR, Omore A, Staal S, Mdoe NSY. Investigation of the risk of exposure to antimicrobial residues present in marketed milk in Tanzania. Journal of Food Protection. 2006;**69**:2487-2492

[50] Sun MM, Ye M, Schwab AP, Li X, Wan JZ, Wei Z, et al. Human migration activities drive the fluctuations of ARGs: Case study of Landfills in Nanjing, Eastern China. Journal of Hazard Materials. 2016;**315**:93-101

[51] World Bank. Drug-Resistant Infections: A Threat to Our Economic Future. Washington, DC: World Bank; 2017

[52] Centers for Disease Control and Prevention. Antibiotic Resistance Threats in the United States. Atlanta: CDC; 2013

[53] World Health Organization. The Evolving Threat of Antimicrobial Resistance: Options for Action. Geneva, Switzerland: World Health Organization; 2012

[54] Van Boeckel TP, Pires J, Silvester R, Zhao C, Song J, Criscuolo NG, et al. Global trends in antimicrobial resistance in animals in low-and middle-income countries. Science. 2019;**365**:eaaw1944

[55] Chatterjee A, Modarai M, Naylor NR, Boyd SE, Atun R, Barlow J, et al. Quantifying drivers of antibiotic resistance in humans: A systematic review. The Lancet Infectious Diseases. 2018;**18**:e368-e378

[56] European Centre for Disease Prevention and Control (ECDC); European Food Safety Authority (EFSA); European Medicines Agency (EMA). ECDC/EFSA/EMA second joint report on the integrated analysis of the consumption of antimicrobial agents and occurrence of antimicrobial resistance in bacteria from humans and food-producing animals: Joint Interagency Antimicrobial Consumption and Resistance Analysis (JIACRA) Report. EFSA Journal. 2017;**15**:e04872

[57] Delepierre A, Gayot A, Carpentier A. Update on counterfeit antibiotics worldwide; public health risks. Médecine et Maladies Infectieuses. 2012;**42**:247-255

[58] Okeke IN, Lamikanra A, Edelman R. Socioeconomic and behavioral factors leading to acquired bacterial resistance to antibiotics in developing countries. Emerging Infectious Diseases. 1999;**5**:18

[59] Byarugaba DK. Antimicrobial resistance in developing countries and responsible risk factors. International Journal of Antimicrobial Agents. 2004;**24**:105-110

[60] Guinovart MC, Figueras A, Llor C. Selling antimicrobials without prescription_ far beyond an administrative problem. Enfermedades Infecciosas Y Microbiologia Clinica (Engl. Ed.). 2018;**36**:290-292

[61] Auta A, Hadi MA, Oga E, Adewuyi EO, Abdu-Aguye SN, Adeloye D,

et al. Global access to antibiotics without prescription in community pharmacies: A systematic review and meta-analysis. The Journal of Infection. 2019;**78**:8-18

[62] Kwena Z, Sharma A, Wamae N, Muga C, Bukusi E. Provider characteristics among staff providing care to sexually transmitted infection self-medicating patients in retail pharmacies in Kibera slum, Nairobi, Kenya. Sexually Transmitted Diseases. 2008;**35**:480-483

[63] Fleming-Dutra KE, Hersh AL, Shapiro DJ, Bartoces M, Enns EA, File TM, et al. Prevalence of inappropriate antibiotic prescriptions among US ambulatory care visits, 2010-2011. JAMA. 2016;**315**:1864-1873

[64] Brink AJ, Messina AP, Feldman C, Richards GA, Becker PJ, Go DA, et al. Antimicrobial stewardship across 47 South African hospitals: An implementation study. The Lancet Infectious Diseases. 2016;**16**:1017-1025

[65] Brink AJ, Messina AP, Feldman C, Richards GA, van den Bergh D. Netcare Antimicrobial Stewardship Study Alliance. From guidelines to practice: A pharmacist-driven prospective audit and feedback improvement model for peri-operative antibiotic prophylaxis in 34 South African hospitals. The Journal of Antimicrobial Chemotherapy. 2017;**72**:1227-1234

[66] Elements of Hosp AMS- Centres for Disease Control and Prevention. Core Elements of Hospital Antibiotic Stewardship Programs; US Department of Health and Human Services. Atlanta, GA, USA: CDC; 2014

[67] World Health Statistics. Monitoring Health for Sustainable Development. World Health Oganisation. Geneva, Switzerland. 2021

[68] Joshua IA, Bauche J, Abdulla S. Managing Antimicrobial Resistance from Medical and Veterinary Health Systems Perspectives to Achieving Universal Health Coverage in the African Region (Review Article). SEEJPH 2021, posted: 18 May 2021. DOI: 10.11576/seejph-4446

[69] Laxminarayan R, Chaudhury RR. Antibiotic resistance in India: Drivers and opportunities for action. PLoS Medicine. 2016;**13**:e1001974

[70] Aidara-Kane A, Angulo FJ, Conly JM, Minato Y, Silbergeld EK, McEwen SA, et al. World Health Organization (WHO) guidelines on use of medically important antimicrobials in food-producing animals. Antimicrobial Resistance and Infection Control. 2018;7:7

[71] World Health Organization. Ten Threats to Global Health in 2019. Geneva, Switzerland: World Health Organization; 2019

[72] Coulter S, Merollini K, Roberts JA, Graves N, Halton K. The need for cost-effectiveness analyses of antimicrobial stewardship programmes: A structured review. International Journal of Antimicrobial Agents. 2015;**46**:140-149

Chapter 5

Molecular Tools for the Study of Resistance to Disinfectants

Samantha Mc Carlie, Gunther Staats, Bernadette Belter, Boudine Van Der Walt and Robert Bragg

Abstract

Disinfectants, antiseptics, and sanitizers are crucial for hygiene standards and disease control, as recently emphasized by the SARS-CoV-2 (COVID-19) pandemic. With the foreshadowing of antibiotic resistance, new cutting-edge technologies and innovative methodology need to be applied to prevent the latest emerging antimicrobial resistance crisis, resistance to disinfectants. Disinfectant resistance is a relatively novel field of study, and although some molecular mechanisms have been elucidated, little is known about complex mechanisms, cross-resistance with antibiotics, and the existence of resistance biomarkers. Fortunately, great advances have been made in the field of sequencing technology and bioinformatics. Although there are many limitations to this technology, various "omics" approaches to disinfectant resistance will be crucial in directing environment-specific disinfection programs. In addition, the vast amounts of data generated by sequencing technologies can be applied by artificial intelligence (AI) models to identify key disinfectant resistance markers and in the surveillance of disinfectant resistance genes. A combination of these approaches will be crucial in identifying new disinfectant resistance mechanisms, in monitoring resistant populations, and in identifying cellular targets for new disinfectant formulations. These molecular tools will be vital in the battle against disinfectant resistance, the latest development in the antimicrobial resistance crisis.

Keywords: biomarkers, antimicrobial resistance, biocide resistance, omics, artificial intelligence

1. Introduction

The SARS-CoV-2 (COVID-19) pandemic has highlighted our reliance on disinfectants, antiseptics, and sanitizers. These products are used extensively in the agricultural, food, and beverage industries, as well as in veterinary and medical environments. Disinfectants play a crucial role in biosecurity as a preventative measure in hygiene and disease control. A great deal of research has been carried out on antimicrobial resistance (AMR) and antibiotic resistance in particular, however little is known about resistance to disinfectants. Disinfectant resistance is a relatively novel field of study; however, this phenomenon is emerging at a troubling rate.

IntechOpen

Quaternary ammonium compounds (QACs) are one of the most widely-used disinfectants globally and are the best studied in terms of disinfectant resistance. In general, the underlying basis of resistance is a decrease in the intracellular concentration of the disinfectant within the microbial cell [1]. Resistance mechanisms may include changes in cell membrane structures, biofilm formation, efflux pumps, enzymatic activity and metabolism, and degradation of these compounds [1, 2]. These properties may be selected for and proliferate under stress conditions (such as QAC exposure), and some may be transferred on mobile genetic elements to other organisms (also to/from other species) [3]. Nonspecific resistance mechanisms, such as multidrug efflux pumps, can result in cross-resistance to several antimicrobials, including resistance to antibiotics, disinfectants, and antiseptics simultaneously [3, 4].

Although advances have been made in the study of disinfectant resistance, the literature is vastly less than that on antibiotic resistance. One main difference between antibiotics and disinfectants is that antibiotics often have one or two cellular targets, whereas disinfectants have multiple cellular targets to bring about the microbicidal effect [1, 5]. As a result, the antimicrobial effect of disinfectants is much more complex and so are the resistance mechanisms. Therefore, it may become crucial to use various "omics" methods in the study of disinfectant resistance that can reveal the full extent of what is happening inside a cell. This includes whole-genome sequencing, metagenomics, transcriptomics, proteome analysis, and metabolome research. This technology has become crucial in the study of disinfectant resistance and may be applied to discover new antimicrobial compounds.

In addition, to sequencing technology, recent advances in artificial intelligence (AI) have resulted in models that can trace and predict antimicrobial resistance patterns [6]. AI models together with disinfectant resistance biomarkers will be integral in the tracking of resistant populations and directing disinfection programs. Although extensive work is still required to develop these techniques and the database they rely on, these pose a promising alternative to studying the rapidly emerging disinfectant resistance crisis.

2. Mechanisms of disinfectant resistance

The biocidal activity of QACs has been attributed to their cationic charge interacting with the anionic charge cell wall of microorganisms and diffuse binding to the cytoplasmic membrane resulting in the formation of an electrostatic bond [5, 7, 8]. QACs cause damage by disrupting the cell membrane, distorting the permeability of the cell wall, loss of osmoregulation, disrupting the flow of nutrients into the cell, leakage of intracellular molecules, protein denaturation, and degradation of nucleic acids; ultimately resulting in cell death [8, 9].

While QACs and their active concentrations vary based on target organisms, they are typically utilized in concentrations below 1000 ppm. Nevertheless, the inappropriate use of disinfectants exposing microbes to sub-lethal concentrations can facilitate tolerance, reduced susceptibility, and resistance to these compounds [10, 11]. Furthermore, the release of QACs in run-off into the environment from food, healthcare, and animal production industries further poses a risk by exposing potential pathogens to sub-inhibitory concentrations of QACs. Bacterial resistance to biocides may develop by several mechanisms.

2.1 Cell membrane/wall alterations

While many studies have examined the development and spread of antibiotic resistance, the rise of disinfectant resistance further threatens biosecurity. While it has been shown that QACs act primarily by disrupting the cellular membranes, some microorganisms have intrinsic resistance provided by their phenotypic and physiological characteristics that challenge the penetration of the QACs [1, 2]. The phenotypic traits that facilitate inherent resistance to QACs often involve sophisticated membrane lipid permeability barriers, reducing the penetration of these compounds. The unique outer membrane, rich in lipopolysaccharides (LPS), phospholipids, and lipoproteins, that surrounds the cellular membrane of Gram-negative bacteria, makes them less susceptible than Gram-positive bacteria [10]. In addition, slime layers and cell walls rich in complex lipid molecules may confer tolerance to QACs based on physiological traits [1].

Apart from intrinsic resistance, reduced susceptibility to QACs can be induced over time by exposure to sub-inhibitory concentrations [12]. This change can be in the form of acquired resistance through the reduction of membrane permeability by changes in the fatty acid and phospholipid composition, and LPS [13, 14]. These alterations result in the cellular membrane becoming more negatively charged and hydrophobic, limiting the diffusion of QACs into the cell via the membrane [2]. Another mechanism to avoid QAC penetration involves density reduction, changes in the composition of porins, and protein composition of the outer membrane [15].

2.2 Biofilms

Exposure to QACs at sub-lethal concentrations enhances biofilm formation [16]. The physiological adaptation of certain bacteria to biofilms aids in their survival as these cells embed in the biofilm polysaccharide matrix and form part of microenvironments [16]. Within the biofilm, any antimicrobial treatment is hindered due to a lack of cell penetration and lower intracellular inhibitory concentrations [10, 16].

2.3 Efflux pumps

Inherent resistance to QACs may also result from the basic activity of broad-spectrum chromosomally encoded efflux pumps [7, 17]. These are transmembrane proteins that may provide resistance to numerous antimicrobial agents, including antibiotics and QACs [17]. While their main physiological purpose includes the transport of natural substances, it has been found that the resistance nodulation division (RND) family, the major facilitator (MF) superfamily, the small multidrug-resistance (SMR) family, and the multidrug and toxic compound extrusion (MATE) family of efflux pumps can expel antimicrobials from cells [7, 17, 18].

QAC resistance mediated through the action of efflux pumps has received considerable attention due to its genetic origin, ability to confer co-resistance to both antibiotics and other antimicrobials, and the ability to be transferred across microbial species via horizontal gene transfer (HGT) [17, 18]. QAC resistance may be induced by the overexpression of these pumps following exposure to QACs [12]. QAC resistance genes for efflux pumps can also be acquired, such as *qacE*, *qacF*, *qacG*, *qacH*, *qacI*, *qacJ*, and *qacZ*, which form part of the SMR efflux family [2, 18]. These

QAC resistance genes have mainly been found on mobile genetic elements, including transposons, plasmids, integrons, and integrative and conjugative elements (ICEs) allowing for HGT [3, 19].

2.4 Degradation and metabolism of QACs

Various studies have suggested an alternative fate of QACs that includes, degradation and metabolism [20, 21]. Some microorganisms have demonstrated the ability to degrade QACs under aerobic conditions, as a result of exposure to a range of sub-inhibitory concentrations [22]. QAC degradation has been found in various microbial communities, where microbes have been able to utilize QACs, (benzalkonium chloride (BC) and dodecyl dimethylammonium chloride (DDAC)) as their sole carbon and energy source [20, 22, 23].

2.5 Mobile genetic elements

Apart from intrinsic resistance mechanisms, the acquisition of mobile genetic elements (MGEs) is an important method for the attainment of resistance genes. Multidrug-resistant microorganisms develop via the acquisition of resistance determinants that exist in the global microbial gene pool [24]. The selective pressure of QACs may enhance the transfer of MGEs, such as ICEs, plasmids, insertion sequences, integrons, and transposons. Furthermore, the movement of these elements facilitates HGT and leads to the rise of antimicrobial resistance as a result of the acquisition and spread of resistance genes [25].

The exposure of bacteria to disinfectants may result in nonoptimal gene expression, which potentially reduces susceptibility through altered gene expression and mutations [26]. Hence, any microbe can possess resistance genes and while these may not always be expressed, they may also be constitutively expressed or induced, resulting in optimal expression in response to environmental changes.

2.6 Linking antibiotic and disinfectant resistance

The mobilization of resistance genes through intracellular mechanisms allows multiple resistance genes to cluster together, forming a single genetic unit. This means that bacteria may acquire resistance to multiple compounds (including disinfectants and antibiotics) simultaneously through one conjugation event [24, 27, 28]. This type of co-resistance to disinfectants and antibiotics has already been observed in multiple bacterial groups and co-resistance plasmids with similar structural arrangements have been isolated from various groups of unrelated bacteria [27, 29, 30].

Multidrug efflux pumps play important roles in linking antibiotic and disinfectant resistance, as they are effective against quaternary ammonium compounds along with various antibiotics. Additionally, disinfectant resistance genes (*qac*) are often found alongside antibiotic resistance genes on plasmids, further increasing the significance and complexity of resistance [30–32].

3. Innovative and cutting-edge solutions to counter disinfectant resistance

3.1 Biomarkers: a new way to track disinfectant resistance

The infectious disease poses a great threat to global healthcare systems and is estimated to cause annual mortality of over 17 million worldwide according to the

World Health Organization [33]. This estimate can be expected to rise due to AMR becoming more prevalent. Antimicrobial resistance arises from the overuse and abuse of antibiotics and other antimicrobial chemicals [2, 34]. The ability to accurately diagnose infections will guide clinicians to select the correct treatment strategy while maintaining sensible use of antimicrobials to prevent further proliferation of AMR. Moreover, scheduled surveying of the microbial population and AMR status of the population within a given setting can provide vital information to attempt to control resistance development.

To appropriately respond to AMR either in a clinical sense or as a public health threat, rapid detection of multidrug-resistant (MDR) isolates will be required. Conventional susceptibility testing, such as agar dilution, disk diffusion, gradient diffusion, or broth macrodilution techniques, has the limitation of only providing results 24 h after bacterial isolation [35]. To overcome this limitation the combination of biological marker indicators, cost-effective and time-efficient techniques could provide the solution.

Biological markers or biomarkers are indicators of a particular disease state or other physiological states of an organism. These indicators can include genes, proteins, genetic variations, and/or differences in metabolic expression [36]. The presence or absence of certain biomarker indicators can provide insight into the physiological state of the cell. Recent developments in biomarker identification and research have focused on antibiotic resistance and virulence biomarkers [37]. However, some techniques and workflows allow for elucidation of novel resistance mechanisms, which could be used to expand the currently small pool of disinfectant biomarkers [38].

A functional metagenomic workflow was used to rapidly identify and validate AMR biomarkers from clinical isolates for three antibiotics, namely tobramycin, ciprofloxacin, and trimethoprim-sulfamethoxazole [38]. This technique was functionally verified for known AMR biomarkers but of substantial interest would be the applicability to elucidate potentially novel resistance mechanisms. While the focus of AMR remains on antibiotics due to the connectivity associated with patient utility and infection, disinfectant resistance is an emerging issue and needs to be approached in the same manner.

Currently, one of the key tools that can be used to combat AMR is rapid detection and diagnostic evaluation of infections. Rapid elucidation of the AMR status of infections can improve the antimicrobial treatment prescribed and affect the outcome of patients. Machine learning approaches have successfully used genomic and transcriptomic data to distinguish and provide antibiotic resistance capabilities for the bacterial pathogen *Pseudomonas aeruginosa* [39]. Genetic features, such as gene expression, gene presence/absence, and single nucleotide polymorphisms (SNPs), were analyzed to categorize the isolates as susceptible or resistant. Interestingly, the resistance predictions were heavily dependent on the specific antimicrobial investigated. However, contrasting prediction criteria are present in other bacterial families, such as the Enterobacteriaceae, where the presence of resistance-conferring genes is sufficient for susceptibility predictions [39].

Another rapid technique for the detection of MDR isolates that have been implemented in clinical microbiology laboratories is matrix-assisted laser desorption ionization-time of flight (MALDI-TOF) mass spectrometry (MS). This technique is used to either replace or assist conventional phenotypic identification for many bacterial isolates [40]. MADLI-TOF MS can rapidly provide the simultaneous detection of multiple resistance characteristic peaks in the spectra generated when the identification of a specific strain occurs. Although the detection of the peaks responsible for resistance is feasible, the identification of the protein responsible for the peak is lacking [41]. Even though the proteins responsible for resistance cannot be identified, the presence of resistance can be identified, this could aid in the potential of MADLI-TOF MS to detect resistance presence or absence as potential biomarker candidates.

In the case of the detection of disinfectant resistance biomarkers, the need for urgency and rapid results is not as pressing as with antibiotic resistance markers present in hospitalized patients. The concept of tracking disinfectant resistance using biomarkers does not require as much speed, because rapid treatment is not of utmost importance. However, being able to elucidate what disinfectant resistance is present is necessary to maintain sanitary surfaces within hospitals, food production, or animal husbandry facilities. Once the disinfectant resistance biomarkers have been detected, follow-up procedures, such as susceptibility testing, can be conducted to appropriately correlate the MADLI-TOF MS peaks with distinctive minimum inhibitory concentration results [41].

3.2 Expanded surveillance of disinfectant resistance

With the growing threat of antimicrobial resistance and emerging pathogens, the need for effective surveillance and tracking of the spread of antimicrobial resistance and possible determinants is a new avenue for possible disease prevention [42]. Resistant determinants include both resistance genes and mutations that provide microbes with the ability to resist the effects, typically biocidal, of antimicrobials or other drugs. As with antibiotic resistance, disinfectant resistance may also be intrinsic or acquired via HGT on plasmids or other mobile genetic elements [25, 42].

Continued misuse and overuse of current essential antibiotics have resulted in less effective therapeutic options and a push into a post-antibiotic era [34, 43, 44]. While much research is underway in developing novel antimicrobials, there is also a need to establish effective strategies and preventative measures to reduce antimicrobial resistance. As disinfection is the main form of biosecurity in many human and animal environments, this has quickly become a concern as these treatments are becoming less reliable [45].

Some advances in molecular typing methods and the availability of genome information for various microorganisms provide tools and insights for further understanding molecular epidemiology, genetic content, and the spread of antimicrobial resistance [19, 46]. Hegstad and coworkers [19] examined mobile genetic elements and their contribution to the resistant Enterococcus species and described specific resistance and virulence determinants. They saw certain genes, transposons, and plasmids may confer certain phenotypic characteristics such as resistance to specific antibiotics. In addition, the development of simplistic typing methods allowed for an assessment of the relative contributions of microbes, such as enterococci, to the spread of defined resistance phenotypes [19]. These typing methods also allowed elucidation of the potential risk for the transfer of conjugative elements to other bacteria genera. In some cases, R-plasmids transferred antibiotic resistance without the selection pressure present. It was also suggested that genetically manipulated CRISPR interference mechanisms may be useful in limiting the spread of antibiotic-resistant enterococci [19].

Investigation of potential pathogens requires the knowledge of pre-established resistance determinants such as antibiotic resistance genes (β-lactam) or disinfectant resistance genes (*qacH*, *qacA*, *qacB*, and *qacC*) [32]. With the use of these known resistance determinants, microbes can be isolated from areas such as those observed in healthcare, food, and animal production industries, and genomic screening of the isolates can be done to elucidate their susceptibility to antibiotics or disinfectants. Such screening was done by Zmantar and coworkers [47] through the examination of *Staphylococcus aureus* associated with dental infections to monitor the epidemiology and spread of the multi-drug resistant staphylococci. In this study, *S. aureus* strains were isolated from bloodstream infections and identified using specific primers, and minimal inhibitory concentrations of BAC and antibiotics were determined. Approximately 50% of the isolated strains were resistant to BAC and harbored efflux-mediated resistance genes [47]. The *qacA/B* and other *qac* genes were found following molecular analysis, these are typically found on pSK1 plasmids and could be why such a high proportion of this population exhibited resistance to BAC [48, 49].

The tracking of resistance determinants provides surveillance of potential pathogens and resistance genes, however, the prevention of the spread of these determinants is a major concern as the control of microbes in many environments is difficult [50]. It may be possible to enforce proper control and sterilization or disinfection procedures in controlled environments such as healthcare and food production industries. Through the screening of nosocomial pathogens in hospitals the possible resistance profile of these pathogens may be elucidated, and help to tailor individual antimicrobial strategies, including changes in terms of QAC concentrations or the use of alternative disinfectants [50].

4. "Omics" methods to study disinfectant resistance

The omics revolution stems from recent significant advances in sequencing technology and bioinformatics. From the beginning of the central dogma, whole-genome sequencing can identify which genes are present in a cell or population. Thereafter, transcriptomics can be used to study how these genes are transcribed and under what conditions. Furthermore, proteomics can be used to reveal which proteins have matured, if any post-translational modifications have occurred, and how the protein profile of a cell can change due to environmental factors. Lastly, metabolomics has the power to elucidate the impact of a changing proteome on metabolic pathways, and which metabolic pathways are key in a microbial response to antimicrobials.

Pan-genome analysis and metagenomics are powerful tools that can reveal any resistance determinants present in the genome of a microorganism. Together with bioinformatics, this approach can be applied to determine what is present in a resistant strain that is absent in a related susceptible strain and therefore infers the cause of the resistance phenotype. By comparison of susceptible against resistant strains, complex resistance mechanisms can be elucidated, and the additive effect of hundreds of genes throughout the genome can be revealed. Bland and coworkers [51] used whole-genome sequencing to gain insight into sanitizer tolerance amongst *Listeria monocytogenes* isolates exhibiting different resistance profiles. The genomic relatedness of the isolates was analyzed, and the origin and dissemination of these populations were tracked throughout the facility where they were isolates [51]. In addition, genetic elements were found associated with decreased susceptibility to QAC-based disinfectants, this included bcrABC efflux cassette and four *Listeria* pathogenicity islands (LIPI-4, [51]). This application of whole-genome sequencing and data mining shows that this methodology can be used to reveal known disinfectant resistance determinants and be applied in the surveillance and tracking of persistent resistance populations. Metagenomics in this approach can reveal the impact of a bacterial community and population dynamics that lead to the development of resistant populations [46].

An additional advantage of whole-genome sequencing is an analysis of full genome sequences and the comparison of the role of the core genome and accessory genome in the resistance phenotype. A great deal can be revealed by focusing on acquired resistance determinants; however, recently it has been discovered that the core genome can play a much greater role than initially anticipated [52]. Gallagher and coworkers [52] used whole-genome and transposon sequencing to reveal that elements of the core genome of *A. baumanii* play an instrumental role in the extreme antibiotic resistance phenotype and not the accessory elements. Research into resistance to disinfectants has focused mainly on accessory genome elements that are acquired through horizontal gene transfer. This has led to the discovery of efflux pumps (*smr* and *qac* genes) harbored on mobile genetic elements and numerous plasmids responsible for resistance to disinfectants [53–55]. However, the core genome may present an equally important untapped refuge for disinfectant resistance determinants. Future work should include a combination of these strategies, the comparison of susceptible and resistant strains together with metagenomics and pan-genome sequencing to elucidate novel molecular mechanisms of resistance in both the accessory and core genomes of bacteria eliciting resistance to disinfectants and antiseptics.

Whole-genome sequencing has played an integral role in identifying genetic elements responsible for decreased susceptibility to antimicrobials. However, simply because a resistance gene is present does not mean it plays a role in the resistance phenotype. Gene activation and regulation of expression can be crucial in resistance to an antimicrobial [12, 56]. Transcriptomics has the power to link the genotype of resistance to the resultant phenotype exhibited. This can be particularly useful when genotypic data needs to be linked to a phenotype, or when a resistance phenotype is exhibited but no obvious antimicrobial resistance genes can be found.

A few methods to analyze transcriptomics are by using Microarray, Real-Time PCR, and RNA-Sequencing technology. Microarray and real-time PCR can show differential expression in response to antimicrobial treatment. However, certain genes need to be targeted, and therefore the mechanisms and gene sequence needs to be known and characterized. Whereas, RNA-Sequencing requires no prior knowledge, and allows for a full view of the core and accessory genomes. In this way, novel mechanisms of resistance can be found more easily, and differential expression can be seen genome-wide.

The transcriptomic analysis allows for the numeration of differentially expressed genes and quantitative data on what fold change is the down or upregulation of expression. This allows for a generation of a complete differential expression profile for thousands of genes for an isolate when exposed to certain environmental conditions. This means that a "screenshot" of gene expression can be taken at a certain time under specific conditions. This gene expression profile can then be compared at the same time under different conditions or under the same conditions but over time. As an application of this information, entire networks and metabolic pathways of hundreds of genes can be characterized. In addition, the impact of complex interconnected pathways can be mapped, and regulatory circuits can be revealed. This is an important application when a resistant isolate does not harbor any attributable resistance genes, the resistance phenotype could be elicited by overexpression of non-resistant "housekeeping" genes or those yet to be annotated as resistance genes [57].

Kim and coworkers [4, 12] combined a genomic and transcriptomic approach to study bacterial resistance to the disinfectant benzalkonium chloride (BC). In a population exposed to BC long-term, RNA-Sequencing revealed an upregulation in efflux pump genes, down-regulation of porins, and a reduced growth rate [12]. In addition, mutations in the *pmrB* genes and upregulation of spermidine synthase genes affected the charge on the cell membrane of bacteria in the resistant population, resulting in a hindrance of BC uptake [12]. These mutations and differential expression, identified by a combination of genomic and transcriptomic methods, work synergistically to reduce the intracellular concentration of BC. The use of both methods revealed a complex multifaceted approach used by bacterial populations to reduce susceptibility to disinfectants. Due to this discovery, the mutation in the *pmrB* gene could be an attractive biomarker for BC resistance moving forward.

Although transcriptomics can give insight and detail into gene function and regulation, a great deal can change after expression. Resistance profiles can be affected during the protein synthesis stage in the form of post-translational modifications, protein maturation, and combinatory effects. Proteomics can also reveal intercellular changes due to antimicrobial treatment. A protein profile includes any extracellular proteins and those that are secreted to form part of the intercellular microenvironment. Proteomics can be studied by Liquid chromatography-mass spectrometry (LC-MS) and two-dimensional difference in gel electrophoresis (DIGE) and generates an overall protein profile of the cell [46]. This includes both qualitative and quantitative data so not only can we see which proteins are present but in what number. Similarly, to transcriptomics, this protein profile can be generated to reveal protein changes in different conditions or can be monitored over time. Proteomics is best used in combination with real-time PCR to determine if regulation of protein expression occurs at the level of transcription, translation, or post-translational modification [46]. This information can tell us how the proteome changes in response to different antimicrobials and what cellular structures are affected. Zhang and coworkers [58] used proteomic analysis to gain insight into the effect of environmental conditions on an increased tolerance of *P. aeruginosa* to monochloramine disinfection in drinking water. The proteomic profile was compared under different environmental conditions and revealed that stress conditions (starvation and low temperatures) significantly aided in tolerance to monochloramine disinfection [58]. This decrease in susceptibility occurred by triggering oxidative stress defense, dormancy, osmotic stress response, and the stringent response; these responses have been shown to play a vital role in reduced susceptibility to disinfectants [58]. Therefore, proteomic signatures can be generated including protein networks for different antimicrobials, and thus provide insight into how resistance to these antimicrobials develops. Additionally, protein profiles can be used to identify antimicrobial targets and therefore in the design of new antimicrobials [46].

Finally, the metabolome is a new area of study in terms of antimicrobial resistance. Metabolomics reveals a profile of metabolites in a system at a specific time under certain conditions. As with proteomics, this is not limited to the cellular metabolome but includes the intercellular metabolome as well (secreted metabolites). The metabolome can be studied through LC-MS and Nuclear Magnetic Resonance (NMR) [46]. Metabolomics is integral in identifying which metabolic pathways are affected by which antimicrobials, to create a metabolic profile for certain antimicrobial treatments. Metabolomic profiles can be created for individual antimicrobials or antimicrobials used in combination. This will provide insight into the mode of action for antimicrobials, any combinatory effects and reveal how resistance mechanisms develop. In turn, metabolomics can also create a metabolic profile of resistant organisms under different stress conditions over time and identify important metabolic pathways in resistance. Lin and coworkers [59] used comparative metabolomics on susceptible and MDR *Escherichia coli* strains to identify and characterize 273 differing metabolites between the susceptible and resistant strains. Bioinformatics analyses revealed that the resistant strains all had enriched biosynthesis of amino acids, biosynthesis of phenylpropanoids, and purine metabolism while the susceptible strains did not [59]. This study represents the first step in the prediction and characterization of metabolic pathways crucial in multidrug-resistance profiles in bacteria.

The "omics" methods mentioned generate immense amounts of data. In particular, transcriptomics, proteomics, and metabolomics provide a screenshot of what is happening within a cell at a certain time under certain conditions. These methods can also be used to study how the response of an organism changes over time to a certain antimicrobial, or how the response of the microbe changes between different antimicrobials. However, these methods are all limited by what can be done with the data generated.

5. Artificial intelligence for the prediction of antimicrobial resistance profiles and directing of antimicrobial treatment

Artificial intelligence (AI) has been used to predict antimicrobial resistance profiles based on sequence data. This is done using several models including random forests, naïve Bayes, decision trees, artificial neural networks, and support vector machines [6]. Whole-genome sequencing is a technology that is readily available and becoming more cost-effective year on year, in turn, a massive amount of genomic data now exists. This genomic information has been stored on various databases (NCBI, GenBank, etc.) and is widely available. Artificial intelligence models use this data to identify multiple biomarkers of resistance, in turn, these biomarkers allow for the generation of a predicted resistance profile. This is done by searching the genome for the presence of resistance determinants and labeling them as biomarkers for associated predicted resistance phenotypes.

The Naïve Bayes method has been used to identify resistance determinants and build a resistance profile of biomarkers [60]. In addition, this method has been used in to determine the probability of effective antibiotic treatment when not targeted to a specific pathogen [61]. The support vector machines model has been used to label susceptible or resistant isolates, when applied to *E. coli* for antibiotic resistance it correctly predicted the susceptibility profile with a 95% accuracy [62]. This model can be applied to antimicrobial resistance surveillance and tracking as well as directing treatment for clinical pathogens [6]. Random forest is an algorithm that can direct antibiotic combinations to find synergistic properties and lower total dosage given over time to patients [6]. The decision tree model has been used to estimate the impact of antimicrobial resistance and strategize a proportional response for the allocation of medical resources [63, 64]. This model has also been applied to direct antibiotic use to shorten treatment time [65, 66]. Finally, artificial neural network models have been used to identify new antimicrobial compounds and methods to modify existing antimicrobials to increase effectiveness [6, 67].

The resultant resistance profile generated can determine which genes, mutations, and resistance mechanisms exist and therefore which antimicrobials will be least likely to be effective [61, 62]. In turn, this resistance profile can be used to determine which resistance biomarkers are not present and therefore which antimicrobials the isolate is most likely susceptible [62]. This generates a list of potential antimicrobials that will be effective against a particular pathogen based on its genomic characteristics and biomarkers present in the genome [60].

This technology can be used for an individual isolate, an infection caused by multiple microbes (as part of a biofilm) or an environmental population, as the total DNA can be extracted, and pan-genome metagenomics is used to screen multiple genomes for biomarkers [62]. This may be of importance when analyzing population dynamics in environments like hospitals where multiple MDR isolates could be harbored together, each exhibiting a different resistance profile. These techniques can be used to determine which antimicrobial will be effective against all microbes present in the population to ensure that one or two do not persist and give rise to a new resistant population.

The resistance profile for an isolate can be used to direct individual targeted antimicrobial treatment. Currently, this method is still in development and being directed mainly to antibiotic treatment of persistent infections [62]. However, this technology has a wide range of applications including chemical treatment of microorganisms by disinfection. Disinfectant resistance is emerging at an alarming rate and some molecular mechanisms of resistance to disinfectants have been discovered [3]. If the molecular resistance mechanisms are known, these resistance genes or mutations that give rise to the resistance phenotype can be flagged as biomarkers of disinfectant resistance. A resistance profile can be generated to determine which active ingredient is most likely to be effective, based on a lack of resistance biomarkers for that specific compound. From there, chemical treatment by certain disinfectants can be recommended based on their active ingredient. Molecular mechanisms of resistance to quaternary ammonium compound (QAC) disinfectants are well characterized. The *qac* gene family plays a direct role in resistance to QAC-based disinfectants as well as smr efflux pump genes [3]. These genes can be flagged as potential resistance markers for QAC-based disinfectants.

This technology can also be applied to direct day-to-day antibiotic treatment by predicting which drug combinations will give the best treatment against a certain pathogen [6]. AI models have been used to elucidate which drug combinations could work synergistically to amplify the antibiotic effect, minimize patient side effects, and prevent the development of antibiotic resistance [6, 65, 66]. This methodology can be applied to antimicrobial chemical treatments (such as disinfectants), which can be applied in combination to create a synergistic effect and prevent the development of resistance to disinfectants. For example, the presence of *smr* efflux pump genes in a microbial population can bring about resistance to a variety of antimicrobial compounds [53]. However, this effect can be negated by the addition of efflux pump inhibitors [68, 69]. This information can be used to direct the treatment of a resistant population by adding efflux pump inhibitors to be used in combination synergistically with disinfectants.

6. Measures to counter resistant organisms once identified

Alternative control methods have been suggested for many years to overcome MDR bacteria, including bacteriocins, essential oils, bacteriophage therapies, nanotherapeutics, antibodies, and more recently quorum sensing inhibitors [44, 70]. Many of these methods currently suffer from their inability to be stand-alone replacements for antibiotics due to their infancy. Additionally, these alternative methods will most likely be classified as supplementary options in addition to traditional antibiotic or antimicrobial treatments. Suggesting that combination therapy could be the best way forward to combat the global threat of MDR.

Combination therapy involves the coadministration of antibiotics/antimicrobial agents with other chemicals that lack antimicrobial properties or different antibiotics/antimicrobials with differing modes of action [71, 72]. The advantage of combination therapy is that further resistance development can be hindered but the susceptibility of MDR bacteria to treatment can also be restored [73]. The success of combination therapy is shadowed by theoretical predictions that can cause unexpected administration results. The interaction between the different chemicals administered with one another and/or the environment could be antagonistic rather than synergistic, resulting in failure of combinational treatment [74]. Emphasizing that solutions to MDR bacteria are not simple but rather require a great degree of complexity to ensure that the trend of resistance spread does not continue.

There are a plethora of resistance mechanisms present in bacteria, some of the most prevalent include low outer membrane permeability, production of degradation enzymes, efflux pumps, and target modification [75]. Efflux pumps possibly provide the most versatile mechanism to both provide resistance to a broad range of antimicrobial compounds while simultaneously providing additional characteristics, such as increased virulence [76]. Multidrug transporters are efflux pumps with the capability to recognize a wide variety of dissimilar substrates and these types of transporters are often key in MDR bacteria [76].

Since the discovery of MDR caused by efflux pumps the development of efflux pump inhibitors (EPIs) as a strategy to combat this resistance has been considered. The combination of antibiotics/antimicrobials and EPIs could allow the return of certain antibiotics/antimicrobial chemicals that have lost functionality in clinical practice. Additionally, the spectrum of usable compounds could be broadened by the addition of EPIs to allow antimicrobials to be able to adequately target Gram-positive and more naturally resistant Gram-negative bacteria [77]. The inhibition of efflux pumps can have dual purposes, of increasing the susceptibility to some of the antimicrobial substrates which would normally be resisted and potentially some attenuation of the virulence that is connected to efflux pump expression.

Many EPIs have been elucidated, some having specific inhibitory activity against select efflux pumps and others having broader inhibitory effects [77]. Some examples of EPIs include verapamil, reserpine, phenylalanine-arginine beta-naphthylamide (PAβN), 1-(1-naphthylmethyl)-piperazine (NMP), and carbonyl cyanide m-chlorophenyl hydrazone (CCCP). The inhibition of efflux pumps that have substrates of high clinical significance appears to be a very attractive approach. However, this approach will only prove productive for bacterial populations with prevalent efflux-mediated resistance. EPI use has been shown to reduce the frequency of emergence of bacterial strains with clinically relevant levels of resistance to certain antibacterial chemicals [78]. However, the introduction of EPIs as a viable form of combination therapy suffers from several downfalls, which prevent its immediate introduction into clinical settings. When considering EPIs that function to inhibit efflux by specific interaction with the efflux protein via competitive inhibition. Some substrates of the same efflux pump might have different binding sites that are not inhibited by the competitive binding inhibition of the EPI. Highlighting the complexity of interactions between antibacterial compounds and efflux pumps, and how unique combinations of antimicrobials and EPIs will have to be identified to provide desired inhibition effects [78, 79].

EPIs with broad inhibitory capabilities, such as CCCP, inhibit efflux by targeting the energy production of the cell [80]. In both nonspecific and specific inhibition, the end result is to increase the susceptibility of bacterial isolates to antibacterial compounds by increasing intracellular accumulation. The major downfall of EPIs is related to the stringent requirements that these chemicals need to obtain to be successfully classified and used. These compounds must not have inherent antibacterial properties, which could potentiate the problem of MDR bacteria, and must be selective for bacterial efflux pumps, without interaction with eukaryotic efflux systems. Additionally, they must fulfill certain pharmacological characteristics, most notably non-toxicity, and must be economically feasible at commercial production levels [81]. To date, no EPIs have reached clinical use, as none can successfully meet all the requirements to be regarded safe for combination therapy in humans. However, the utility of EPIs could be beneficial for studying the prevalence and contribution that efflux plays in acquired and intrinsic resistance to antibacterial compounds within clinical bacterial isolates to allow better understating of efflux-mediated MDR in the clinical setting. Increased study into alternatives to antibiotics has put a spotlight on treatment options, such as EPIs, further research will be required to get EPIs into clinical use as a combinational therapy option. EPIs can also be used in combination with disinfectants and antiseptics where efflux pumps contribute to reduced susceptibility of these compounds. The synergistic use of EPIs together with antimicrobials may be a crucial step in the fight against AMR in the near future.

7. Conclusions

The antibiotic resistance crisis is a foreshadowing of an equally troubling phenomenon, resistance to disinfectants. Although we have elucidated molecular mechanisms of disinfectant resistance there is still a great deal we do not understand. Complex mechanisms of disinfectant resistance are poorly characterized, little to no research has been done on disinfectant resistance biomarkers and surveillance of disinfectant resistant populations is not a priority. Our current disinfectants need to be safeguarded and the search for new disinfectant formulas must become a priority. Advances in sequencing technologies in the form of omics, biomarkers, and AI will be key in the battle against emerging disinfectant resistance and will go far in characterizing synergistic treatment options. Disinfectants are heavily relied on for hygiene purposes and infection control in the agricultural, food and beverage industries, and so designing a disinfection program that is effective is vital. Additionally, characterizing disinfectant resistance profiles in environments such as the food industry and medical environments will be key to effective control ever changing persistent resistant populations. Advances in sequencing and antimicrobial resistance research have never been so important, as this methodology can be applied to disinfectant resistance to avoid another worldwide resistance crisis.

Acknowledgements

The authors would like to acknowledge the National Research Foundation (NRF) of South Africa for funding and support.

Conflict of interest

The authors declare no conflict of interest.

Author details

Samantha Mc Carlie, Gunther Staats, Bernadette Belter, Boudine Van Der Walt
and Robert Bragg*
University of the Free State, Bloemfontein, South Africa

*Address all correspondence to: braggrr@ufs.ac.za

References

[1] McDonnell G, Russell A. Antiseptics and disinfectants: Activity , action, and resistance. Clinical Microbiology Reviews. 1999;**12**(1):147-179

[2] Bragg R, Jansen A, Coetzee M, van der Westhuizen W. Bacterial resistance to quaternary ammonium compounds (QAC) disinfectants. Advances in Experimental Medicine and Biology. 2014;**808**(1):1-13. DOI: 10.1007/978-81-322-1774-9

[3] Mc Carlie S, Boucher C, Bragg R. Molecular basis of bacterial disinfectant resistance. Drug Resistance Updates. 2020;**48**:100672. Available from: http://www.ncbi.nlm.nih.gov/pubmed/31830738

[4] Kim M, Weigand MR, Oh S, Hatt JK, Krishnan R, Tezel U, et al. Widely used benzalkonium chloride disinfectants can promote antibiotic resistance. Applied and Environmental Microbiology. 2018;**84**(17):e01201-18. DOI: 10.1128/AEM.01201-18

[5] Yoshimatsu T, Hiyama KI. Mechanism of the action of didecyldimethylammonium chloride (DDAC) against *Escherichia coli* and morphological changes of the cells. Biocontrol Science. 2007;**12**(3):93-99. Available from: http://joi.jlc.jst.go.jp/JST.Journalarchive/bio1996/12.93?from=CrossRef

[6] Lv J, Deng S, Zhang L. A review of artificial intelligence applications for antimicrobial resistance. Biosafety and Health. 2021;**3**(1):22-31

[7] Jansen A, Boucher CE, Coetsee E, Kock J, van Wyk P, Swart H, et al. The influence of didecyldimethylammonium chloride on the morphology and elemental composition of *Staphylococcus aureus* as determined by NanoSAM. Scientific Research and Essays. 2013;**8**(3):152-160

[8] Gerba CP. Quaternary ammonium biocides: Efficacy in application. Applied and Environmental Microbiology. 2015;**81**(2):464-469. DOI: 10.1128/AEM.02633-14

[9] Denyer SP, Stewart GSAB. Mechanisms of action of disinfectants. International Biodeterioration & Biodegradation. 1998;**41**(3-4):261-268. Available from: https://linkinghub.elsevier.com/retrieve/pii/S0964830598000237

[10] Tezel U, Pavlostathis SG. Quaternary ammonium disinfectants: Microbial adaptation, degradation and ecology. Current Opinion in Biotechnology. 2015;**33**:296-304. Available from: https://linkinghub.elsevier.com/retrieve/pii/S0958166915000622

[11] Tezel U, Pavlostathis SG. Role of quaternary ammonium compounds on antimicrobial resistance in the environment. In: Keen P, Mantforts M, editors. Antimicrobial Resistance in the Environment. 1st ed. Hoboken, NJ: Wiley-Blackwell; 2012. pp. 349-387

[12] Kim M, Hatt JK, Weigand MR, Krishnan R, Pavlostathis SG, Konstantinidis KT. Genomic and transcriptomic insights into how bacteria withstand high concentrations of benzalkonium chloride biocides. Applied and Environmental Microbiology. 2018;**84**(12):1-15

[13] Boeris PS, Liffourrena AS, Salvano MA, Lucchesi GI. Physiological role of phosphatidylcholine in the *pseudomonas putida* a ATCC 12633 response

to tetradecyltrimethylammonium bromide and aluminium. Letters in Applied Microbiology. 2009;**49**(4):491-496. DOI: 10.1111/j.1472-765X.2009.02699.x

[14] Poursat BAJ, van Spanning RJM, de Voogt P, Parsons JR. Implications of microbial adaptation for the assessment of environmental persistence of chemicals. Critical Reviews in Environmental Science and Technology. 2019;**49**(23):2220-2255. DOI: 10.1080/10643389.2019.1607687

[15] Ishikawa S, Matsumura Y, Yoshizako F, Tsuchido T. Characterization of a cationic surfactant-resistant mutant isolated spontaneously from *Escherichia coli*. Journal of Applied Microbiology. 2002;**92**(2):261-268. DOI: 10.1046/j.1365-2672.2002.01526.x

[16] Pagedar A, Singh J, Batish VK. Adaptation to benzalkonium chloride and ciprofloxacin affects biofilm formation potential, efflux pump and haemolysin activity of *Escherichia coli* of dairy origin. The Journal of Dairy Research. 2012;**79**(4):383-389. Available from: https://www.cambridge.org/core/product/identifier/S0022029912000295/type/journal_article

[17] Poole K. Efflux pumps as antimicrobial resistance mechanisms. Annals of Medicine. 2007;**39**(3):162-176. DOI: 10.1080/07853890701195262

[18] Braga T, Marujo P, Pomba C, Lopes M. Involvement and dissemination of the enterococcal small multidrug resistance transporter QacZ in resistance to quaternary ammonium compounds. The Journal of Antimicrobial Chemotherapy. 2011;**66**:283-286

[19] Hegstad K, Mikalsen T, Coque TM, Werner G, Sundsfjord A. Mobile genetic elements and their contribution to the emergence of antimicrobial resistant *Enterococcus faecalis* and *Enterococcus faecium*. Clinical Microbiology and Infection. 2010;**16**(6):541-554. Available from: http://www.ncbi.nlm.nih.gov/pubmed/20569265

[20] Dean-Raymond D, Alexander M. Bacterial metabolism of quaternary ammonium compounds. Applied and Environmental Microbiology. 1977;**33**(5):1037-1041. Available from: https://aem.asm.org/content/33/5/1037

[21] van Ginkel CG, van Dijk JB, Kroon AG. Metabolism of hexadecyltrimethylammonium chloride in *pseudomonas* strain B1. Applied and Environmental Microbiology. 1992;**58**(9):3083-3087. Available from: https://aem.asm.org/content/58/9/3083

[22] van Ginkel CG. Complete degradation of xenobiotic surfactants by consortia of aerobic microorganisms. Biodegradation. 1996;**7**(2):151-164. DOI: 10.1007/s002030050390

[23] Patrauchan MA, Oriel PJ. Degradation of benzyldimethylalkylammonium chloride by *Aeromonas hydrophila* sp. K. Journal of Applied Microbiology. 2003;**94**(2):266-272. DOI: 10.1046/j.1365-2672.2003.01829.x

[24] Frost LS, Leplae R, Summers AO, Toussaint A. Mobile genetic elements: The agents of open source evolution. Nature Reviews. Microbiology. 2005;**3**(9):722-732. Available from: http://www.nature.com/articles/nrmicro1235

[25] Van Der Meer JR. Environmental pollution promotes selection of microbial degradation pathways. Frontiers in Ecology and the Environment. 2006;**4**(1):35-42

[26] Palmer AC, Chait R, Kishony R. Nonoptimal gene expression creates latent potential for antibiotic resistance.

Csaba P, editor. Molecular Biology and Evolution 2018;**34**(11):2669-2684. 10.1093/molbev/msy163/5085501

[27] Barlow M. What Antimicrobial Resistance Has Taught Us About Horizontal Gene Transfer. Berlin/Heidelberg, Germany: Springer; 2009. pp. 397-411. DOI: 10.1007/978-1-60327-853-9_23

[28] Carattoli A. Plasmids and the spread of resistance. International Journal of Medical Microbiology. 2013;**303**(6-7):298-304. Available from: https://linkinghub.elsevier.com/retrieve/pii/S1438422113000167

[29] Poirel L, Cattoir V, Nordmann P. Is plasmid-mediated quinolone resistance a clinically significant problem? Clinical Microbiology and Infection. 2008;**14**(4):295-297. Available from: https://linkinghub.elsevier.com/retrieve/pii/S1198743X14627993

[30] Russel A. Plamids and bacterial resistance to biocides. Journal of Applied Microbiology. 1997;**83**(2):155-165

[31] Langsrud S, Sundheim G, Brogman-Stahsen R. Intrinsic and acquired resistance to quaternary ammonium compounds in food-related *pseudomonas* spp. Journal of Applied Microbiology. 2003;**95**:874-882

[32] Jaglic Z, Cervinkova D. Genetic basis of resistance to quaternary ammonium compounds & the *qac* genes and their role: A review. Veterinární Medicína (Praha). 2012;**57**(6):275-281. Available from: http://www.agriculturejournals.cz/web/vetmed.htm?volume=57&firstPage=275&type=publishedArticle

[33] Bloom DE, Cadarette D. Infectious disease threats in the twenty-first century: Strengthening the global response. Frontiers in Immunology. 2019;**28**:10. DOI: 10.3389/fimmu.2019.00549/full

[34] Laxminarayan R, Duse A, Wattal C, Zaidi AKM, Wertheim HFL, Sumpradit N, et al. Antibiotic resistance—The need for global solutions. The Lancet Infectious Diseases. 2013;**13**(12):1057-1098. Available from: https://linkinghub.elsevier.com/retrieve/pii/S1473309913703189

[35] Hindler JA, Humphries RM. Colistin MIC variability by method for contemporary clinical isolates of multidrug-resistant gram-negative *bacilli*. Journal of Clinical Microbiology. 2013;**51**(6):1678-1684. DOI: 10.1128/JCM.03385-12

[36] Huss R. Chapter 19 Biomarkers. Translational Regenerative Medicine. Amsterdam, Netherlands: Elsevier Inc.; 2015. p. 235-241

[37] Russo TA, Olson R, Fang C-T, Stoesser N, Miller M, MacDonald U, et al. Identification of biomarkers for differentiation of hypervirulent *Klebsiella pneumoniae* from Classical *K. pneumoniae*. Journal of Clinical Microbiology. 2018;**56**(9):e00776-18. Available from: http://www.ncbi.nlm.nih.gov/pubmed/29925642

[38] Volz C, Ramoni J, Beisken S, Galata V, Keller A, Plum A, et al. Clinical resistome screening of 1,110 *Escherichia coli* isolates efficiently recovers diagnostically relevant antibiotic resistance biomarkers and potential novel resistance mechanisms. Frontiers in Microbiology. 2019;**13**:10. Available from: https://www.frontiersin.org/article/10.3389/fmicb.2019.01671/full

[39] Khaledi A, Weimann A, Schniederjans M, Asgari E, Kuo T, Oliver A, et al. Predicting antimicrobial resistance in *Pseudomonas aeruginosa* with machine learning-enabled molecular diagnostics. EMBO Molecular Medicine. 2020;**12**(3):e10264. DOI: 10.15252/emmm.201910264

[40] Bizzini A, Greub G. Matrix-assisted laser desorption ionization time-of-flight mass spectrometry, a revolution in clinical microbial identification. Clinical Microbiology and Infection. 2010;**16**(11):1614-1619. Available from: https://linkinghub.elsevier.com/retrieve/pii/S1198743X1460553X

[41] Burckhardt I, Zimmermann S. Susceptibility testing of bacteria using Maldi-Tof mass spectrometry. Frontiers in Microbiology. 2018;**6**:9. DOI: 10.3389/fmicb.2018.01744/full

[42] Meade E, Slattery MA, Garvey M. Biocidal resistance in clinically relevant microbial species: A major public health risk. Pathogens. 2021;**10**(5):598. Available from: https://www.mdpi.com/2076-0817/10/5/598

[43] Allen HK, Trachsel J, Looft T, Casey TA. Finding alternatives to antibiotics. Annals of the New York Academy of Sciences. 2014;**1323**(1): 91-100. DOI: 10.1111/nyas.12468

[44] Bragg RR, Meyburgh CM, Lee JY, Coetzee M. Potential treatment options in a post-antibiotic era. In: Advances in Experimental Medicine and Biology. Vol. 1052. Singapore: Springer; 2018. pp. 51-61. DOI: 10.1007/978-981-10-7572-8_5

[45] Hulme PE. One biosecurity: A unified concept to integrate human, animal, plant, and environmental health. Emerging Topics in Life Sciences. 2020;**4**(5):539-549. Available from: http://www.ncbi.nlm.nih.gov/pubmed/33111945

[46] Chernov VM, Chernova OA, Mouzykantov AA, Lopukhov LL, Aminov RI. Omics of antimicrobials and antimicrobial resistance. Expert Opinion on Drug Discovery. 2019;**14**(5):455-468. DOI: 10.1080/17460441.2019.1588880

[47] Zmantar T, Kouidhi B, Hentati H, Bakhrouf A. Detection of disinfectant and antibiotic resistance genes in *Staphylococcus aureus* isolated from the oral cavity of Tunisian children. Annales de Microbiologie. 20122;**62**(1):123-128. DOI: 10.1007/s13213-011-0236-3

[48] Littlejohn TG, DiBerardino D, Messerotti LJ, Spiers SJ, Skurray RA. Structure and evolution of a family of genes encoding antiseptic and disinfectant resistance in *Staphylococcus aureus*. Gene. 1991;**101**(1):59-66. Available from: http://linkinghub.elsevier.com/retrieve/pii/037811199190224Y

[49] Noguchi N. Antiseptic susceptibility and distribution of antiseptic-resistance genes in methicillin-resistant *Staphylococcus aureus*. FEMS Microbiology Letters. 1999;**172**(2):247-253. DOI: 10.1016/S0378-1097(99)00041-5

[50] Liu S, Wang M, Zheng L, Guan W. Antimicrobial resistance profiles of nosocomial pathogens in regional China: A brief report from two tertiary hospitals in China. Medical Science Monitor. 2018;**24**:8602-8607. Available from: https://www.medscimonit.com/abstract/index/idArt/911229

[51] Bland RN, Johnson JD, Waite-Cusic JG, Weisberg AJ, Riutta ER, Chang JH, et al. Application of whole genome sequencing to understand diversity and presence of genes associated with sanitizer tolerance in listeria monocytogenes from produce handling sources. Foods (Basel, Switzerland). 2021;**10**(10):2454. Available from: http://www.ncbi.nlm.nih.gov/pubmed/34681501

[52] Gallagher LA, Lee SA, Manoil C. Importance of Core genome functions for an extreme antibiotic resistance trait. mBio. 2017;**8**(6):e01655-17. Available from: http://www.ncbi.nlm.nih.gov/pubmed/29233894

[53] Bjorland J, Sunde M, Waage S. Plasmid-borne *smr* gene causes resistance to quaternary ammonium compounds in bovine *Staphylococcus aureus*. Journal of Clinical Microbiology. 2001:3999-4004

[54] Heir E, Sundheim G, Holck A. The *qacG* gene on plasmid pST94 confers resistance to quaternary ammonium compounds in *staphylococci* isolated from the food industry. Journal of Applied Microbiology. 1999;**86**:378-388

[55] Bjorland J, Steinum T, Sunde M, Waage S, Heir E. Novel plasmid borne gene *qacJ* mediates resistance to quaternary ammonium compounds in equine *Staphylococcus aureus, Staphylococcus simulans,* and *Staphylococcus intermedius*. Antimicrobial Agents and Chemotherapy. 2003;**47**:3046-3052

[56] Khaledi A, Schniederjans M, Pohl S, Rainer R, Bodenhofer U, Xia B, et al. Transcriptome profiling of antimicrobial resistance in *Pseudomonas aeruginosa*. Antimicrobial Agents and Chemotherapy. 2016;**60**(8):4722-4733

[57] Tansirichaiya S, Reynolds LJ, Cristarella G, Wong LC, Rosendahl K, Roberts AP. Reduced susceptibility to antiseptics is conferred by heterologous housekeeping genes. Microbial Drug Resistance. 2018;**24**(2):105-112. DOI: 10.1089/mdr.2017.0105

[58] Zhang S-P, Feng H-Z, Wang Q, Quan S-W, Yu X-Q, Tao X, et al. Proteomic analysis reveals the mechanism of different environmental stress-induced tolerance of *Pseudomonas aeruginosa* to monochloramine disinfection. Journal of Hazardous Materials. 2021;**417**:126082. Available from: https://linkinghub.elsevier.com/retrieve/pii/S0304389421010463

[59] Lin Y, Li W, Sun L, Lin Z, Jiang Y, Ling Y, et al. Comparative metabolomics shows the metabolic profiles fluctuate in multi-drug resistant *Escherichia coli* strains. Journal of Proteomics. 2019;**207**:103468. Available from: https://linkinghub.elsevier.com/retrieve/pii/S1874391919302404

[60] Rezaei-hachesu P, Samad-Soltani T, Yaghoubi S, GhaziSaeedi M, Mirnia K, Masoumi-Asl H, et al. The design and evaluation of an antimicrobial resistance surveillance system for neonatal intensive care units in Iran. International Journal of Medical Informatics. 2018;**115**:24-34. Available from: https://linkinghub.elsevier.com/retrieve/pii/S1386505618304106

[61] Choisy M, Van Cuong N, Bao TD, Kiet BT, Hien BV, Thu HV, et al. Assessing antimicrobial misuse in small-scale chicken farms in Vietnam from an observational study. BMC Veterinary Research. 2019;**15**(1):206. DOI: 10.1186/s12917-019-1947-0

[62] Her H-L, Wu Y-W. A pan-genome-based machine learning approach for predicting antimicrobial resistance activities of the *Escherichia coli* strains. Bioinformatics. 2018;**34**(13):i89-i95. Available from: https://academic.oup.com/bioinformatics/article/34/13/i89/5045729

[63] Naylor NR, Zhu N, Hulscher M, Holmes A, Ahmad R, Robotham JV. Is antimicrobial stewardship cost-effective? A narrative review of the evidence. Clinical Microbiology and Infection. 2017;**23**(11):806-811. Available from: https://linkinghub.elsevier.com/retrieve/pii/S1198743X17303300

[64] Naylor NR, Atun R, Zhu N, Kulasabanathan K, Silva S, Chatterjee A, et al. Estimating the burden of antimicrobial resistance: A systematic literature review. Antimicrobial Resistance and Infection

Control. 2018;**7**(1):58. DOI: 10.1186/s13756-018-0336-y

[65] Voermans AM, Mewes JC, Broyles MR, Steuten LMG. Cost-effectiveness analysis of a procalcitonin-guided decision algorithm for antibiotic stewardship using real-world U.S. hospital data. OMICS: A Journal of Integrative Biology. 2019;**23**(10):508-515. DOI: 10.1089/omi.2019.0113

[66] Schuetz P, Amin DN, Greenwald JL. Role of procalcitonin in managing adult patients with respiratory tract infections. Chest. 2012;**141**(4):1063-1073. Available from: https://linkinghub.elsevier.com/retrieve/pii/S0012369212602311

[67] Stokes JM, Yang K, Swanson K, Jin W, Cubillos-Ruiz A, Donghia NM, et al. A deep learning approach to antibiotic discovery. Cell. 2020;**180**(4):688-702.e13. Available from: https://linkinghub.elsevier.com/retrieve/pii/S0092867420301021

[68] Wang Y, Mowla R, Guo L, Ogunniyi AD, Rahman T, De Barros Lopes MA, et al. Evaluation of a series of 2-napthamide derivatives as inhibitors of the drug efflux pump AcrB for the reversal of antimicrobial resistance. Bioorganic & Medicinal Chemistry Letters. 2017;**27**(4):733-739. DOI: 10.1016/j.bmcl.2017.01.042

[69] Lowrence RC, Subramaniapillai SG, Ulaganathan V, Nagarajan S. Tackling drug resistance with efflux pump inhibitors: From bacteria to cancerous cells. Critical Reviews in Microbiology. 2019;**45**(3):334-353. DOI: 10.1080/1040841X.2019.1607248

[70] Zaidi S, Misba L, Khan AU. Nano-therapeutics: A revolution in infection control in post antibiotic era. Nanomedicine Nanotechnology, Biology and Medicine. 2017;**13**(7):2281-2301. Available from: https://linkinghub.elsevier.com/retrieve/pii/S1549963417301235

[71] Tamma PD, Cosgrove SE, Maragakis LL. Combination therapy for treatment of infections with gram-negative bacteria. Clinical Microbiology Reviews. 2012;**25**(3):450-470. DOI: 10.1128/CMR.05041-11

[72] Paul M, Lador A, Leibovici L. Beta lactam antibiotic monotherapy versus beta lactam- aminoglycoside antibiotic combination therapy for sepsis (review). The Cochrane Library. 2014;**1**:1-147

[73] Mouton JW. Combination therapy as a tool to prevent emergence of bacterial resistance. Infection. 1999;**27**(S2):S24-S28. DOI: 10.1007/BF02561666

[74] Brochado AR, Telzerow A, Bobonis J, Banzhaf M, Mateus A, Selkrig J, et al. Species-specific activity of antibacterial drug combinations. Nature. 2018; **559**(7713):259-263. Available from: http://www.nature.com/articles/s41586-018-0278-9

[75] Nikaido H. Multidrug Resistance in Bacteria. Annual Review of Biochemistry. 2009;**78**(1):119-146. DOI: 10.1146/annurev.biochem.78.082907.145923

[76] Alcalde-Rico M, Hernando-Amado S, Blanco P, Martínez JL. Multidrug efflux pumps at the crossroad between antibiotic resistance and bacterial virulence. Frontiers in Microbiology. 2016;**21**:7. DOI: 10.3389/fmicb.2016.01483/abstract

[77] Lomovskaya O, Watkins W. Inhibition of efflux pumps as a novel approach to combat drug resistance in bacteria. Journal of Molecular Microbiology and Biotechnology. 2001;**3**(2):225-236. Available from:

http://www.ncbi.nlm.nih.gov/pubmed/11321578

[78] Lomovskaya O, Bostian KA. Practical applications and feasibility of efflux pump inhibitors in the clinic—A vision for applied use. Biochemical Pharmacology. 2006;**71**(7):910-918. Available from: https://linkinghub.elsevier.com/retrieve/pii/S0006295205008154

[79] Dreier J, Ruggerone P. Interaction of antibacterial compounds with RND efflux pumps in *Pseudomonas aeruginosa*. Frontiers in Microbiology. 2015;**6**:1-21

[80] Mallea M, Chevalier J, Bornet C, Eyraud A, Davin-Regli A, Bollet C, et al. Porin alteration and active efflux: Two in vivo drug resistance strategies used by *Enterobacter aerogenes*. Microbiology. 1998;**144**(11):3003-3009. DOI: 10.1099/00221287-144-11-3003

[81] Bhardwaj AK, Mohanty P. Bacterial efflux pumps involved in multidrug resistance and their inhibitors: Rejuvinating the antimicrobial chemotherapy. Recent Patents on Anti-Infective Drug Discovery. 2012;7(1):73-89. Available from: http://www.eurekaselect.com/openurl/content.php?genre=article&issn=1574-891X&volume=7&issue=1&spage=73

Chapter 6

Worldwide Colistin Use and Spread of Resistant-*Enterobacteriaceae* in Animal Production

Carla Miranda, Gilberto Igrejas, Rosa Capita, Carlos Alonso-Calleja and Patrícia Poeta

Abstract

Colistin has been administrated for a long time in both human and veterinary medicine. Since the detection of the colistin resistance gene in animals, the increased concern about the impact on public health of colistin resistance has been evident, and several measures have been implemented. Some countries banned colistin use in food-producing animals, however, other countries continue the animal administration of colistin without restrictions. Consequently, colistin resistance originated on animal production can be transmitted to humans through the food chain or the contaminated environment. Nowadays, this antibiotic was considered as the last resort for the treatment of multidrug-resistant Gram-negative infections or patients with fibrosis cystic. For these reasons, this review aimed to summarize the trend of antimicrobial use in livestock and aquaculture production, as well as, colistin-resistant bacteria in these animals, and the impact of its resistance on human health and the environment. In general, consumption and colistin use in livestock production have shown to decrease worldwide. In animal production, the detection of mcr genes, is well documented, demonstrating global dissemination of colistin resistance in *Enterobacteriaceae* isolates and the emergence of novel colistin-resistant genes. Moreover, identification of these genes has also been reported in animal food, humans and the environment.

Keywords: antibiotics, antimicrobial resistance, colistin resistance, *mcr*-1, *Enterobacteriaceae*, livestock, aquaculture

1. Introduction

In veterinary medicine, colistin or polymyxin E has been administrated for a long time [1], mainly for therapeutic and prophylactic purposes in food animals. In pets, colistin is administered alone or in combination with other antibiotics for eye application and eardrops [2]. Colistin belongs to the antimicrobial class of polymyxins and currently was classified as the highest priority critically important antimicrobials for human medicine, according to WHO CIA list [3], and included in the category 2 (restrict), according to the EU antimicrobial advice ad hoc expert group (AMEG) classification. Polymyxins, as well as, fluoroquinolones and 3rd- or

4th-generation cephalosporins are considered the classes of antimicrobial agents that required the most urgent measures to tackle risks of the antimicrobial resistance (AMR) [4]. Therefore, this agent is used in both people and food animals for the treatment and prevention of *Enterobacteriaceae* infections, which include *Klebsiella* spp., *Escherichia coli*, *Acinetobacter* spp. and *Pseudomonas aeruginosa*. Colistin is highly used as the last resource in critical care settings or tackling multidrug-resistant (MDR) microorganisms, such as *Enterobacteriaceae*, in humans. Taking into account the increase of MDR *Enterobacteriaceae* prevalence in human infections, the amount of colistin administration can be considerable worldwide [1, 3]. In addition, this antimicrobial showed lower toxicity and more efficiency compared to beta-lactams and fluoroquinolones used as alternatives [5].

The detection of bacteria resistance to colistin has worried the scientific community, due to the presence of mobile-colistin-resistance (*mcr*) genes that are on plasmids and can be transferred among bacteria conferring resistance to colistin via the food chain and/or the environment through the contaminated manure, soil, air or water [1, 3, 6]. Furthermore, the scarce of alternatives against MDR Gram-negative bacteria infections can constitute an emerging threat to animal and human health. In addition, data regarding the presence of colistin resistance in livestock animals and animal products are dispersed. These factors, such as the detection of colistin-resistant genes in humans, animals and food, and the increase of colistin use to combat multi-resistant bacteria in human infections, implied international and national strategies to determine the risk and limit the colistin use in veterinary medicine [5]. For these reasons, this review summarizes to trend of antimicrobial use in food-producing animals, as well as, colistin-resistant bacteria in these animals, and the impact of its resistance on public health.

2. Colistin worldwide use in animal production

Antibiotics, including colistin, are required in pets but mainly in aquaculture and livestock to maintain animal welfare, reproductive performance or as growth-promoting, and food security, although all antimicrobial growth promoters (AGPs) for food animal production are prohibited in Sweden, Denmark and Europe since 1986, 2000 and 2006, respectively (**Table 1**). Recently, China has banned colistin use and all AGPs in animal production in 2017 and 2020, respectively [6]. Of the 146 member countries, 86 (59%) have not authorized any antimicrobial drugs for growth promotion in animals since 2016, according to the World Organization for Animal Health (OIE) annual report [21]. In 2022, the EU will implement new legislation (Regulation (EU) 2019/6), prohibiting all forms of routine and prophylactic farm antibiotic use and banning the importation of all animal food produced with AGPs. Although some countries, such as the United States, Canada, Australia and New Zealand, have banned the antibiotic use classified as medically important, continue to administer other antibiotics as growth promoters like bacitracin [19, 20].

In animal production, estimates more than 57 million kilograms of antibiotics are annually used worldwide [22]. Antibiotic consumption in food-producing animals was globally estimated to increase by 67% between 2010 and 2030 [22], and 11.5% between 2017 and 2030 [23]. In parallel, a recent study reported the gap of register and control by the antibiotics sales online, showing the ease availability of veterinary antibiotics for purchase without a prescription, the factor that can raise the risk of resistant microorganisms [24].

Year	Country	Antibiotic use	Colistin use	Reference
2011–2018	Europe	↓ 34%	All AGPs banned (2006), Colistin sales ↓ 70%	[7]
2011–2018	Norway	—	No polymyxins sales	[7]
2011–2018	Finland	—	No polymyxins sales	[7]
2011–2018	Iceland	—	No polymyxins sales	[7]
2015–2019	UK	↓ 45%	Colistin sales ↓ 99.9%	[8]
2018–2019	Netherlands	↓ 16%	Colistin sales ↑ 13%	[9]
2010–2019	Sweden	↓ 32%	All AGPs banned (1986), Polymyxins sales ↑ 73%	[10]
2015–2016	Denmark	↓ 5%	All AGPs banned (2000), Colistin sales ↑ 4%	[5]
2011–2018	France	↓ 48%	Colistin sales ↓ 76%	[11]
2018–2019	Belgium	↓ 8%	Polymyxin sales ↓ 11%	[12]
2013–2017	China	↓ 46%	Colistin and all AGPs banned	[6]
2015–2017	Japan	Trend to ↑	Colistin 0.3–1.6% (2005–2010)	[13, 14]
2015–2018	USA	↓ 38%	MIA banned (2017), no colistin products sales	[15]
2019	India	—	Colistin banned and no colistin products sales	[16]
2016–2017	Canada	↓ 12%	MIA banned, no colistin products sales	[17]
2016	Brazil	—	Colistin banned as AGPs	[18]
2008–2010	Australia	↑ 11%	Some AGPs banned, no colistin products sales	[19]
2016–2017	New Zealand	↑ 3%	—	[20]

AGPs, antimicrobial growth promoters; MIA, medically important antimicrobials.

Table 1.
The sales of antimicrobial agents and colistin for use in livestock animals.

In veterinary medicine, colistin is mainly used to treat gastrointestinal infections caused by *Escherichia coli* in ruminants (cattle, sheep and goats), pigs, rabbits and poultry, and endotoxemia caused by other Gram-negative bacteria (**Figure 1**). In pig production, is often used as medicated premixes after weaning for the treatment and prevention of colibacillosis, diarrhea and septicemia caused by *E. coli* mainly in piglets [25, 26]. Colistin can be administrated via oral through the feed or drinking water, parental and intra-mammary in the form of colistin sulfate, for metaphylactic, prophylactic, growth promotion and therapeutic purposes [2].

2.1 European countries

In European countries, the total sales for veterinary antimicrobial agents, 99% were used for food-producing animals in 2016 and 2018 (approximately 8 and 6.4 tones, respectively), according to the annuals European Surveillance of Veterinary Antimicrobial Consumption (ESVAC) report [4, 7]. The sales of polymyxins, including colistin only, for animal production, represented 5% in 2016, decreasing its sales by 40% in 25 countries of which three countries like Norway, Finland and Iceland not reported polymyxin sales from 2011 to 2018 [4]. In general, the antibiotic use between 2011 and 2018 in European animals has been declined at 34%, in particular the antimicrobial classes used in the treatment of severe human infections, such as

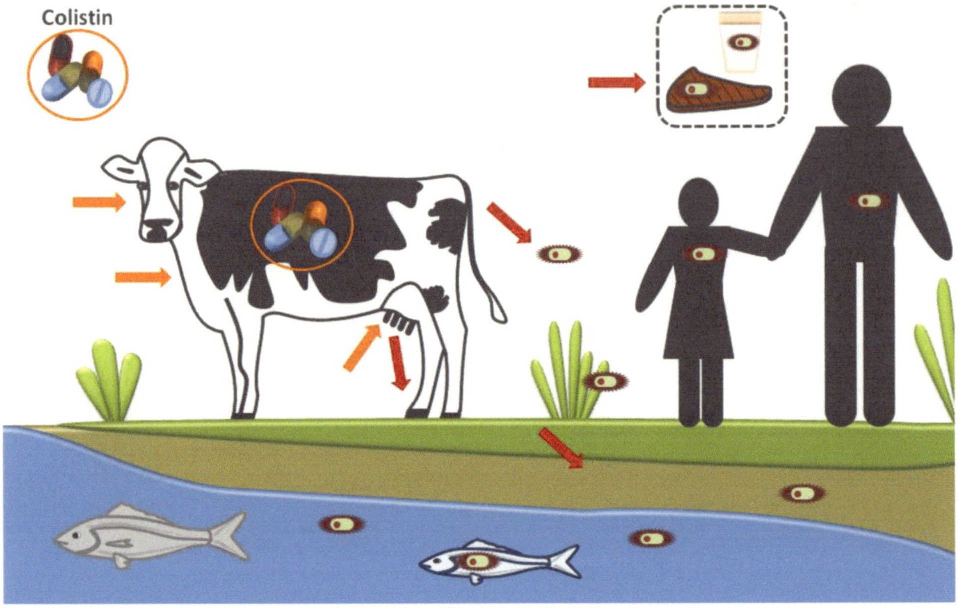

Figure 1.
The possible pathways of the colistin administration into a food-producing animal (orange arrows) and consequently, the spreading of the colistin-resistant bacteria to humans, other animals and the environment (red arrows).

polymyxins, 3rd- and 4th-generation cephalosporins and fluoroquinolones dropped by 70%, 24% and 4%, respectively [7]. However, antibiotic consumption is high and increasing in some countries because of the growing demands by consumers for animal proteins and depending on animal production composition and epidemiological situations of the production systems in each country [4, 22]. In Europe, the countries with the lowest antibiotic use were Norway, Iceland and Sweden, while Cyprus, Italy and Spain showed the highest farm antibiotic use in 2019 [7, 20].

In the United Kingdom (UK), the sold veterinary antibiotics for administration in food-producing animals showed a relevant reduction (45%) in 2015 but an increase (5%) in 2018 in comparison to 2019 (**Table 1**). Since 2015, sales of antimicrobial classes classified as a highest priority on the WHO CIA list decrease 74%, in particular, colistin showing a reduction of 99.9%. In 2019, colistin was sold 1.2 kg only for UK animal production and 8.8 kg was exported as medicated feed [8]. In the Netherlands, antibiotic sales decreased by 70% for farm animals including pigs, veal calves, broilers and dairy cattle, between 2009 and 2019, resulting from efforts of the veterinarians, livestock sectors and the authorities. Additionally, the polymyxins are considered as third choice agents and the use of colistin increased by 13% in 2019, mainly in weaned piglets and poultry (excepting broilers and turkey), although cattle farms not have prescriptions since 2015, according to the annual NethMap/MARAN 2020 report [9]. In Sweden, all antimicrobial classes showed a reduction notably in sales (32%) since 2010, leveled between 2014 and 2018, and in 2019 decreased 8% comparatively to 2018. The sales of polymyxins have decreased by 73%, between 2010 and 2019. In addition, no colistin is used in poultry [10]. In France, the sales of veterinary medicinal products showed a reduction by 48% between 2011 and 2018, in which for all species was observed a reduction of the administered antimicrobials. Taking into account the aim of the EcoAntibio 2017–2021 plan to reduce by 50% the colistin use in food-producing animals, in which for pigs has been achieved with a decreasing of 63%, while for cattle and poultry production only decreasing 48% and 49%, respectively, in 2018 [11].

In Denmark, the DANMAP program has allowed monitoring the use of antimicrobial agents in both animals and humans since 1995. Since then, the consumption of antimicrobial for animals has been reduced, due to the limits to its use to veterinarians and the national measures implemented of veterinary preventive medicine. Moreover, the total antimicrobial use for animals decreased by 5% in Denmark, between 2015 and 2016, in particular for pig (75%) and cattle (12%) production, and the aquaculture industry. Although the antimicrobial use classified as critically important remained low in food production, the colistin use that represented 1% for pigs in 2016, showing an increase (40 kg) mainly for the gastroenteritis treatment in weaners, between 2015 and 2016. This increase of the colistin use for pigs has been demonstrated since 2009, reporting a high increase, from 407 kg to 864 kg, between 2009 and 2016 respectively [5]. According to the annual BelVet-SAC report, the antimicrobial use in animals mainly veal calves, poultry and pigs, showed a relevant decrease by 40% and 8% in 2019 relative to 2011 and 2018, respectively, in Belgium. The polymyxin use also decreased by 66% and 11% in 2019 when compared to 2013 and 2018, respectively [12].

2.2 Other countries

In the United States (US), the total sales of medically important antibiotics approved for food-producing animals reduced by 21% and 38% from 2009 and 2015 through 2018, respectively, however, there was an increase by 9% from 2017, according to the report on antimicrobials sold or distributed for use in food-producing animals [15]. In 2018, these sales were estimated that 42%, 39%, 11% and 4% were intended for administration in cattle, swine, turkeys and chickens, respectively. In addition, colistin products have never been sold for animal use in the US, only products containing polymyxin B also belonging to the antimicrobial class of polymyxins were approved for ophthalmic infections in animals [2, 15]. In Canada, of which antimicrobials were distributed in 2014, 73% were in the same antimicrobial classes used in humans, and in 2017, 77% of distributed antimicrobials were intended for food-producing animals and 20% for humans. Moreover, antimicrobials sales showed a decrease by 12% mainly in pig and broiler chicken farms between 2016 and 2017, due to the implementation of measures for medically important antimicrobials banning them to use for growth promotion, according to the Canadian Animal Health Institute and the Public Health Agency of Canada. However, antibiotic sales for animal production increased by 6% between 2017 and 2018. Moreover, there are no colistin products approved for animal use and the predominant antimicrobial classes sold were tetracyclines and beta-lactams [17, 27]. The government of Brazil after the emergence of colistin resistance in animals and humans banned colistin use as AGPs in 2016 [18].

Australia showed one of the lowest levels of antibiotic resistance, although the Australian Pesticides and Veterinary Medicines Authority (APVMA) reported that 98% of the overall antibiotics sold for animal use between 2005 and 2010 were administered in food-producing animals, showing an increase of 11% in 2010 compared with 2008. Of which 43% were used for therapeutic purposes in poultry (49%), pigs (36%) and cattle (15%) mainly by feed (76%) and water (18%); and 6% were used as growth promoters. In addition, colistin was only reported for human use in

Australia [19]. In New Zealand, the antibiotic sold for farm animals increased by 4% and 3% between 2014 and 2017, and between 2016 and 2017, respectively [20].

The overall antimicrobial sales for animals in Japan showed a trend to decrease from 2001 to 2014 and since 2015 there was an upward trend until 2017, according to the Japanese Veterinary Antimicrobial Resistance Monitoring System (JVARM) that was implemented in 1999 [13]. From 2005 to 2010, the sales of polymyxins (colistin) for farm animals were 0.3–1.6%, with therapeutic use in beef cattle, pigs and broiler chickens [14]. China, the country that is expected to be the largest antimicrobial consumer worldwide in livestock animals in 2030, showed a reduction of 46% in antibiotic consumption from 2013 to 2017 [23]. With the emergence of colistin-resistant bacteria in China, colistin use was banned as a feed additive in animal production in 2016. Additionally, from the annual production of 12,000 tons of colistin in China, their use was estimated at 8000 tons per year in this country. The remains were exported to South Korea, Vietnam and India [28]. Similar to China, India is also one of the top antibiotic consumers for farm animals. However, the manufacture, sale and distribution of colistin, as well as, its formulations applied for food-producing animals, aquafarming and animal feed supplements were banned by the Indian Ministry of Health and Family welfare, since July 2019 [16].

3. Dissemination of colistin-resistant *Enterobacteriaceae* in animal production

In food-producing animals, the widely antibiotic use constitutes a key factor to the development of resistant microorganisms in humans, these animals work as potential reservoirs of antibiotic-resistant bacteria transferable to them through food supply [29, 30]. The second joint interagency antimicrobial consumption and resistance analysis (JIACRA) report in bacteria from food-producing animals and humans produced by the European centre for disease prevention and control (ECDC), European food safety authority (EFSA) and European medicines agency (EMA) showed that the average antimicrobial consumption (AMC) was lower in humans (124 mg/kg) than in animals (152 mg/kg), however, the median AMC was higher in humans (118 and 67 mg/kg, respectively), in 2014. In addition, the results demonstrated statistically-significant associations between antimicrobial consumption and resistance for 3rd- and 4th-generation cephalosporins and *Escherichia coli* in humans, for tetracyclines and polymyxins and *Escherichia coli* in animals, and fluoroquinolones and *Escherichia coli* in food-producing animals and humans. For carbapenems and polymyxins in *Klebsiella pneumoniae* from humans, there was also a significant association between AMC and AMR [30].

Colistin, an old antibiotic, used for the treatment of *E. coli* infections, was reintroduced as last resort for the treatment of emergent multi-resistant Gram-negative infections [26, 31]. In 2015, the *mcr*-1 gene responsible for the plasmid-mediated (transferrable) colistin resistance was detected in *E. coli* isolates from animals, foods and patients in China [32]. With the emergence of a colistin resistance mechanism including the *mcr*-1 gene, first detected in China and remaining spreading worldwide (**Figure 2**); the concerns around colistin use in animals and humans have increased [32, 33]. Since then, several other countries worldwide have been reported the presence of colistin-resistant genes in Gram-negative bacteria isolated from food animals or products animals, even before 2015.

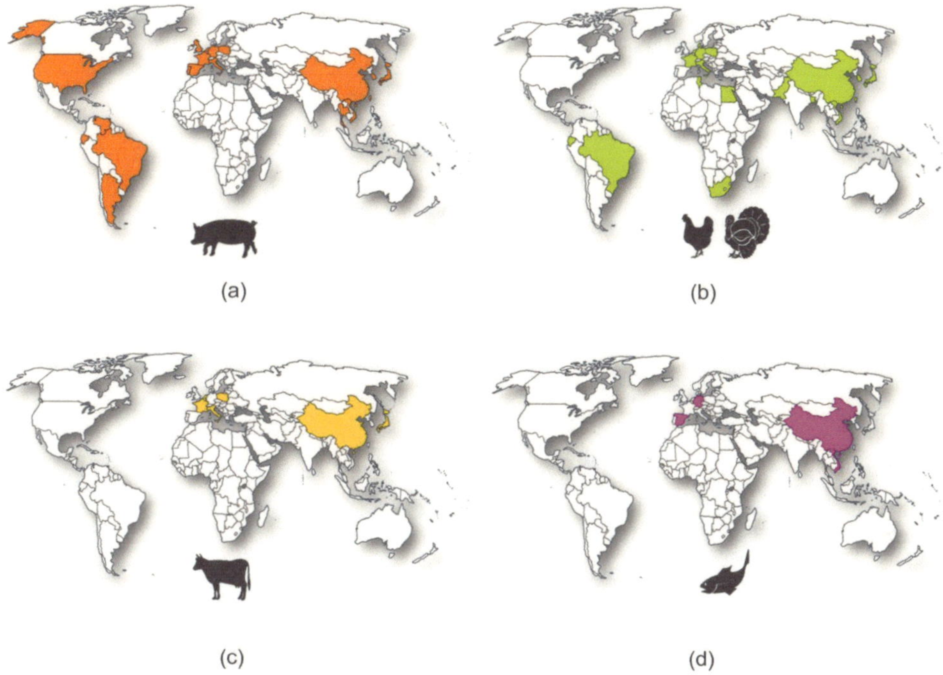

Figure 2.
Geographical distribution of the spreading colistin-resistant genes detected in worldwide animal production. Orange, presence of colistin-resistant genes in pigs; green, presence of colistin-resistant genes in poultry; yellow, presence of colistin-resistant genes in cattle; Purple, presence of colistin-resistant genes in fish.

3.1 Colistin-resistant *E. coli*

The occurrence of colistin resistance in indicator *E. coli* from poultry (turkeys and broilers) of 27 analyzed countries in 2014 and pigs for 18 countries in 2015, were significantly positive for the consumption of polymyxins [30]. Migura-Garcia et al. [34] reported the circulation of *mrc*-1 in *E. coli* isolated from Spanish pig farms since 2005 (**Table 2**). In Great Britain, one *E. coli* carrying the *mrc*-1 gene was also identified from a pig farm [41]. However, the UK reported that in 2015, 2017 and 2019, there was no resistance to colistin in *E. coli* from pigs and chickens [8]. In Denmark, *mcr*-1 gene was identified in *E. coli* from imported poultry meat [37]. In France, the presence of the *mcr*-1 gene from *E. coli* strains isolated from diseased pigs was confirmed by sequencing between 2009 and 2013 [42]. In this country, Perrin-Guyomard et al. [43] also reported the *mrc*-1 presence in 0.5% pigs analyzed in 2013, and in 6% turkeys and 2% broilers analyzed in 2014; and Haenni et al. [44] reported that the *mcr*-1 gene was present in 21% of the extended-spectrum-lactamase (ESBL)-producing *E. coli* isolates obtained from bovine calves in 2014 and the *mcr*-3 gene in 3% of ESBL-producing *E. coli* from cattle between 2011 and 2016 [45]. In Germany, also three ESBL-producing *E. coli* that originated from swine were positive for the *mcr*-1 gene during 2010–2011 [46]. In concordance, Roschanski et al. [67] reported the presence of the *mcr*-1 in *E. coli* collected from fecal samples of pig-fattening farms during 2011–2012; whilst *mcr*-2 was not detected in Germany (**Table 2**). In Portugal, two pig farms also showed *mcr*-1 positive *E. coli* isolates, in which this gene was identified on IncHI2, IncP, and IncX4 plasmids. This study

Animal specie	Country	Year of isolation	Isolate strain	Colistin-resistant gene	Reference
Pigs	China	2011–2014	*E. coli*	*mcr*-1	[32]
Chicken	China	2012–2014	*E. coli*	*mcr*-1	[35]
Pigs	China	2015	*E. coli*	*mcr*-3	[36]
Poultry meat (imported)	Denmark	2015	*E. coli*	*mcr*-1	[37]
Pigs and calves	Belgium	2011–2012	*E. coli*	*mcr*-1	[38]
Pigs and Cattles	Belgium	2016	*E. coli*	*mcr*-2	[39]
Pigs	Belgium	2015–2016	*E. coli*	*mcr*-4	[40]
Pigs	Great Britain	2016	*E. coli*	*mcr*-1	[41]
Pigs	France	2009–2013	*E. coli*	*mcr*-1	[42]
Pigs and poultry	France	2013–2014	*E. coli*	*mcr*-1	[43]
Calves, Cattle	France	2014, 2011–2016	*E. coli*	*mcr*-1, *mcr*-3	[44, 45]
Pigs	Germany	2010–2011	*E. coli*	*mcr*-1	[46]
Broilers, veal calves and turkey	Netherlands	2010–2013	*E. coli*	*mcr*-1	[47]
Pigs, veal calves and turkey meat	Netherlands	2019	*E. coli*	*mcr*-1, *mcr*-4	[9]
Poultry, pigs and cattle	Poland	2011–2016	*E. coli*	*mcr*-1	[48]
Pigs	Spain	2005–2014	*E. coli*	*mcr*-1, *mcr*-4	[34]
Pigs	Spain	2015–2016	*E. coli*	*mcr*-4	[40]
Pigs	Italy	2015–2016	*E. coli*	*mcr*-1	[49]
Pigs, turkeys, broilers and cattle	Italy	2014–2015	*E. coli*	*mcr*-1, *mcr*3, *mcr*-4	[50]
Pigs	Portugal	2016	*E. coli*	*mcr*-1	[51, 52]
Poultry and pigs	South Korea	2013–2015	*E. coli*	*mcr*-1	[53]
Pigs	South Korea	2011–2018	*E. coli*	*mcr*-3	[54]
Pigs	Thailand	—	*E. coli*	*mcr*-1, *mcr*-3	[55]
Pigs	USA	2016	*E. coli*	*mcr*-1	[56]
Chicken	Egypt	2010	*E. coli*	*mcr*-1	[35]
Chicken	Bangladesh	2017–2018	*E. coli*	*mcr*-1, *mcr*-2	[57]
Cattle, pigs and broilers	Japan	2008–2014	*E. coli*	*mcr*-1	[58]
Pigs and chickens	Vietnam	2013–2014	*E. coli*	*mcr*-1	[59]
Chickens	Tunisia	2015	*E. coli*	*mcr*-1	[60]
Chickens	Pakistan	2016–2017	*E. coli*	*mcr*-1	[61]
Pigs and chickens	Ecuador	2017	*E. coli*	*mcr*-1	[62]
Pigs and chickens	Brazil	2012–2013	*E. coli*	*mcr*-1	[63]
Pigs	Venezuela	2015	*E. coli*	*mcr*-1	[64]
Pigs	Argentina	2017	*E. coli*	*mcr*-1	[65]
Chickens	South Africa	2015	*E. coli*	*mcr*-1	[66]

Table 2.
Summary of the worldwide distribution of the mcr-positive Escherichia coli, resistance to colistin, from livestock animals.

reported that these food animals had received feed colistin as metaphylaxis during the 6 weeks before sample collection, suggesting a selective pressure. In addition, the presence of extended-spectrum beta-lactamases (ESBL)-encoding genes, such as $bla_{\text{CTX-M-2}}$, $bla_{\text{TEM-1}}$ and chloramphenicol gene, *flo*R, were also identified into same plasmids [51]. Similar results were also obtained by Manageiro et al. [68] and Fournier et al. [52].

In South Korea, the first detection of *mcr*-1 positive *E. coli* isolated from food animals was documented since 2013. Between 2013 and 2015, the *mcr*-1 gene was identified in 12 *E. coli* isolates collected from healthy chicken fecal samples and chicken carcasses at slaughterhouses, and a diseased pig. Additionally, all colistin-resistant *E. coli* strains showed multidrug resistance (≥3 antimicrobial classes) and this gene was not detected in cattle samples [53]. Lima Barbieri et al. [35] reported the identification of the *mcr*-1 gene in diseased poultry fecal *E. coli* from Egypt and China and no *mcr*-1 or -2 gene was detected in the healthy birds. In the United States, two *E. coli* isolates from swine fecal samples at slaughter were positive for *mcr*-1 gene, however, fecal content samples from cattle, chickens and turkeys were negative to this resistance in 2016 [56]. The presence of *mcr*-1 gene was detected in food animals, such as healthy cattle, pigs and broilers since 2008 in Japan [58] and in pigs and chickens during 2013–2014 in Vietnam [59]. In Tunisia, three chicken farms were positive to colistin-resistant *E. coli* carrying the *mcr*-1 and $bla_{\text{CTX-M-1}}$ genes, in which these animals were imported from France [60]. The *mcr*-1 gene was detected in *E. coli* obtained from swine samples in Venezuela [64], from chickens in South Africa [66] and swine and chickens samples collected in Brazil [63] and Ecuador [62]. In Argentina, *mcr*-1-positive and ESBL-producing *E. coli* isolates were identified from healthy fattening pigs and diarrheic piglets [65]. In Pakistan, *mcr*-1 positive *E. coli* were identified in healthy broilers during 2016–2017 [61] and in retail freshwater fish [69].

During 2011–2012, 12% of *E. coli* diarrhea strains originated Belgian calves and piglets were positive for *mcr*-1 [38]. Also, in Belgium, a novel plasmid-mediated colistin-resistant gene, *mcr*-2, was documented in porcine and bovine colistin-resistant *E. coli* in 2016, showing a higher prevalence of *mcr*-2 in porcine colistin-resistant *E. coli* in comparison to *mcr*-1 [39, 70]. In Bangladesh, the *mcr*-1 and *mcr*-2 genes were predominant in chicken gut *Enterobacteriaceae*, including *E. coli*, during 2017–2018 [57]. In China, a novel *mcr* gene, *mcr*-3, was identified in *E. coli* collected from health pigs in 2015 [36]. This gene, *mcr*-3 was also identified in porcine *E. coli* in South Korea [54] and Thailand [55]. Between 2015 and 2016, the *mcr*-4 gene was detected in *E. coli* isolates collected from piglets with diarrhea in Spain and Belgium [40]. In Italy, 72% of the *E. coli* isolates from pigs with post-weaning diarrhea were positive for *mcr*-1 from 2015 to 2016 [49]. Another Italian study identified *E. coli* carrying the *mcr*-1, -3 and -4 genes in fattening turkeys, broilers, pigs and bovines during 2014–2015 [50]. In the Netherlands, the *mcr*-1 gene was identified in *E. coli* from fecal samples obtained in slaughter pigs and white veal calves in 2019, representing less than 1%. Additionally, the *mcr*-4 was detected in white veal calves at a low level (2%). In turkey meat, colistin-resistant *E. coli* was identified (13%). However, *E. coli* isolated from broilers and chicken meat no *mcr* genes were identified [9]. Veldman et al. [47] also report *mcr*-1-positive *E. coli* from broilers, veal calves and turkeys from 2010 to 2013, in the Netherlands. In Poland, 62% of *E. coli* isolated from fecal samples of several food-producing animals, mainly turkeys, following broilers, laying hens, pig and bovine were present for *mcr*-1 gene, although no *mcr*-2 to -5 were detected during 2011–2016 [48].

3.2 Colistin-resistant *Salmonella* spp

Salmonella Typhimurium carrying the *mrc*-1 gene was identified from a pig in Great Britain [41]. More recently, two *Salmonella* strains (serovars *S.* Dublin and *S.* Bovismorbificans) isolated from pig farm were resistant to colistin in 2019, although neither isolate had known transferable colistin resistance genes detected (**Table 3**). In ad dition, *S.* Dublin showed a degree of intrinsic colistin resistance and *S.* Bovismorbificans sequencing demonstrated the presence of chromosomal mutations conferring colistin resistance [8]. Four *S. enterica* (serovar Typhimurium and Rissen) collected from swine farms were positive for *mcr*-1 in Spain [71]. In Taiwan, the colistin resistance gene *mcr*-1 was identified in two *Salmonella* serovars (Typhimurium and Anatum) isolated from food animals with diarrhea, pigs and chickens, during 2012–2015. However, no gene was identified in all colistin-resistant strains tested phenotypically [75]. *Salmonella* Typhimurium carrying the *mrc*-1 gene, and *flo*R and *oqx*AB genes was identified from a pig in China, during 2013–2014 [76] and the *mcr*-1 gene in *S. enterica* from diseased chickens during 2014–2015 [77]. In Japan, *S. enterica* serovar Typhimurium carrying the *mcr*-1 gene was also detected in swine [78]. In Bangladeshi chicken, the *mcr*-1 and *mcr*-2 were isolated from *Salmonella* spp. [57].

A study that screened the *mcr*-1 positive *Salmonella* spp. in European food-producing animals, reported three isolates from pigs and cattle located in France and Germany from 2004 to 2014 [73]. During 2012–2015, 56% of the colistin-resistant *Salnomella* isolates from swine and 15% from poultry showed the presence of *mcr*-1 gene in Italy, as well as, in food samples of pork [72]. Another Italian study only identified *Salmonella* spp. carrying the *mcr*-1 gene in fattening turkeys in 2014 [50]. In 2013, a novel colistin-resistant gene, *mcr*-4, was detected in *S. enterica* serovar Typhimurium obtained from the pig fecal sample at Italian slaughter [40]. German sample collection of *Salmonella* Paratyphi B isolated in 2012 from poultry, carried a novel colistin-resistant gene, *mcr*-5 [74].

3.3 Colistin-resistant *Klebsiella pneumoniae*

In Portugal, *Klebsiella pneumoniae* isolated from swine farms were positive for *mcr*-1 gene (**Table 3**), carrying on IncHI2 and IncP plasmids, collected in 2016 [51, 52]. In Bangladesh, *Klebsiella* spp. isolated from chicken was positive to *mcr*-1 and *mcr*-2 [57]. In China, *K. pneumoniae* isolates harboring the *mcr*-1 gene were collected from pigs, cattle and chickens in 2016 [79] and harboring the *mcr*-3 gene were detected in chicken samples during 2016–2017 [80]. Other study reported the coexistence in *K. pneumoniae* isolates of the carbapenemase-encoding gene $bla_{CTX\text{-}M\text{-}55}$ and *mcr*-7 from chickens [81], and the carbapenemase-encoding gene bla_{NDM} and *mcr*-8 from pigs and chicken in China [82].

Recently, studies showed that the mechanism responsible for colistin resistance in *K. pneumoniae* resulted in inactivation of the *mgr*B gene through the upregulation of the PhoPQ system and consequently the overexpression of the *pmr*HFIJKLM operon [83, 89]. In France, the *mgr*B gene was identified in *K. pneumoniae* isolates from bovine mastitis [83].

3.4 Other colistin-resistant *Enterobacteriaceae* and non-*Enterobacteriaceae* bacteria

In China, the *mcr*-1 gene was detected in *Citrobacter freundii* isolates collected from pigs in 2012 (**Table 3**) [84] and *Citrobacter braakii* isolates collected from chickens in

Animal specie	Country	Year of isolation	Isolate strain	Colistinresistant gene	Reference
Pigs	Spain	2010–2011	*S.* Typhimurium and *S.* Rissen	*mcr*-1	[71]
Pigs	Great Britain	2016	*S.* Typhimurium	*mcr*-1	[41]
Pigs	UK	2019	*S.* Dublin and *S.* Bovismorbificans	*mcr*-1	[8]
Pigs and poultry	Italy	2012–2015	*Salmonella* spp.	*mcr*-1	[72]
Pigs	Italy	2013	*S.* Typhimurium	*mcr*-4	[40]
Turkeys	Italy	2014	*Salmonella* spp.	*mcr*-1	[50]
Pigs and cattle	France	2005–2010	*Salmonella* spp.	*mcr*-1	[73]
Pigs	Germany	2010	*Salmonella* spp.	*mcr*-1	[73]
Poultry	Germany	2012	*Salmonella* Paratyphi B	*mcr*-5	[74]
Pigs and chickens	Taiwan	2012–2015	*S.* Typhimurium and *S.* Anatum	*mcr*-1	[75]
Pigs	China	2013–2014	*S.* Typhimurium	*mcr*-1	[76]
Chickens	China	2014–2015	*S.* enterica	*mcr*-1	[77]
Pigs	Japan	2013	*S.* Typhimurium	*mcr*-1	[78]
Chicken	Bangladesh	2017–2018	*Salmonella* spp.	*mcr*-1, *mcr*-2	[57]
Pigs, chickens and cattle	China	2016	*K. pneunomiae*	*mcr*-1	[79]
Chickens	China	2016–2017	*K. pneunomiae*	*mcr*-3	[80]
Chickens	China	2010–2015	*K. pneunomiae*	*mcr*-7	[81]
Pigs and chickens	China	2015–2017	*K. pneunomiae*	*mcr*-1, *mcr*-8	[82]
Cattle	France	2015	*K. pneunomiae*	*mgr*B	[83]
Pigs	Portugal	2016	*K. pneunomiae*	*mcr*-1	[51, 52]
Chicken	Bangladesh	2017–2018	*Klebsiella* spp.	*mcr*-1, *mcr*-2	[57]
Chickens	China	2012	*Citrobacter freundii*	*mcr*-1	[84]
Pigs	China	2015	*Citrobacter braakii*	*mcr*-1	[85]
Chicken	Bangladesh	2017–2018	*Proteus* spp.	*mcr*-1, *mcr*-2	[57]
Chicken	Bangladesh	2017–2018	*Shigella* spp.	*mcr*-1	[57]
Chicken	Bangladesh	2017–2018	*Enterobacter* spp.	*mcr*-1	[57]
Turkey	German	2012	*Aeromonas media*	*mcr*-3	[86]
Pigs	China	2018	*Acinetobacter baumannii*	*mcr*-4	[87]
Pigs	Great Britain	2014–2015	*Moraxella pluranimalium*	*mcr*-6	[88]

Table 3.
Summary of the worldwide distribution of the colistin-resistant Enterobacteriaceae (Salmonella, Klebsiella and Citrobacter) and non-Enterobacteriaceae (Aeromonas, Acinetobacter and Moraxella) in livestock animals.

2015 [85]. In Bangladeshi chicken, *mcr*-1 gene was identified from *Proteus* spp., *Shigella* spp. and *Enterobacter* spp., and the *mcr*-2 gene was detected in *Proteus* spp. [57].

Furthermore, an *Aeromonas media* carrying the *mcr*-3 was detected in German turkey in 2012 [86]. The *mcr*-4 gene was identified in *Acinetobacter baumannii* isolated from fecal pig samples at a Chinese slaughter in 2018 [87]. From healthy pigs in Great Britain, a novel mobile colistin resistance gene, *mcr*-6, was identified in *Moraxella pluranimalium* (**Table 3**) [88].

4. Colistin resistance in aquaculture production

In aquaculture production, information about the presence of colistin resistance is scarce. This ecosystem is a possible reservoir of antibiotic resistance and it can contribute to the transfer of colistin-resistant genes to humans through the food chain and to aquatic environments [90, 91], requiring strict sanitary control by the public health authorities [92].

Hassan et al. [90] reported the identification of *E. coli* strains positive for *mcr*-1 isolated from Rainbow trout guts in Lebanon (**Table 4**). Moreover, isolates were multidrug-resistant to bla_{TEM-1}, *tet*A and *str*A genes. In Vietnam, a AmpC-producing *E. coli* isolated from a fish sample was positive to *mcr*-1 in 2014 [93] and two ESBL-producing *E. coli* harboring *mcr*-1 and *sul* genes were detected from wild fish and striped catfish during 2014–2015 [94]. In China, *mcr*-1 gene was detected in seven *E. coli* from retail grass carp collected from fish markets in 2016 [69], and in two *E. coli* isolated from farmed fish in 2017 [96]. In same country, ESBL-producing *E. coli* strains carrying the *mcr*-1, bla_{CTX-Ms} and *fos*A3 genes were isolated from shrimp purchased in markets and supermarkets during 2015–2016 [95].

Salmonella enterica serovar Rissen strains obtained from mussel samples carried the *mcr*-1 gene, in Spain from 2012 to 2016, and their genotypic profile showed other antibiotic resistance, such as *bla*TEM-1B, *aac*(6′)-Iaa, *tet*(A), *cml*A1, *aad*A1, *aad*A2, *dfr*A1, *sul*1 and *sul*3 [92].

An *Aeromonas* carrying the *mcr*-3 gene was isolated from a Chinese market sample in 2017 [96]. In Germany, three *Aeromonas* isolates were positive for *mcr*-3 genes recovered from ornamental fish, such as koi carp (*Aeromonas jandaei* and *Aeromonas hydrophila*) and golden orfe (*Aeromonas allosacharophila*), between 2005 and 2008, suggesting that the *mcr*-3 genes have circulated in European food-producing aquaculture animals at least 10 years before the first *mcr*-1-positive isolates from China [86].

Recently, two new no mobile genetic genes encoding phosphoethanolamine transferase, *ept*Av3 and *ept*Av7 were detected in *Aeromonas jandaei* strain isolated from retail fish in China. These genes demonstrated a high amino acid identity (80 and 79.9%) relative to *mcr*-3.1 and *mcr*-7.1 mobile genes, respectively. The strain demonstrated high-level colistin resistance, suggesting that *Aeromonas* can emerge and constitute as reservoir for *mcr*-3 and -7 genes [97]. Although the colistin-resistance prevalence in fish products is low, the antibiotic use in aquacultural activities can be a key factor for the rapid generation and dissemination of colistin-resistant bacteria in terrestrial and aquatic ecosystems [98]. In addition, antibiotic use for therapeutic and metaphylaxis purposes should be regulated and better assessed in aquaculture since there is antimicrobial consumption trend to increasing 33% between 2017 and 2030, suggesting that intensity use of antibiotics in some fish species can surpass the level of consumption in humans and terrestrial animals [98, 99].

Animal specie	Country	Year of isolation	Isolate strain	Colistin-resistant gene	Reference
Rainbow trout	Lebanon	—	*E. coli*	*mcr*-1	[90]
Fish	Vietnam	2014	*E. coli*	*mcr*-1	[93]
Wild fish and striped catfish	Vietnam	2014–2015	*E. coli*	*mcr*-1	[94]
Shrimp	China	2015–2016	*E. coli*	*mcr*-1	[95]
Grass carp	China	2016	*E. coli*	*mcr*-1	[69]
Fish	China	2017	*E. coli*	*mcr*-1	[96]
Mussels	Spain	2012–2016	*Salmonella enterica* Rissen	*mcr*-1	[92]
Ornamental fish	German	2005–2008	*Aeromonas jandaei, A. hydrophila* and *A. allosacharophila*	*mcr*-3	[86]
Fish	China	2017	*Aeromonas* spp.	*mcr*-3	[96]
Fish	China	—	*Aeromonas jandaei*	*ept*Av3, *ept*Av7	[97]

Table 4.
Summary of the worldwide distribution of the colistin-resistant Enterobacteriaceae and non-Enterobacteriaceae in aquaculture production.

5. Global impact of colistin-resistant *Enterobacteriaceae* to public health

Inappropriate antibiotic use and its administration in food animals is a public health concern, contributing to the emergence of resistant bacteria in animals that can be transmitted to humans by ingestion of contaminated food or proximity with animals and their environment or other contaminated vehicles [5, 27]. In addition, characterization of colistin-resistance genes allows understanding of the molecular basis and evolution of colistin resistance, facilitating new interventions, such as the development of diagnostic tests [18]. Therefore, restriction of colistin use is essential to prevent the transmission of resistant genes, such as *mcr*-1, to other bacteria in the same or different animals, to the food supply and human community [53]. Moreover, the colistin-resistant genes have been identified in a variety of plasmids, such as IncHI2, IncP, Incl2 and IncX4, which are associated with the dissemination of other antibiotic resistance, in particular for β-lactams [34, 51].

In past years, the colistin administration in animal production has been decreasing. However, there is evidence that colistin resistance can increase due to the use of other antibiotics, in particular bacitracin, in animal production. Bacitracin is used in human medicine and veterinary medicine as a growth promoter in some countries, although the US is not marketed for food animals as a growth promoter for instance [20]. The bacitracin use in livestock production cans facilitate the transmission of bacitracin-resistant bacteria to agricultural ecosystems and humans by the food chain. Xu et al. [100] highlighted the extensive and imprudent usage of bacitracin in food animals to mitigate the spread of colistin resistance, since they observed that *mcr-1* gene confers cross-resistance to bacitracin, serving as a risk factor for the plasmid-mediated transmissible colistin resistance. This finding can explain the study performed by Lentz et al. [101] that reported the *mcr*-1 positive *E. coli* isolates from chickens that had not been fed with colistin antibiotic but had been fed with bacitracin and other antibiotics.

In some countries, high administration of colistin has been reported in food animals [2]. Touati et al. [102] demonstrated that farms are an important reservoir of colistin resistance genes, as well as, other antibiotic resistance genes. These genes are transferred from manure and slurry animal directly to soil and water used for irrigation, allowing their dissemination in the environment, which posture an emergent threat to public health. In Estonia, ESBL-producing *E. coli* isolates obtained from swine slurry samples carried the *mcr*-1 gene on a plasmid of the IncX4 group [103]. In Algeria, *E. coli* isolates carrying the *mcr*-1 and *mcr*-3 genes were isolated from bovine manure, agricultural soil and irrigation water [102]. In addition, the colistin resistance gene has also been identified in bacteria from other different environment niches and commodities, like vegetables, wells, rivers and lakes and coastal water or beaches and vegetables [104–106]. Moreover, wild animals can contribute to the dissemination of colistin resistance since these animals feed on landfills of urban household waste. Migura-Garcia et al. [34] reported colistin-resistant *E. coli* isolated from white storks (Ciconia ciconia) in Spain. Fernandes et al. [105] highlighted a potential new environmental reservoir in the case of public urban beaches, where colistin-resistant *E. coli* isolates carrying the *mcr*-1 and $bla_{\text{CTX-M}}$ genes were isolated and there is a high tourist turnover, facilitating the rapid spread of antimicrobial resistance. Zhao et al. [107] reported the presence of *mcr*-1 *K. pneumoniae* in hospital sewage.

Although there is a high rate of widespread dissemination of colistin-resistant bacteria in livestock, the rate of colistin resistance in healthy animals and animal products is low [53]. However, several studies have been reported the detection of *mcr-1* in food that originated from animals, confirming the risk of consuming these products colonized with resistant bacteria and one via transmission of colistin-resistant genes to humans [6]. For instance, *mcr*-1 positive *E. coli* isolates from lean ground beef, chicken meat samples and raw milk cheese were collected from markets of Canada, Brazil and Egypt, respectively [108–110]. Shen et al. [96] reported the aquaculture role as a reservoir for colistin resistance, suggesting the transmission of *mcr* genes via the aquatic food chain.

In humans, the administration of colistin in the hospital sector, mainly in patients in intensive care units and with cystic fibrosis, is increasing [30]. Most colistin-resistant bacteria also showed resistance to several antimicrobial agents available for treatment, leading to antibiotic treatment failure, life-threatening illness and enormous clinical costs 5]***. According to the ECDC report, countries with a high rate of carbapenem resistance also show a high rate of polymyxin-resistant isolates, suggesting loss of effect of antibiotic alternatives for the treatment of the infections caused by Gram-negative bacteria. In addition, colistin resistance was common in *K. pneumoniea*, *Acinetobacter* spp., *E. coli* and *Pseudomonas aeruginosa* [111, 112]. It is urgent to control the risk and rate of the spreading of colistin resistance worldwide, for this, it is necessary to work together in a One Health approach, according to Liu and Liu [112]. The data analysis of colistin resistance detecting in animals and humans is essential, as well as, motoring the epidemiological distribution of colistin-resistant Gram-negative bacteria in animals [112].

6. Conclusions

The immensely and novel diversity of colistin-resistant *Enterobacteriaceae* isolates is well documented in various ecosystems, where the occurrence of *mcr* genes with livestock, food, human and environmental origin have been detected. Here, this information was summarized to understand the global dissemination of colistin resistance, particular in food-producing animals, and to highlight the impact of colistin use in livestock and aquaculture production, and consequently public health. Although some countries had banned colistin use in food animals, in others colistin is administrated without restrictions. This review verified that the circulation of colistin-resistant genes started about 10 years before the first detection in 2015. Thereby, identification and continuous monitoring of colistin-resistance genes, as well as, prudent colistin use in both animals and humans are vital for tracking the colistin resistance threat and the associated socioeconomic costs.

Acknowledgements

This work was supported by the Associate Laboratory for Green Chemistry—LAQV which is financed by national funds from FCT/MCTES (UIDB/50006/2020 and UIDP/50006/2020). This work was founded by the Ministerio de Ciencia, Innovación y Universidades (Spain, Project RTI2018-098267-R-C33) and the Junta de Castilla y León (Consejería de Educación, Spain, Project LE018P20). This work was supported by the project UIDB/CVT/00772/2020 funded by the Portuguese Foundation for Science and Technology (FCT).

Conflict of interest

The authors declare no conflict of interest.

Author details

Carla Miranda[1,2,3,†], Gilberto Igrejas[3,4,5,†], Rosa Capita[6,7], Carlos Alonso-Calleja[6,7] and Patrícia Poeta[1,3,8,9*]

1 Microbiology and Antibiotic Resistance Team (MicroART), University of Trás-os-Montes and Alto Douro, Vila Real, Portugal

2 Department of Sciences, University Institute of Health Sciences (IUCS), CESPU, CRL, Gandra, Portugal

3 Associated Laboratory for Green Chemistry (LAQV-REQUIMTE), University NOVA of Lisboa, Lisboa, Caparica, Portugal

4 Department of Genetics and Biotechnology, University of Trás-os-Montes and Alto Douro, Vila Real, Portugal

5 Functional Genomics and Proteomics Unit, University of Trás-os-Montes and Alto Douro, Vila Real, Portugal

6 Department of Food Hygiene and Technology, Veterinary Faculty, University of León, León, Spain

7 Institute of Food Science and Technology, University of León, León, Spain

8 Department of Veterinary Sciences, University of Trás-os-Montes and Alto Douro, Vila Real, Portugal

9 Veterinary and Animal Research Centre, Associate Laboratory for Animal and Veterinary Science (AL4AnimalS), University of Trás-os-Montes and Alto Douro (UTAD), Vila Real, Portugal

*Address all correspondence to: ppoeta@utad.pt

† These authors have contributed equally to this work.

References

[1] Kempf I, Jouy E, Chauvin C. Colistin use and colistin resistance in bacteria from animals. International Journal of Antimicrobial Agents. 2016;**48**(6):598-606. DOI: 10.1016/j.ijantimicag.2016.09.016

[2] EMA/AMEG. Updated Advice on the Use of Colistin Products in Animals Within the European Union: Development of Resistance and Possible Impact on Human and Animal Health. 2016. Available from: http://www.ema.europa.eu/docs/en_GB/document_library/Scientific_guideline/2016/07/WC500211080.pdf [Accessed: February 11, 2021]

[3] World Health Organization. WHO list of Critically Important Antimicrobials for Human Medicine (WHO CIA list). 6th revision 2018. 2019. Available from: www.who.int/foodsafety/publications/antimicrobials-sixth/en/ [Accessed: February 11, 2021]

[4] European Medicines Agency (EMA), European Surveillance of Veterinary Antimicrobial Consumption (ESVAC). Sales of veterinary antimicrobial agents in 30 European countries in 2016. 2018. Eighth ESVAC report (EMA/275982/2018)

[5] Borck Høg B, Korsgaard HB, Wolff Sönksen U, et al., editors. DANMAP 2016—Use of Antimicrobial Agents and Occurrence of Antimicrobial Resistance in Bacteria from Food Animals, Food and Humans in Denmark. Statens Serum Institut, National Veterinary Institute, Technical University of Denmark National Food Institute, Technical University of Denmark; 2017. ISSN: 1600-2032. Available from: https://orbit.dtu.dk/en/publications/danmap-2016-use-of-antimicrobial-agents-and-occurence-of-antimic [Accessed: February 10, 2021]

[6] Ma F, Xu S, Tang Z, Li Z, Zhang L. Use of antimicrobials in food animals and impact of transmission of antimicrobial resistance on humans. Biosafety and Health. 2021;**3**(1):32-38. DOI: 10.1016/j.bsheal.2020.09.004

[7] EMA. European Medicines Agency, European Surveillance of Veterinary Antimicrobial Consumption, 2020. Sales of Veterinary Antimicrobial Agents in 31 European Countries in 2018. Tenth ESVAC Report (EMA/24309/2020). 2020

[8] UK-VARSS. UK Veterinary Antibiotic Resistance and Sales Surveillance Report (UK-VARSS 2019). New Haw, Addlestone: Veterinary Medicines Directorate; 2020. Available from: https://www.gov.uk/government/collections/veterinary-antimicrobialresistance-and-sales-surveillance [Accessed: February 10, 2021]

[9] NethMap-MARAN. Consumption of Antimicrobial Agents and Antimicrobial Resistance among Medically Important Bacteria in the Netherlands. Monitoring of Antimicrobial Resistance and Antibiotic Usage in Animals in the Netherlands in 2019. 2020. Available from: https://swab.nl/en/nethmap [Accessed: February 16, 2021]

[10] Swedres-Svarm 2019. Sales of Antibiotics and Occurrence of Resistance in Sweden. A Report on Swedish Antibiotic Sales and Resistance in Human Medicine (Swedres) and Swedish Veterinary Antibiotic Resistance Monitoring (Svarm). Solna/Uppsala. ISSN1650-6332. 2020. Available from: https://www.sva.se/media/0hihej1c/swedres-svarm-2019.pdf [Accessed: February 10, 2021]

[11] ANSES. Sales Survey of Veterinary Medicinal Products Containing Antimicrobials in France in 2018. Annual Report. French Agency for Food, Environmental and Occupational Health & Safety (ANSES) and Agency for Veterinary Medicinal Products (ANMV); 2019. Available from: https://www.anses.fr/en/system/files/ANMV-Ra-Antibiotiques2018EN.pdf [Accessed: February 10, 2021]

[12] BelVet-SAC. Belgian Veterinary Surveillance of Antibacterial Consumption National Consumption Report 2019. 2020. Available from: https://belvetsac.ugent.be/BelvetSac_report_2019.pdf [Accessed: February 10, 2021]

[13] JVARM. Report on Japanese Veterinary Antimicrobial Resistance Monitoring System (JVARM) 2016-2017. National Veterinary Assay Laboratory, Ministry of Agriculture, Forestry and Fisheries; 2020. Available from: https://www.maff.go.jp/nval/yakuzai/pdf/200731_JVARMReport_2016-2017.pdf [Accessed: February 10, 2021]

[14] Hosoi Y, Asai T, Koike R, Tsuyuki M, Sugiura K. Sales of veterinary antimicrobial agents for therapeutic use in food-producing animal species in Japan between 2005 and 2010. Revue Scientifique et Technique. 2014;**33**(3):1007-1015. DOI: 10.20506/rst.33.3.2337

[15] FDA. US Food Drug Administration, Summary Report on Antimicrobials Sold or Distributed for Use in Food-Producing Animals, 2018. 2019. Available from: https://www.fda.gov/media/133411/download [Accessed: February 10, 2021]

[16] DHFW. Department of Health and Family Welfare. Notification. New Delhi, the 19th July, 2019. 2019. Available from: http://www.egazette.nic.in/WriteReadData/2019/207345.pdf [Accessed: February 10, 2021]

[17] CIPARS. Canadian Integrated Program for Antimicrobial Resistance Surveillance 2017: Integrated Findings. Guelph, Ontario: Public Health Agency of Canada; 2020. Available from: HP2-4-2017-eng-2.pdf (publications.gc.ca [Accessed: February 10, 2021]

[18] Sun J, Zeng X, Li XP, Liao XP, Liu YH, Lin J. Plasmid-mediated colistin resistance in animals: current status and future directions. Animal Health Research Reviews. 2017;**18**(2):136-152. DOI: 10.1017/S1466252317000111

[19] APVMA. Australian Pesticides and Veterinary Medicines Authority. Quantity of Antimicrobial Products Sold For Veterinary Use in Australia, 2005-2010. 2014. Available from: https://apvma.gov.au/sites/default/files/images/antimicrobial_sales_report_march-2014.pdf [Accessed: February 10, 2021]

[20] Nunan C. Farm Antibiotics and Trade Deals—Could UK Standards be Undermined? Alliance to Save Our Antibiotics. 2020. Available from: https://www.saveourantibiotics.org/media/1864/farm-antibiotics-and-trade-could-uk-standards-be-undermined-asoa-nov-2020.pdf [Accessed: February 10, 2021]

[21] OIE Annual Report on the Use of Antimicrobial Agents Intended for Use in Animals. Better Understanding of the Global Situation. Second Report. 2017. Available from: https://www.oie.int/fileadmin/home/eng/our_scientific_expertise/docs/pdf/amr/annual_report_amr_2.pdf [Accessed: February 16, 2021]

[22] Van Boeckel TP, Brower C, Gilbert M, Grenfell BT, Levin SA, Robinson TP, et al. Global trends in antimicrobial use in food animals. Proceedings of the National

Academy of Sciences of the United States of America. 2015;**112**:5649-5654. DOI: 10.1073/pnas.1503141112

[23] Tiseo K, Huber L, Gilbert M, Robinson TP, Van Boeckel TP. Global trends in antimicrobial use in food animals from 2017 to 2030. Antibiotics. 2020;**9**(12):918. DOI: 10.3390/antibiotics9120918

[24] Garcia JF, Diez MJ, Sahagun AM, Diez R, Sierra M, Garcia JJ, et al. The online sale of antibiotics for veterinary use. Animals. 2020;**10**(3):503. DOI: 10.3390/ani10030503

[25] Kempf I, Fleury MA, Drider D, Bruneau M, Sanders P, Chauvin C, et al. What do we know about resistance to colistin in *Enterobacteriaceae* in avian and pig production in Europe? International Journal of Antimicrobial Agents. 2013;**42**(5):379-383. DOI: 10.1016/j.ijantimicag.2013.06.012

[26] Rhouma M, Beaudry F, Letellier A. Resistance to colistin: What is the fate for this antibiotic in pig production? International Journal of Antimicrobial Agents. 2016;**48**(2):119-126. DOI: 10.1016/j.ijantimicag.2016.04.008

[27] PHAC. Public Health Agency of Canada. Canadian Antimicrobial Resistance Surveillance System—Report 2016. 2016. Available from: https://www.canada.ca/en/public-health/services/publications/drugs-health-products/canadian-antimicrobial-resistance-surveillance-system-report-2016.html [Accessed: February 11, 2021]

[28] Davies M, Walsh TR. A colistin crisis in India. The Lancet Infectious Diseases. 2018;**18**(3):256-257. DOI: 10.1016/S1473-3099(18)30072-0

[29] Marshall BM, Levy SB. Food animals and antimicrobials: Impacts on human health. Clinical Microbiology Reviews. 2011;**24**(4):718-733. DOI: 10.1128/CMR.00002-11

[30] JIACRA. Joint Interagency Antimicrobial Consumption and Resistance Analysis Report. ECDC/EFSA/EMA second joint report on the integrated analysis of the consumption of antimicrobial agents and occurrence of antimicrobial resistance in bacteria from humans and food-producing animals. EFSA Journal. 2017;**15**(7):4872, p. 135. DOI: 10.2903/j.efsa.2017.4872

[31] Theuretzbacher U, Van Bambeke F, Cantón R, Giske CG, Mouton JW, Nation RL, et al. Reviving old antibiotics. The Journal of Antimicrobial Chemotherapy. 2015;**70**(8):2177-2181. DOI: 10.1093/jac/dkv157

[32] Liu YY, Wang Y, Walsh TR, Yi LX, Zhang R, Spencer J, et al. Emergence of plasmid-mediated colistin resistance mechanism MCR-1 in animals and human beings in China: A microbiological and molecular biological study. The Lancet Infectious Diseases. 2016;**16**(2):161-168. DOI: 10.1016/S1473-3099(15)00424-7

[33] Bialvaei AZ, Samadi KH. Colistin, mechanisms and prevalence of resistance. Current Medical Research and Opinion. 2015;**31**(4):707-721. DOI: 10.1185/03007995.2015.1018989

[34] Migura-Garcia L, González-López JJ, Martinez-Urtaza J, Aguirre Sánchez JR, Moreno-Mingorance A, Perez de Rozas A, et al. mcr-colistin resistance genes mobilized by IncX4, IncHI2, and IncI2 plasmids in *Escherichia coli* of pigs and white stork in Spain. Frontiers in Microbiology. 2020;**10**:3072. DOI: 10.3389/fmicb.2019.03072

[35] Lima Barbieri N, Nielsen DW, Wannemuehler Y, Cavender T, Hussein A,

Yan SG, et al. *mcr*-1 identified in avian pathogenic *Escherichia coli* (APEC). PLoS ONE. 2017;**12**(3):e0172997. DOI: 10.1371/journal.pone.0172997

[36] Yin W, Li H, Shen Y, Liu Z, Wang S, Shen Z, et al. Novel plasmid-mediated colistin resistance gene *mcr*-3 in *Escherichia coli*. MBio. 2017;**8**:e00543-e00517. DOI: 10.1128/mBio.00543-17

[37] Hasman H, Hammerum AM, Hansen F, Hendriksen RS, Olesen B, Agersø Y, et al. Detection of mcr-1 encoding plasmid-mediated colistin-resistant *Escherichia coli* isolates from human bloodstream infection and imported chicken meat, Denmark 2015. Eurosurveillance. 2015;**20**(49):30085

[38] Malhotra-Kumar S, Xavier BB, Das AJ, Lammens C, Butaye P, Goossens H. Colistin resistance gene *mcr*-1 harboured on a multidrug resistant plasmid. The Lancet Infectious Diseases. 2016;**16**(3):283-284. DOI: 10.1016/S1473-3099(16)00012-8

[39] Xavier BB, Lammens C, Ruhal R, Kumar-Singh S, Butaye P, Goossens H, et al. Identification of a novel plasmid-mediated colistin-resistance gene, *mcr*-2, in *Escherichia coli*, Belgium, June 2016. Euro Surveillance. 2016;**21**(27):30280. DOI: 10.2807/1560-7917.ES.2016.21.27.30280

[40] Carattoli A, Villa L, Feudi C, Curcio L, Orsini S, Luppi A, et al. Novel plasmid-mediated colistin resistance *mcr*-4 gene in *Salmonella* and *Escherichia coli*, Italy 2013, Spain and Belgium, 2015 to 2016. Euro Surveillance. 2017;**22**(31):30589. DOI: 10.2807/1560-7917.ES.2017.22.31.30589

[41] Anjum MF, Duggett NA, AbuOun M, Randall L, Nunez-Garcia J, Ellis RJ, et al. Colistin resistance in *Salmonella* and *Escherichia coli* isolates from a pig farm in Great Britain. The Journal of Antimicrobial Chemotherapy. 2016;**71**(8):2306-2313. DOI: 10.1093/jac/dkw149

[42] Delannoy S, Le Devendec L, Jouy E, Fach P, Drider D, Kempf I. Characterization of colistin-resistant *Escherichia coli* isolated from diseased pigs in France. Frontiers in Microbiology. 2017;**8**:2278. DOI: 10.3389/fmicb.2017.02278

[43] Perrin-Guyomard A, Bruneau M, Houee P, Deleurme K, Legrandois P, Poirier C, et al. Prevalence of mcr1 in commensal *Escherichia coli* from French livestock, 2007 to 2014. Euro Surveillance. 2016;**21**(6):30135. DOI: 10.2807/1560-7917.ES.2016.21.6.30135

[44] Haenni M, Metayer V, Gay E, Madec J-Y. Increasing trends in mcr-1 prevalence among extended-spectrum-beta-lactamaseproducing *Escherichia coli* isolates from French calves despite decreasing exposure to colistin. Antimicrobial Agents and Chemotherapy. 2016;**60**:6433-6434. DOI: 10.1128/AAC.01147-16

[45] Haenni M, Beyrouthy R, Lupo A, Châtre P, Madec JY, Bonnet R. Epidemic spread of *Escherichia coli* ST744 isolates carrying mcr-3 and blaCTX-M-55 in cattle in France. The Journal of Antimicrobial Chemotherapy. 2018;**73**(2):533-536. DOI: 10.1093/jac/dkx418

[46] Falgenhauer L, Waezsada SE, Yao Y, Imirzalioglu C, Käsbohrer A, Roesler U, et al. Colistin resistance gene *mcr*-1 in extended-spectrum β-lactamase-producing and carbapenemase-producing Gram-negative bacteria in Germany. The Lancet Infectious Diseases. 2016;**16**(3):282-283. DOI: 10.1016/S1473-3099(16)00009-8

[47] Veldman K, van Essen-Zandbergen A, Rapallini M, Wit B, Heymans R, van Pelt W, et al. Location of colistin resistance gene *mcr*-1 in *Enterobacteriaceae* from livestock and meat. The Journal of Antimicrobial Chemotherapy. 2016;**71**(8):2340-2342. DOI: 10.1093/jac/dkw181

[48] Zajac M, Sztromwasser P, Bortolaia V, Leekitcharoenphon P, Cavaco LM, Zietek-Barszcz A, et al. Occurrence and characterization of *mcr*-1-positive *Escherichia coli* isolated from food-producing animals in Poland, 2011-2016. Frontiers in Microbiology. 2019;**10**:1753. DOI: 10.3389/fmicb.2019.01753

[49] Curcio L, Luppi A, Bonilauri P, Gherpelli Y, Pezzotti G, Pesciaroli M, et al. Detection of the colistin resistance gene *mcr*-1 in pathogenic *Escherichia coli* from pigs affected by post-weaning diarrhoea in Italy. Journal of Global Antimicrobial Resistance. 2017;**10**:80-83. DOI: 10.1016/j.jgar.2017.03.014

[50] Alba P, Leekitcharoenphon P, Franco A, Feltrin F, Ianzano A, Caprioli A, et al. Molecular epidemiology of mcr-encoded colistin resistance in *Enterobacteriaceae* from food-producing animals in Italy revealed through the EU harmonized antimicrobial resistance monitoring. Frontiers in Microbiology. 2018;**9**:1217. DOI: 10.3389/fmicb.2018.01217

[51] Kieffer N, Aires-de-Sousa M, Nordmann P, Poirel L. High rate of MCR-1-PRODUCING *Escherichia coli* and *Klebsiella pneumoniae* among pigs, Portugal. Emerging Infectious Diseases. 2017;**23**(12):2023-2029. DOI: 10.3201/eid2312.170883

[52] Fournier C, Aires-de-Sousa M, Nordmann P, Poirel L. Occurrence of CTX-M-15- and MCR-1-producing Enterobacterales in pigs in Portugal: Evidence of direct links with antibiotic selective pressure. International Journal of Antimicrobial Agents. 2020;**55**(2):105802. DOI: 10.1016/j.ijantimicag.2019.09.006

[53] Lim SK, Kang HY, Lee K, Moon DC, Lee HS, Jung SC. First detection of the *mcr*-1 gene in *Escherichia coli* isolated from livestock between 2013 and 2015 in South Korea. Antimicrobial Agents and Chemotherapy. 2016;**60**:6991-6993. DOI: 10.1128/AAC.01472-16

[54] Mechesso AF, Moon DC, Kang HY, Song HJ, Kim SJ, Choi JH, et al. Emergence of *mcr*-3 carrying *Escherichia coli* in Diseased Pigs in South Korea. Microorganisms. 2020;**8**(10):1538. DOI: 10.3390/microorganisms8101538

[55] Khine NO, Lugsomya K, Kaewgun B, Honhanrob L, Pairojrit P, Jermprasert S, et al. Multidrug resistance and virulence factors of *Escherichia coli* harboring plasmid-mediated colistin resistance: *mcr*-1 and *mcr*-3 genes in contracted pig farms in Thailand. Frontiers in Veterinary Science. 2020;**7**:582899. DOI: 10.3389/fvets.2020.582899

[56] Meinersmann RJ, Ladely SR, Plumblee JR, Cook KL, Thacker E. Prevalence of *mcr*-1 in the cecal contents of food animals in the United States. Antimicrobial Agents and Chemotherapy. 2017;**61**(2):e02244-e02216. DOI: 10.1128/AAC.02244-16

[57] Islam S, Urmi UL, Rana M, Sultana F, Jahan N, Hossain B, et al. High abundance of the colistin resistance gene *mcr*-1 in chicken gut-bacteria in Bangladesh. Scientific Reports. 2020;**10**(1):17292. DOI: 10.1038/s41598-020-74402-4

[58] Kawanishi M, Abo H, Ozawa M, Uchiyama M, Shirakawa T, Suzuki S, et al.

Prevalence of colistin resistance gene *mcr*-1 and absence of *mcr*-2 in *Escherichia coli* isolated from healthy food-producing animals in Japan. Antimicrobial Agents and Chemotherapy. 2016;**61**(1): e02057-e02016. DOI: 10.1128/AAC.02057-16

[59] Nguyen NT, Nguyen HM, Nguyen CV, Nguyen TV, Nguyen MT, Thai HQ, et al. Use of colistin and other critical antimicrobials on pig and chicken farms in Southern Vietnam and its association with resistance in commensal *Escherichia coli* bacteria. Applied and Environmental Microbiology. 2016;**82**(13):3727-3735. DOI: 10.1128/AEM.00337-16

[60] Grami R, Mansour W, Mehri W, Bouallègue O, Boujaâfar N, Madec JY, et al. Impact of food animal trade on the spread of *mcr*-1-mediated colistin resistance, Tunisia, July 2015. Euro Surveillance. 2016;**21**(8):30144. DOI: 10.2807/1560-7917.ES.2016.21.8.30144

[61] Lv J, Mohsin M, Lei S, Srinivas S, Wiqar RT, Lin J, et al. Discovery of a *mcr*-1-bearing plasmid in commensal colistin-resistant *Escherichia coli* from healthy broilers in Faisalabad, Pakistan. Virulence. 2018;**9**(1):994-999. DOI: 10.1080/21505594.2018.1462060

[62] Yamamoto Y, Calvopina M, Izurieta R, Villacres I, Kawahara R, Sasaki M, et al. Colistin-resistant *Escherichia coli* with *mcr* genes in the livestock of rural small-scale farms in Ecuador. BMC Research Notes. 2019;**12**(1):121. DOI: 10.1186/s13104-019-4144-0

[63] Fernandes MR, Moura Q, Sartori L, Silva KC, Cunha MP, Esposito F, et al. Silent dissemination of colistin-resistant *Escherichia coli* in South America could contribute to the global spread of the *mcr*-1 gene. Euro Surveillance. 2016;**21**(17):30214. DOI: 10.2807/1560-7917.ES.2016.21.17.30214

[64] Delgado-Blas JF, Ovejero CM, Abadia-Patiño L, Gonzalez-Zorn B. Coexistence of *mcr*-1 and *bla*NDM-1 in *Escherichia coli* from Venezuela. Antimicrobial Agents and Chemotherapy. 2016;**60**(10):6356-6358. DOI: 10.1128/AAC.01319-16

[65] Faccone D, Moredo FA, Giacoboni GI, Albornoz E, Alarcón L, Nievas VF, et al. Multidrug-resistant *Escherichia coli* harbouring *mcr*-1 and *bla*CTX-M genes isolated from swine in Argentina. Journal of Global Antimicrobial Resistance. 2019;**18**:160-162. DOI: 10.1016/j.jgar.2019.03.011

[66] Perreten V, Strauss C, Collaud A, Gerber D. Colistin resistance gene *mcr*-1 in avian-pathogenic *Escherichia coli* in South Africa. Antimicrobial Agents and Chemotherapy. 2016;**60**(7):4414-4415. DOI: 10.1128/AAC.00548-16

[67] Roschanski N, Falgenhauer L, Grobbel M, Guenther S, Kreienbrock L, Imirzalioglu C, et al. Retrospective survey of *mcr*-1 and *mcr*-2 in German pig-fattening farms, 2011-2012. International Journal of Antimicrobial Agents. 2017;**50**(2):266-271. DOI: 10.1016/j.ijantimicag.2017.03.007

[68] Manageiro V, Clemente L, Romão R, Silva C, Vieira L, Ferreira E, et al. IncX4 plasmid carrying the new mcr-1.9 gene variant in a CTX-M-8-producing *Escherichia coli* isolate recovered from swine. Frontiers in Microbiology. 2019;**10**:367. DOI: 10.3389/fmicb.2019.00367

[69] Lv L, Cao Y, Yu P, Huang R, Wang J, Wen Q, et al. Detection of *mcr*-1 gene among *Escherichia coli* isolates from farmed fish and characterization of *mcr*-1-bearing IncP plasmids. Antimicrobial

Agents and Chemotherapy. 2018;**62**(3): e02378-e02317. DOI: 10.1128/AAC.02378-17

[70] Al-Tawfiq JA, Laxminarayan R, Mendelson M. How should we respond to the emergence of plasmid-mediated colistin resistance in humans and animals? International Journal of Infectious Diseases. 2017;**54**:77-84. DOI: 10.1016/j.ijid.2016.11.415

[71] Quesada A, Ugarte-Ruiz M, Iglesias MR, Porrero MC, Martínez R, Florez-Cuadrado D, et al. Detection of plasmid mediated colistin resistance (MCR-1) in *Escherichia coli* and *Salmonella enterica* isolated from poultry and swine in Spain. Research in Veterinary Science. 2016;**105**:134-135. DOI: 10.1016/j.rvsc.2016.02.003

[72] Carnevali C, Morganti M, Scaltriti E, Bolzoni L, Pongolini S, Casadei G. Occurrence of *mcr*-1 in colistin-resistant *Salmonella enterica* isolates recovered from humans and animals in Italy, 2012 to 2015. Antimicrobial Agents and Chemotherapy. 2016;**60**(12):7532-7534. DOI: 10.1128/AAC.01803-16

[73] El Garch F, Sauget M, Hocquet D, LeChaudee D, Woehrle F, Bertrand X. mcr-1 is borne by highly diverse *Escherichia coli* isolates since 2004 in food-producing animals in Europe. Clinical Microbiology and Infection. 2017;**23**(1):51.e1-51.e4. DOI: 10.1016/j.cmi.2016.08.033

[74] Borowiak M, Fischer J, Hammerl JA, Hendriksen RS, Szabo I, Malorny B. Identification of a novel transposon-associated phosphoethanolamine transferase gene, *mcr*-5, conferring colistin resistance in d-tartrate fermenting *Salmonella enterica* subsp. *enterica* serovar Paratyphi B. The Journal of Antimicrobial Chemotherapy. 2017;**72**(12):3317-3324. DOI: 10.1093/jac/dkx327

[75] Chiou CS, Chen YT, Wang YW, Liu YY, Kuo HC, Tu YH, et al. Dissemination of mcr-1-carrying plasmids among colistin-resistant *Salmonella* strains from humans and food-producing animals in Taiwan. Antimicrobial Agents and Chemotherapy. 2017;**61**(7):e00338-e00317. DOI: 10.1128/AAC.00338-17

[76] Yi L, Wang J, Gao Y, Liu Y, Doi Y, Wu R, et al. mcr-1-harboring *Salmonella enterica* serovar typhimurium sequence type 34 in pigs, China. Emerging Infectious Diseases. 2017;**23**(2):291-295. DOI: 10.3201/eid2302.161543

[77] Yang YQ, Zhang AY, Ma SZ, Kong LH, Li YX, Liu JX, et al. Co-occurrence of mcr-1 and ESBL on a single plasmid in *Salmonella enterica*. The Journal of Antimicrobial Chemotherapy. 2016;**71**(8):2336-2338. DOI: 10.1093/jac/dkw243

[78] Suzuki S, Ohnishi M, Kawanishi M, Akiba M, Kuroda M. Investigation of a plasmid genome database for colistin-resistance gene *mcr*-1. The Lancet Infectious Diseases. 2016;**16**(3):284-285. DOI: 10.1016/S1473-3099(16)00008-6

[79] Wang R, Liu Y, Zhang Q, Jin L, Wang Q, Zhang Y, et al. The prevalence of colistin resistance in *Escherichia coli* and *Klebsiella pneumoniae* isolated from food animals in China: Coexistence of mcr-1 and blaNDM with low fitness cost. International Journal of Antimicrobial Agents. 2018;**51**(5):739-744. DOI: 10.1016/j.ijantimicag.2018.01.023

[80] Xiang R, Ye X, Tuo H, Zhang X, Zhang A, Lei C, et al. Co-occurrence of mcr-3 and blaNDM-5 genes in multidrug-resistant *Klebsiella pneumoniae* ST709 from a commercial chicken farm in China. International

Journal of Antimicrobial Agents. 2018;**52**(4):519-520. DOI: 10.1016/j.ijantimicag.2018.07.007

[81] Yang YQ, Li YX, Lei CW, Zhang AY, Wang HN. Novel plasmid-mediated colistin resistance gene *mcr*-7.1 in *Klebsiella pneumoniae*. The Journal of Antimicrobial Chemotherapy. 2018;**73**(7):1791-1795. DOI: 10.1093/jac/dky111

[82] Wang X, Wang Y, Zhou Y, Li J, Yin W, Wang S, et al. Emergence of a novel mobile colistin resistance gene, *mcr*-8, in NDM-producing *Klebsiella pneumoniae*. Emerging Microbes & Infections. 2018;**7**(1):122. DOI: 10.1038/s41426-018-0124-z

[83] Kieffer N, Poirel L, Nordmann P, Madec J-Y, Haenni M. Emergence of colistin resistance in *Klebsiella pneumoniae* from veterinary medicine. The Journal of Antimicrobial Chemotherapy. 2015;**70**(4):1265-1267. DOI: 10.1093/jac/dku485

[84] Li XP, Fang LX, Jiang P, Pan D, Xia J, Liao XP, et al. Emergence of the colistin resistance gene *mcr*-1 in *Citrobacter freundii*. International Journal of Antimicrobial Agents. 2017;**49**(6):786-787. DOI: 10.1016/j.ijantimicag.2017.04.004

[85] Liu J, Yang Y, Li Y, Liu D, Tuo H, Wang H, et al. Isolation of an IncP-1 plasmid harbouring *mcr*-1 from a chicken isolate of *Citrobacter braakii* in China. International Journal of Antimicrobial Agents. 2018;**51**(6):936-940. DOI: 10.1016/j.ijantimicag.2017.12.030

[86] Eichhorn I, Feudi C, Wang Y, Kaspar H, Feßler AT, Lübke-Becker A, et al. Identification of novel variants of the colistin resistance gene *mcr*-3 in *Aeromonas* spp. from the national resistance monitoring programme GERM-Vet and from diagnostic submissions. The Journal of Antimicrobial Chemotherapy. 2018;**73**(5):1217-1221. DOI: 10.1093/jac/dkx538

[87] Ma F, Shen C, Zheng X, Liu Y, Chen H, Zhong L, et al. Identification of a novel plasmid carrying *mcr*-4.3 in an *Acinetobacter baumannii* strain in China. Antimicrobial Agents and Chemotherapy. 2019;**63**(6):e00133-e00119. DOI: 10.1128/AAC.00133-19

[88] AbuOun M, Stubberfield EJ, Duggett NA, Kirchner M, Dormer L, Nunez-Garcia J, et al. *mcr*-1 and *mcr*-2 variant genes identified in *Moraxella* species isolated from pigs in Great Britain from 2014 to 2015. The Journal of Antimicrobial Chemotherapy. 2017;**72**(10):2745-2749. DOI: 10.1093/jac/dkx286

[89] Poirel L, Jayol A, Bontron S, Villegas MV, Ozdamar M, Türkoglu S, et al. The mgrB gene as a key target for acquired resistance to colistin in *Klebsiella pneumoniae*. The Journal of Antimicrobial Chemotherapy. 2015;**70**:75-80. DOI: 10.1093/jac/dku323

[90] Hassan J, Eddine RZ, Mann D, Li S, Deng X, Saoud IP, et al. The mobile colistin resistance gene, mcr-1.1, is carried on IncX4 plasmids in multidrug resistant *E. coli* isolated from rainbow trout aquaculture. Microorganisms. 2020;**8**(11):1636. DOI: 10.3390/microorganisms8111636

[91] Shen Y, Zhang R, Schwarz S, Wu C, Shen J, Walsh TR, et al. Farm animals and aquaculture: significant reservoirs of mobile colistin resistance genes. Environmental Microbiology. 2020;**22**(7):2469-2484. DOI: 10.1111/1462-2920.14961

[92] Lozano-Leon A, Garcia-Omil C, Dalama J, Rodriguez-Souto R,

Martinez-Urtaza J, Gonzalez-Escalona N. Detection of colistin resistance *mcr*-1 gene in *Salmonella enterica* serovar Rissen isolated from mussels, Spain, 2012- to 2016. Euro Surveillance. 2019;**24**(16):1900200. DOI: 10.2807/1560-7917.ES.2019.24.16.1900200

[93] Yamaguchi T, Kawahara R, Harada K, Teruya S, Nakayama T, Motooka D, et al. The presence of colistin resistance gene *mcr*-1 and -3 in ESBL producing *Escherichia coli* isolated from food in Ho Chi Minh City, Vietnam. FEMS Microbiology Letters. 2018;**365**(11). DOI: 10.1093/femsle/fny100

[94] Hoa TTT, Nakayama T, Huyen HM, Harada K, Hinenoya A, Phuong NT, et al. Extended-spectrum beta-lactamase-producing *Escherichia coli* harbouring sul and mcr-1 genes isolates from fish gut contents in the Mekong Delta, Vietnam. Letters in Applied Microbiology. 2020;**71**(1):78-85. DOI: 10.1111/lam.13222

[95] Liu X, Li R, Zheng Z, Chen K, Xie M, Chan EW, et al. Molecular characterization of *Escherichia coli* isolates carrying *mcr*-1, *fos*A3, and extended-spectrum-β-lactamase genes from food samples in China. Antimicrobial Agents and Chemotherapy. 2017;**61**(6):e00064-e00017. DOI: 10.1128/AAC.00064-17

[96] Shen Y, Lv Z, Yang L, Liu D, Ou Y, Xu C, et al. Integrated aquaculture contributes to the transfer of *mcr*-1 between animals and humans via the aquaculture supply chain. Environment International. 2019;**130**:104708. DOI: 10.1016/j.envint.2019.03.056

[97] Liu D, Song H, Ke Y, Xia J, Shen Y, Ou Y, et al. Co-existence of two novel phosphoethanolamine transferase gene variants in *Aeromonas jandaei* from retail fish. International Journal of Antimicrobial Agents. 2020;**55**(1):105856. DOI: 10.1016/j.ijantimicag.2019.11.013

[98] Cabello FC, Tomova A, Ivanova L, Godfrey HP. Aquaculture and *mcr* colistin resistance determinants. MBio. 2017;**8**(5):e01229-e01217. DOI: 10.1128/mBio.01229-17

[99] Schar D, Klein EY, Laxminarayan R, Gilbert M, Van Boeckel TP. Global trends in antimicrobial use in aquaculture. Scientific Reports. 2020;**10**(1):21878. DOI: 10.1038/s41598-020-78849-3

[100] Xu F, Zeng X, Hinenoya A, Lin J. MCR-1 confers cross-resistance to bacitracin, a widely used in-feed antibiotic. mSphere. 2018;**3**(5): e00411-e00418. DOI: 10.1128/mSphere.00411-18

[101] Lentz SA, de Lima-Morales D, Cuppertino VM, Nunes Lde S, da Motta AS, Zavascki AP, et al. Letter to the editor: *Escherichia coli* harbouring *mcr*-1 gene isolated from poultry not exposed to polymyxins in Brazil. Euro Surveillance. 2016;**21**(26):30267. DOI: 10.2807/1560-7917.ES.2016.21.26.30267

[102] Touati M, Hadjadj L, Berrazeg M, Baron SA, Rolain JM. Emergence of *Escherichia coli* harbouring *mcr*-1 and *mcr*-3 genes in North West Algerian farmlands. Journal of Global Antimicrobial Resistance. 2020;**21**: 132-137. DOI: 10.1016/j.jgar.2019.10.001

[103] Brauer A, Telling K, Laht M, Kalmus P, Lutsar I, Remm M, et al. Plasmid with colistin resistance gene *mcr*-1 in extended-spectrum-β-lactamase-producing *Escherichia coli* strains isolated from pig slurry in Estonia. Antimicrobial Agents and Chemotherapy. 2016;**60**(11):6933-6936. DOI: 10.1128/AAC.00443-16

[104] Zurfuh K, Poirel L, Nordmann P, Nüesch-Inderbinen M, Hächler H, Stephan R. Occurrence of the plasmid-borne mcr-1 colistin resistance gene

in extended-spectrum-β-lactamase-producing *Enterobacteriaceae* in river water and imported vegetable samples in Switzerland. Antimicrobial Agents and Chemotherapy. 2016;**60**(4):2594-2595. DOI: 10.1128/AAC.00066-16

[105] Fernandes MR, Sellera FP, Esposito F, Sabino CP, Cerdeira L, Lincopan N. Colistin-resistant *mcr*-1-positive *Escherichia coli* on public beaches, an infectious threat emerging in recreational waters. Antimicrobial Agents and Chemotherapy. 2017;**61**(7): e00234-e00217. DOI: 10.1128/AAC.00234-17

[106] Sun P, Bi Z, Nilsson M, Zheng B, Berglund B, Stålsby Lundborg C, et al. Occurrence of blaKPC-2, blaCTX-M, and *mcr*-1 in *Enterobacteriaceae* from Well Water in Rural China. Antimicrobial Agents and Chemotherapy. 2017;**61**(4): e02569-e02516. DOI: 10.1128/AAC.02569-16

[107] Zhao F, Feng Y, Lü X, McNally A, Zong Z. IncP plasmid carrying colistin resistance gene *mcr*-1 in *Klebsiella pneumoniae* from hospital sewage. Antimicrobial Agents and Chemotherapy. 2017;**61**(2):e02229-e02216. DOI: 10.1128/AAC.02229-16

[108] Yao X, Doi Y, Zeng L, Lv L, Liu J-H. Dissemination of the *mcr*-1 colistin resistance gene. The Lancet. 2016;**16**:288

[109] Monte DF, Mem A, Fernandes MR, Cerdeira L, Esposito F, Galvão JA, et al. Chicken meat as a reservoir of colistin-resistant *Escherichia coli* strains carrying *mcr*-1 genes in South America. Antimicrobial Agents and Chemotherapy. 2017;**61**(5):e02718-e02716. DOI: 10.1128/AAC.02718-16

[110] Hammad AM, Hoffmann M, Gonzalez-Escalona N, Abbas NH, Yao K, Koenig S, et al. Genomic features of colistin resistant *Escherichia coli* ST69 strain harboring *mcr*-1 on IncHI2 plasmid from raw milk cheese in Egypt. Infection, Genetics and Evolution. 2019;**73**:126-131. DOI: 10.1016/j.meegid.2019.04.021

[111] ECDC. European Centre for Disease Prevention and Control. Antimicrobial Resistance Surveillance in Europe 2015. Annual Report of the European Antimicrobial Resistance Surveillance Network (EARS-Net). 2017. Available from: https://ecdc.europa.eu/en/publications-data/antimicrobial-resistance-surveillance-europe-2015 [Accessed: February 11, 2021]

[112] Liu Y, Liu JH. Monitoring colistin resistance in food animals, an urgent threat. Expert Review of Anti-Infective Therapy. 2018;**16**(6):443-446. DOI: 10.1080/14787210.2018.1481749

Chapter 7

Use of Humic Substances from Vermicompost in Poultry

Jesús A. Maguey-González, Sergio Gómez-Rosales, María de Lourdes Ángeles and Guillermo Téllez-Isaías

Abstract

In recent years, there has been a surge in antibiotic resistance in both humans and animals, as well as increased public concern over medication residues in animal products. As a result, the use of antibiotics as growth promoters in chicken has been banned in the European Union, and consumer pressure is likely to lead to their removal in other countries. More recently, the United States of America adopted the same restriction in 2017. Different alternatives to antibiotics have been proposed as a measure to eliminate pathogens or to improve growth and feed conversion in poultry, such as probiotics, enzymes, bacteriophages and antimicrobial peptides, herbal compounds and organic acids. These substances exert their effects on the gastrointestinal biota and digestion processes, directly or indirectly. Humic substances (HS) in animal applications have shown improved live-weight, growth rates and feed intakes by improving immune functions and gut health. In poultry nutrition as an alternative to growth-promoting has been proven with promising results on the growth and health of birds. Additional research suggests that HS can increase gut integrity and performance when combined with good nutrition, management, and biosecurity policies. Therefore, recent results of HS extracted from vermicompost in poultry will be described in this chapter.

Keywords: humic substances, vermicompost, gut health, poultry

1. Introduction

During the previous decade, the poultry business has been the most active and fastest-growing sector in the worldwide market for the production of meat directed to human consumption. The world's population is forecasted to double by 2050, and agricultural production is expected to double as well [1]. The supply of protein of animal origin in countries like Mexico is insufficient to fulfill the population's demand; nevertheless, the OECD-FAO Agricultural Outlook 2016–2025 projects that chicken meat output will expand at a pace of 2.9 per cent per year on average between average 2016 and 2025. As a result, national output and consumption will continue to rise [2].

Increased livestock output must be done with the most efficient use of resources as possible, without compromising environmental integrity or animal or human health. For achieving this, one of the most significant obstacles is the restriction to the use of growth-promoting antibiotics (GPA) as feed additives due to concerns about pathogenic bacteria developing antibiotic resistance, which represents a threat to human and animal health [3]. As a result, non-pharmacological growth promoters have been studied for decades in an attempt to replace or alternate the application of conventional promoters [4].

Humic substances (HS) are one of the alternative additives that has been studied for several years as a means to promote animal health. HS has been employed in agriculture for many years to enrich the soils, but because of their many qualities, they have been recently resurrected in environmental sciences and the human biomedical sector [5, 6]. In veterinary medicine, HS have been used as antidiarrheal, analgesic, immunostimulatory, and antibacterial agents [7]. When HS has been combined with the correct ration, husbandry, and biosecurity procedures, has been shown to enhance intestinal integrity and performance in chickens [8–11].

The majority of HS used in these studies were commercial products generated from mineral sources such as lignites and leonardite Vermicompost, from which compost and leachates are produced, is one of the first HS sources employed by Gomez and Angeles [11, 12], they have been reported to have promising advantages in poultry production as growth promoters.

2. Humic substances

HS are organic macromolecules that play an important role in biochemistry; they are a fraction of soil organic matter and have the highest density in soil and composts. They are produced by the biodegradation of organic matter, which involves physical, chemical, and microbiological processes [5], in which eukaryotes (worms and fungi) and prokaryotes (aerobic bacteria) further decompose organic matter [11]. HS are a natural component of streams, lakes, and oceans, containing the majority of the nutrients in the soil, accounting for approximately 80% of the carbon in soils and 60% of the dissolved carbon in the aquatic environment [13].

Based on their solubility, HS are classified as humic acids (HA), fulvic acids (FA) and humin. It has been stated that the isolating and characterizing of the organic components of soils is challenging due to the breakdown products of organic matter associated with other minerals [5]. For the first time, Senn and Kingman [14] characterized the molecular structure of HS, determining that the oxidized sites provide the molecule with a negative charge, allowing it to attach to mineral ions.

HA molecules have a wide range of weight and size, ranging from hundreds to thousands of atomic mass units, and are constituted of aromatic units, units linked by oxygen and nitrogen, functional groups mostly linked to carboxylic acids, phenols, and hydroxyl alcohol, ketone and quinone groups [15]. These chemical properties provide HA with the capacity to serve as a surfactant, binding to a variety of substances and generating hydrophobic and hydrophilic chemical complexes [16]. HS are excellent at transporting and binding organic and inorganic agents in the environment because of this function, which is combined with their colloidal characteristics [17].

The great electron transport capability of HS in oxidation–reduction processes is their principal attraction [6, 18]. HS also can develop bindings with ions such as Mg^{2+}, Ca^{2+}, Fe^{2+}, and Fe^{3+} due to the presence of carboxylic groups and phenolates. Creating chelate compounds with one or more of these ions and controlling metal ion bioavailability [19].

2.1 Use of humic substances in animals

HS have been used as an antidiarrheal, analgesic, immunostimulatory and antimicrobial agent in veterinary practices in Europe [7]. Furthermore, they have been shown to have a strong affinity for binding to heavy metals, mutagens, minerals, bacteria, and aflatoxins [20]. HS can be used orally in horses, ruminants, pigs and poultry for the treatment of diarrhoeas, dyspepsia and acute intoxications following the recommendations of the Committee for Veterinary Medicinal Products of the European Agency for the Evaluation of Medicinal Products [7].

In vitro studies of the antioxidant properties of HS in doses of 0.1 per cent in rat liver, the organ with the highest metabolic function and responsible for the metabolism of pharmacological compounds, found that HS aids in the elimination of free radicals and superoxide radicals, as well as maintaining the balance in the oxidation–reduction reactions of the mitochondria [21]. In rats with 2/3 of their liver removed, the administration of HS resulted in liver regeneration [18].

When rats were administered HS in their drinking water, Yasar [22] discovered a considerable rise in weight, as well as an increase in epithelial surface, intestinal villi length, and crypt depth.

2.2 Use of humic substances in poultry

The effects of different concentrations of HS in the diet on live weight, feed consumption, carcass characteristics, and gastrointestinal characteristics in broilers have been extensively studied [8, 23–26]. The addition of HS in the drinking water or feed improves most of the productive parameters, such as daily weight gain, in addition to enhancing the carcass yield of broilers.

According to Jin [27], the feed conversion does not change at 21 days but improves at 42 days, implying that the increase in weight gain and feed conversion efficiency might be due to the stimulating impact of HS in the digestive system and the nutrient utilization in metabolic processes. Adding HS to laying and broiler chickens can boost profitability by improving the production performance, reducing mortality, lowering feed conversion, and increasing egg output [28].

HS have been used to reduce mycotoxicosis in chicken due to its adsorbent capability [29–31]. Several studies have found that HS reduces ammonia emissions to the environment [9, 32, 33]. Similarly, in farms with a high density of chicken population, HS have been found to have a considerable anti-stress impact, reducing the negative effects of chronic stress in laying hens in production [34].

The mechanism through which HS affects poultry performance remains unclear. According to Shermer [35], HS might affect poultry performance by modifying the microbiota in the gastrointestinal tract, particularly in *Escherichia coli* populations, by altering the pH and promoting a greater activity of intestinal enzymes and feed digestibility. Several trace elements in HS can act as co-factors, increasing the activity of numerous enzymes involved in digestion and metabolic processes [36].

3. Humic substances derived from vermicompost

Vermicomposting is a bio-oxidative process that involves the breakdown of organic materials by litter-dwelling detritivorous earthworm species. Earthworms are important because they fragment organic materials and increase surface area, but microscopic species in the earthworm stomach and their castings are responsible for the actual decomposition [37, 38]. These gut and cast-related processes are considered to have a significant impact on the properties of vermicompost. *Eisenia fetida*, *Eisenia andrei*, and *Dendrobaena veneta* are the earthworm species used in vermicomposting. *E. fetida* is one of the most frequently utilized earthworm species due to its high rate of organic matter digestion, tolerance to environmental conditions, fast reproductive rate and short life cycle, and tolerance to handling [39].

Two phases can be distinguished in vermicomposting: 1) an active phase in which earthworms process the waste through physical comminution, ingestion, and microbial decomposition, and 2) a maturation-like phase in which earthworms move to fresher layers of undigested waste and microbes provide additional decomposition [40].

Bacteria and fungi are the two primary kinds of microorganisms involved in the composting process. Bacteria degrade sugars and other easily accessible organic materials faster than fungus. They are, therefore, crucial in the early phases of composting, when the feedstocks contain large quantities of carbohydrates. More complex compounds, such as hemicellulose, starch, and even lignin, can be degraded by actinomycetes. They are more common in the later stages of composting, after the majority of easily degradable substrates have been used [41].

The fresh organic matter (animal manure, food wastes, green wastes, agricultural leftovers, etc.) is transformed into more stable humus-like substances, nutrients are recycled, and energy is created throughout the composting process [42]. The process should be aerobic, with a portion of it taking place under thermophilic circumstances. It minimizes phytotoxicity, removes pathogens, and stabilizes the material in terms of nitrogen and oxygen demand, avoiding N immobilization by soil biota, which competes with the plant for limited nitrogen resources [43].

3.1 Establishment of vermicompost

The establishment of our vermicompost took place in the greenhouse area, which has concrete and stone walls and a white polyethylene roof (**Figure 1**). The temperature inside the greenhouse is maintained throughout the year, with maximum temperatures of 35°C and minimum temperatures of 15°C, as well as relative humidity of 70–80%. *Eisenia Foetida* is used in vermicompost; and it is composed of 50% pastures from the area/50% sheep manure; with an internal temperature of 15–25°C, a pH of 6.5 to 7.5 and relative humidity of 80–90%.

Vermicompost is made by enclosing a 2 × 5 m rectangular space with bricks and covering it with polypropylene (**Figure 2**). To reduce the bacterial content, the organic material and manure are first pre-composted for a month. The components are then combined and placed in the compost's lower side; worms are then sewn on top, and the compost is watered. Finally, a part of the organic material is coated over the compost to protect it from light rays, covered with black polyethylene (**Figure 3**).

The compost is moistened once a week, vegetable matter is added once a week, and all parameters are checked regularly for upkeep. To collect the compost, the vermicompost must grow for four months. The worm harvest is conducted first, and then the compost is sifted to eliminate any unprocessed organic and inorganic elements. The compost was preserved after being oven-dried for 24 hours at 60°C.

Figure 1.
Greenhouse area.

Figure 2.
Delimitation of the area with bricks and covered with polypropylene.

3.2 Humic substances extraction/isolation

HA, FA, and humin are the three main types of humic compounds found in soils and sediments [5, 13]. A strongly basic aqueous solution of sodium hydroxide or potassium hydroxide is used to extract HS from humus and other solid phases. The HS precipitate in this solution when the pH is adjusted to 1 with hydrochloric or sulfuric acid, leaving the FA in the solution (upper section) [19, 44]. This is the most important distinction: FA are insoluble in alkaline media, while HA are insoluble in acid media, and humin are insoluble at any pH (**Figure 4**).

Figure 3.
Vermicompost is covered with organic matter and polyethylene.

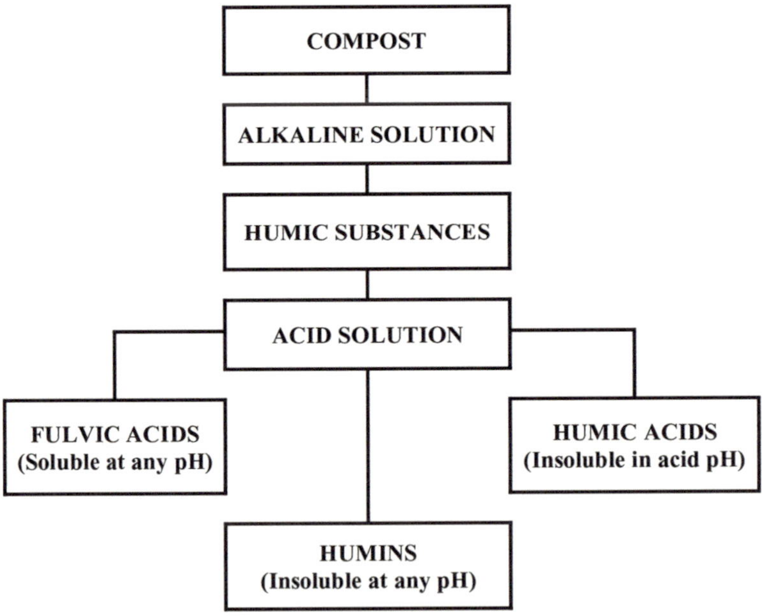

Figure 4.
Fractions of humic substances.

In addition to alkaline solvents, chelating agents, organic solvents, and aqueous saline solutions have been proposed for the extraction of HS. Alkaline solvents are the most efficient and commonly utilized of them [44]. The extraction of HA with NaOH is a standard method for isolating HA, with an extraction efficiency of more than 80% of samples from soils (**Table 1**).

The isolation and extraction of HS from the earthworm compost were carried out using an alkaline extraction process. Sodium hydroxide (0.1 M NaOH) was used in a ratio of 5:1 parts of compost (mL/g); it was left to rest for 24 h at room temperature, and then filtered through a 125 μm mesh; sulfuric acid was added (H2SO4, 10%),

Agent	HS extraction (%)
NaOH	80
Na4P2O7	30
Organic chelators	30
Acetyl-acetone/Hydroxyquinone	30
Formic acid	55

Table 1.
Agents used for the extraction of humic substances.

	Humic acid	Fulvic acid
Formula	$C_{110}H_{105}N_7O_{50}$	$C_{25}H_{17}N_0O_{18}$
Molecular weight	2325.028	619.398
Elemental composition (%)	C (56.82), H (4.55), N (4.22), O (34.41)	C (48.48), H (2.77), N (2.26), O (46.49)
Density, g/cm^3	1.870 ± 0.10	1.935 ± 0.06

Table 2.
Estimated chemical properties of the humic and fulvic acid molecules, extracted from vermicompost.

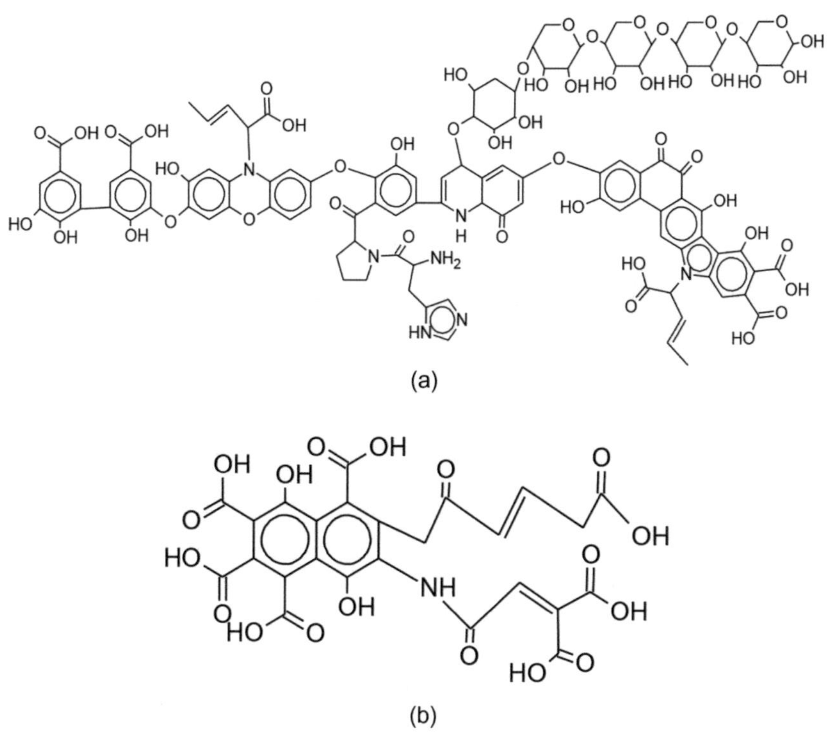

Figure 5.
The flat structure of a humic acid molecule with aromaticity (a) and flat structure of a fulvic acid molecule with aromaticity (b).

rectifying a pH of 2. The solids and liquids were separated by decantation. The precipitate (HA) was washed two times with distilled water to remove sulfuric acid residues and between each washing it was centrifuged for 20 min at 5000 rpm. Then, to remove the remnants of sulfuric acid, in a rotary evaporator, the sample was dried at 60°C until it had a gel consistency. Finally, it was dried in an oven at 60°C. The result was a yellowish-brown powder with a pH of 10.

For the identification of functional groups, the extracted HS were analyzed using an infrared spectrophotometer with Fourier transformation with attenuated total reflectance (FTIR-ATR); the elemental analysis was carried out using energy dispersive X-Ray spectroscopy (EDS); and the crystal types were detected with X-ray diffraction (XRD). These results were used for the calculation of aromaticity and were published in a previous paper [45]. Additionally, the chemical properties (**Table 2**) and the flat structures of the HS molecules with aromaticity (**Figure 5**) using the chemistry software ACD Lab v.12 (Advanced Chemistry Development, Toronto, Canada) were estimated [46].

4. Use of humic substances from vermicompost in poultry

Various *in vitro* and *in vivo* models for evaluating the behavior of different chemicals in chicken supplements have recently been established in our lab. The *in vitro* digestion model replicates broiler body temperature, peristaltic motions, enzymatic and pH conditions in each simulated compartment (crop, proventriculus, and small intestine) [47, 48]. In addition, *in vivo* models of intestinal inflammation in birds have been employed. Non-starch polysaccharide-rich diets [49, 50]; dexamethasone [51]; dextran sodium sulfate (DSS) [52, 53]; and 24 hours of feed restriction [54, 55]. With the help of these research models, we have been able to elucidate some effects of HA.

Firstly, two experiments were conducted to evaluate the effects of HA on recovery of *Salmonella* Enteritidis (SE) [56]; there were no effects of HA on SE recovery in an *in vitro* digestive system, or in Salmonella intestinal colonization, bacterial numbers in ceca, intestinal IgA, or serum FITC-d in neonatal broiler chicks.

Treatment	**Intestinal viscosity ($cP\ Log_{10}$)[b]**	**Serum FITC-d (ng/mL)[c]**	**Liver bacterial translocation (Log_{10} cfu/g)[d]**	
			cfu Log_{10} of liver	**Liver enrichment culture[e]**
Control FR	0.13 ± 0.01^f	828.58 ± 32.85^g	2.83 ± 0.09^g	12/12(100%)
0.2% HA + FR	0.24 ± 0.01^g	544.62 ± 41.84^f	1.60 ± 0.04^f	7/12 (58%)*

[a]*Data expressed as Mean ± SE.*
[b]*Intestinal viscosities evaluated in Log_{10} (in centipoise, cP = 1/100 dyne sec/cm^2), n = 5 chickens/group.*
[c]*Serum (FITC-d) was evaluated in 20 chickens/group.*
[d]*Liver bacterial translocation was evaluated in 12 chickens/group.*
[e]*Data expressed as positive/total chickens (%).*
[f,g]*Superscripts within columns indicate significant difference at $P < 0.05$. *$P < 0.001$.*

Table 3.
Evaluation of intestinal viscosity, serum FITC-d, and bacterial liver translocation in chickens consuming a corn-based diet with or without the inclusion of 0.2% of humic acids following 24 hours of feed restriction (FR) in broiler chickens[a].

Additionally, a second study was carried out [57], to evaluate direct or indirect impacts of HA on intestinal integrity because of their physical and chemical features. Using a 24-hour feed restriction model, the aim was to investigate how HA affected intestinal viscosity, leaky gut, and ammonia excretion in broiler chicks. The experimental group was given 0.2% HA had increased intestinal viscosity and showed lower levels of FITC-d, bacterial liver translocation, and ammonia in the excreta. It was confirming its advantages, enhancing the viscosity and the integrity of the intestine (**Table 3**).

The previous researches were the first to address the effects of HA from vermicompost on bacterial challenges with SE both *in vitro* and *in vivo* in chicks; furthermore, the mechanism of action of HA on the maintenance of intestinal integrity in poultry was demonstrated for the first time. In line with this finding, Mudronová [58] reported that HS positively regulates MUC-2 gene expression.

In further research in broilers fed an extract of HS from vermicompost, increased carcass yield and lactic acid bacteria (LAB) and reduced coccidian oocysts excretion were observed; but increased *Clostridium perfringens* (CP) counts were also seen compared to broilers fed diets supplemented with GPA (**Figure 6**) [45].

In a recent report, broilers kept in floor pens from 1 to 42 days of age and fed increasing levels of HS from vermicompost, showed lower feed intake (FI) and overall mortality, besides, a better feed conversion rate (FCR) compared to negative control birds not supplemented with HS and positive control birds added with antibiotics (**Table 4**) [59]. The greatest benefits of adding HS were observed in the last period of the trial, from 29 to 42; these findings closely resemble the observations in previous research. For the authors, it is unknown whether improved FCR from 29 to 42 days was dependent on, or independent of, the addition of HS from 1 to 14 and 15–28 days. This topic deserves further clarification in future research.

Although, HS have been proven to have prospective benefits in chicken production as growth promoters, data on their antimicrobial properties are inconsistent. Using an *in vitro* chicken digestive system, it was essential to examine the effect of HA isolated from vermicompost on the recovery of *Salmonella Enteritidis* (SE), *E. coli* (EC), *C. perfringens* (CP), *Bacillus subtilis* (BS), and *Lactobacillus salivarius* (LS). In general,

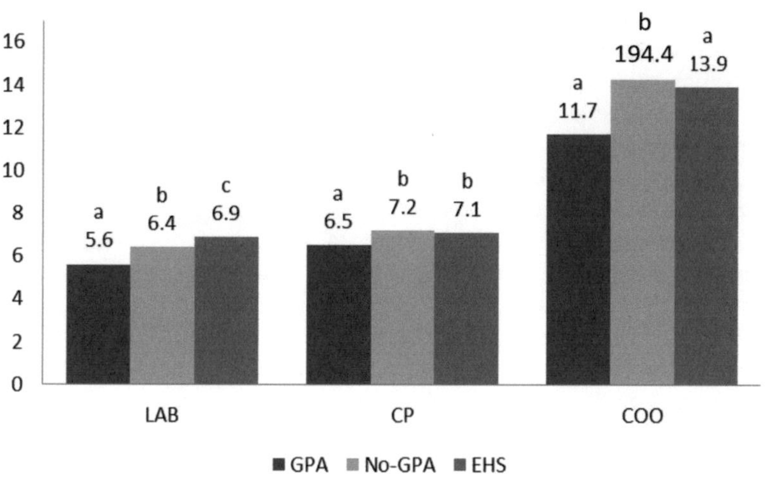

Figure 6.
Counts of lactic acid bacteria (LAB) and Clostridium perfringens (CP), and the number of coccidian oocysts excreted in broilers fed with humic substances.

	Level of humic substances (%)						
	Positive control	0	0.15	0.30	0.45	SEM[a]	p-Value
FI (g/d)	127[b]	125[b,c]	122[c,d]	120[d]	121[d]	1.458	0.01
WG (g/d)	60.8	58.4	59.8	58.3	58.7	0.577	0.29
FCR	2.09[b,c]	2.15[b]	2.04[c]	2.05[c]	2.06[c]	0.019	0.01
Mortality (%)	7.1[c,d]	13.3[c]	5.4[d]	1.7[d]	3.78[d]	2.37	0.05

[a]*SEM = estandard error of the mean.*
[b–d]*Means with a different superscript in the same row differ significantly (P < 0.05).*

Table 4.
Growth performance of broilers from 1 to 42 days of age fed increasing levels of humic substances extracted from vermicompost.

Treatment	Crop counts			Proventriculus counts			Intestine counts		
	SE	EC	CP	SE	EC	CP	SE	EC	CP
Control–	0.00[a]	0.00[a]	0.00[a]	0.00[a]	0.00[a]	0.00[a]	0.00[a]	0.00[a]	0.00[a]
Control+	7.18[b]	7.18[b]	2.19[b]	2.30[b]	2.32[b]	1.12[b]	6.41[b]	6.43[b]	1.61[b]
HA (0.1%)	7.07[b]	7.54[c]	2.16[b]	2.30[b]	2.48[b,c]	1.18[b]	6.82[c]	6.82[c]	1.81[b,c]
HA (0.2%)	7.18[b]	7.52[c]	2.34[b]	2.59[b]	2.84[c]	1.18[b]	7.49[d]	7.24[d]	1.94[c]
HA (0.5%)	6.79[c]	7.72[d]	2.28[b]	3.73[c]	3.48[d]	1.12[b]	8.15[e]	7.68[e]	2.21[d]
HA (1%)	6.89[c]	7.88[d]	2.36[b]	4.72[d]	4.06[d]	1.16[b]	8.15[e]	8.40[f]	2.27[d]
SEM***	0.069	0.104	0.126	0.303	0.189	0.072	0.136	0.112	0.076

**The initial inoculum of SE, EC and CP in the feed was 10^8 CFU/g.*
***Data are expressed as $\log_{10}$ CFU.*
****Standard error of the mean.*
[a–f]*Values in columns with different letters differ significantly (P ≤ 0.0001).*

Table 5.
Effect of humic acids on the counts of Salmonella Enteritidis, Escherichia coli and Clostridium perfringens under an in vitro poultry digestive model**.*

the number of microorganisms counted increased in the final simulation, remarkably when the HA inclusion concentration increased (**Tables 5** and **6**).

HS may be utilized as substrates by bacteria since they are organic sources of carbon, nitrogen, phosphorus, and other nutrients. In this way, they can be used as prebiotics to enhance nutrient digestion.

As previously stated, HS have the potential to chelate minerals; when added to drinking water at concentrations of 345.0, 322.5 and 347.8 g/L, improvements in bone mineralization in broilers at 21 (**Table 7**) and 42d (**Table 8**) of age have been observed [46].

Subsequently, 28-year-old broilers were challenged to a radical change in diet, in addition to inducing immune stress when vaccinated against Newcastle, Infectious bronchitis and Salmonella, with the intention of causing damage to the intestinal mucosa. The results of this research [60] show that HS does not prevent intestinal mucosal atrophy but increase the number of goblet cells, and probably the mucus layer, compared to chickens that received the treatment without GPA.

On the other hand, when compared to a commercial zeolite, the adsorption capacity of HA against AFB1 when added to the feed in a 1% concentration exhibited the maximum effectivity of capture (**Figure 7**).

Treatment	Crop counts		Proventriculus counts		Intestine counts	
	BS	LS	BS	LS	BS	LS
Control-	0.00^{a}	0.00^{a}	0.00^{a}	0.00^{a}	0.00^{a}	0.00^{a}
Control+	6.72^{b}	4.64^{b}	1.46^{b}	2.51^{b}	4.33^{b}	2.51^{b}
HA (0.1%)	$6.69^{b,c}$	4.50^{b}	1.67^{b}	2.45^{b}	4.30^{b}	2.45^{b}
HA (0.2%)	$6.57^{c,d}$	5.13^{c}	1.68^{b}	2.52^{b}	4.40^{b}	2.52^{b}
HA (0.5%)	6.52^{d}	5.12^{c}	$1.70^{b,c}$	2.52^{b}	5.00^{c}	2.52^{b}
HA (1%)	6.71^{b}	5.49^{d}	1.89^{c}	2.49^{b}	5.01^{c}	2.49^{b}
SEM***	0.054	0.140	0.079	0.079	0.065	0.079

**The initial inoculum of BS and LS in the feed was 10^8 CFU/g.*
***Data are expressed as log_{10} CFU.*
****Standard error of the mean.*
$^{a-e}$ Values in columns with different letters differ significantly ($P \leq 0.0001$).

Table 6.
Effect of humic acids on the counts of Bacillus subtilis and Lactobacillus salivarius under an in vitro poultry digestive model**.*

	Humic substances, µg/L of water					
	0	161	322	483	654	SEM[b]
Dry matter (%)	35.24	36.15	35.77	35.87	36.48	0.522
Dry Weight (%)	2.10	1.99	2.07	2.05	2.13	0.060
Ashes (%)	38.11^{d}	40.13^{e}	$39.32^{d,e}$	$39.18^{d,e}$	39.57^{e}	0.539
Ashes (mg)	802.0	797.0	814.8	804.8	845.3	28.075
Ca (%)	$40.20^{f,g}$	36.94^{f}	42.04^{g}	42.20^{g}	36.70^{f}	1.648
Ca (mg)	320.3	298.5	342.3	339.4	307.9	18.551
P (%)	14.02	14.01^{d}	$14.10^{d,e}$	14.15^{e}	14.14^{e}	0.050
P (mg)	112.6	111.8	114.8	113.9	119.4	4.203

$^{a}n = 12$, using six tibias per replicate..
^{b}SEM = standard error of the mean.
^{d,e}Means with a different superscript in the same row differ significantly ($P < 0.05$).
^{f,g}Means with a different superscript in the same row differ significantly ($P < 0.01$).

Table 7.
Dry matter, ashes, calcium and phosphorus content of the tibia of 21 days old broiler chickens added with increasing levels of humic substances in the drinking water[a].

	Humic substances, µg/L of water					
	0	161	322	483	654	SEM[b]
Dry matter (%)	42.25^{d}	$43.06^{d,e}$	$43.16^{d,e}$	43.35^{e}	42.17^{d}	0.440
Dry Weight (%)	7.79	8.32	8.26	8.39	8.21	0.218
Ashes (%)	35.84	35.64	35.11	36.34	35.34	0.596
Ashes (mg)	2785.5	2966.1	2894.4	3046.8	2891.0	73.941
Ca (%)	34.73^{d}	$33.17^{d,e}$	35.81^{d}	35.98^{d}	31.11^{e}	1.242
Ca (mg)	$969.0^{f,g}$	$984.0^{f,g}$	1039.2^{f}	1102.5^{f}	891.9^{g}	45.542

	Humic substances, μg/L of water					
	0	**161**	**322**	**483**	**654**	**SEM** [b]
P (%)	$14.02^{f,g}$	13.98^{f}	14.04^{g}	13.97^{f}	13.97^{f}	0.019
P (mg)	390.6^{d}	$414.5^{d,e}$	$409.9^{d,e}$	425.9^{e}	402.0^{d}	10.187

[a] *n = 12, using six tibias per replicate.*
[b] *SEM = estandard error of the mean.*
[d,e] *Means with different superscript in the same row differ significantly (P < 0.05).*
[f,g] *Means with different superscript in the same row differ significantly (P < 0.01).*

Table 8.
Dry matter, ashes, calcium and phosphorus content of the tibia of 42 days old broiler chickens added with increasing levels of humic substances in the drinking water[a].

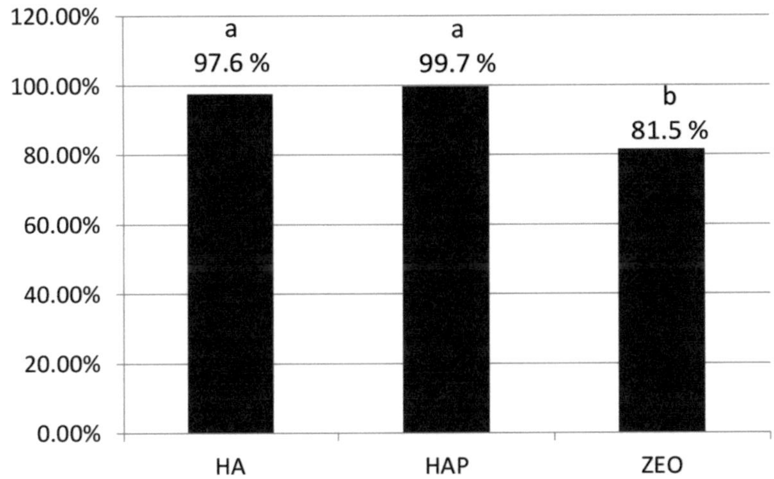

Figure 7.
Adsorption capability of humic acids (HA), humic acids purified (HAP) and zeolite (ZEO) against AFB1 using an in vitro poultry digestive model.

Finally, the addition of HS extracted from vermicompost in piglet feed at weaning [61] improved the weight gain and feed efficiency during 1–42 days post-weaning. The final body weight at 42 days, weight gain and feed efficiency were higher with the inclusion level of 0.50% HS. Most of the productive responses showed an increasing linear pattern as the addition of HS increased.

5. Conclusions

Although, the antibacterial activity of HS/HA is unclear, they might be regarded as viable alternatives to replace or alternate the use of antibiotics in poultry. As was documented, HS extracted from vermicompost improve performance, increase intestinal viscosity and intestinal integrity in poultry. These benefits have been attributed to the macro colloidal structure of HS, which ensures effective protection on gastric and intestinal mucous membranes, as well as the promotion of mucin synthesis in the gut. However, their applications depend on the source and the way they are extracted. HS stimulates the growth of microorganisms, suggesting that they can be used as prebiotics.

In addition, multiple doses were investigated, which might have contributed to the differences in response variables among our investigations. It can also be argued that whether in a more challenging environment, the effects of HS may be better. Finally, our working team is focusing on elucidating the mechanism by which HS exert their effects on the intestinal mucosa and the intestinal microbiota.

Acknowledgements

This research was supported by the Arkansas Biosciences Institute under the project: Development of an avian model for evaluation early enteric microbial colonization on the gastrointestinal tract and immune function.

CONACYT provided funding to carry out several experiments described in this chapter through the project "Efficacy of an extract of humic substances as a growth promoter, reducing infectious diseases and increasing the profitable production of meat of birds and pork" (PDCPN 2017/4777).

Conflict of interest

The authors declare no conflict of interest.

Author details

Jesús A. Maguey-González[1*], Sergio Gómez-Rosales[2], María de Lourdes Ángeles[2] and Guillermo Téllez-Isaías[3]

1 National Autonomous University of Mexico (UNAM), Postgraduate Unit, University City, Coyoacán, Mexico City, Mexico

2 National Center of Disciplinary Research in Animal Physiology and Genetics, National Institute of Research in Forestry, Agriculture and Livestock, Ajuchitlan, Queretaro, Mexico

3 Department of Poultry Science, University of Arkansas, Fayetteville, AR, USA

*Address all correspondence to: magueyjesus@gmail.com

References

[1] La CA, La AY, Todos NP. El Estado Mundial de la Pesca y la Acuicultura. ROMA: FAO; 2016

[2] Fideicomisos Instituidos en Relación con la Agricultura (FIRA). Panorama agroalimentario, avicultura carne; 2016

[3] Van Vuuren M. Antibiotic resistance with special reference to poultry production. Conference OIE. 2001:135-146

[4] Huyghebaert G, Ducatelle R, Van Immerseel F. An update on alternatives to antimicrobial growth promoters for broilers. The Veterinary Journal. 2011;**187**:182-188

[5] Peña-Méndez EM, Havel J, Patočka J. Humic substances-compounds of still unknown structure: Applications in agriculture, industry, environment, and biomedicine. Journal of Applied Biomedicine. 2005;**3**:13-24

[6] Aeschbacher M, Graf C, Schwarzenbach RP, Sander M. Antioxidant properties of humic substances. Environmental Scince Technology. 2012;**46**:4916-4925. DOI: 10.1021/es300039h

[7] EMEA. Committee for Veterinary Medical Products. Humic Acids and their Sodium Salts. 1999. Available from: http:/www.emea.europa.eu/pdfs/vet/mrls/055499en.pdf

[8] Karaoglu M, Macit M, Esenbuga N, Durdag H, Turgut L, Bilgin ÖC. Effect of supplemental humate at different levels on the growth performance, slaughter and carcass traits of broilers. International Journal of Poultry Science. 2004;**3**:406-410

[9] Ji F, McGlone JJ, Kim SW. Effects of dietary humic substances on pig growth performance, carcass characteristics, and ammonia emission. Journal of Animal Science. 2006;**84**:2482-2490

[10] Ipek H, Avci M, Iriadam M, Kaplan O, Denek N. Effects of humic acid on some hematological parameters, total antioxidant capacity and laying performance in Japanese quails. Arch Geflugelk. 2008;**72**:56-60

[11] Gomez-Rosales S, de Angeles ML. Addition of a worm leachate as source of humic substances in the drinking water of broiler chickens. Asian-Australas Journal of Animal Science. 2015;**28**: 215-222. DOI: 10.5713/ajas.14.0321

[12] Gómez-Rosales S, Angeles M, Maguey-González J. Productive responses and nitrogen balance in broilers fed with humic substances in the drinking water. Abanico Veterinario. 2021;**11**:e120. Epub 08 de noviembre de 2021. DOI: 10.21929/abavet2021.30

[13] Lehmann J, Kleber M. The contentious nature of soil organic matter. Nature. 2015;**528**:60-68. DOI: 10.1038/nature16069

[14] Senn TL, Kingman AR. A review of humus and humic acids. Research Series. 1973;**145**:1-5

[15] Saar RA, Weber JH. Complexation of cadmium (II) with water-and soil-derived fulvic acids: Effect of pH and fulvic acid concentration. Canadian Journal of Chemistry. 1979;**57**:1263-1268

[16] Gaffney JS, Marley NA, Clark SB. Humic and fulvic acids and organic colloidal materials in the environment. In: Gaffney JS, Marley NA, Clark SB, editors. Humic and Fulvic Acids. Washington, DC: ACS Publications; 1996. pp. 2-16

[17] Piccolo A. The supramolecular structure of humic substances: A novel understanding of humus chemistry and implications in soil science. Advances in Agronomy. 2002;**75**:57-134. DOI: 10.1016/S0065-2113(02)75003-7

[18] Maśliński C, Fogel WA, Andrzejewski W. The influence of Tołpa peat preparation (TPP) on rat liver regeneration. Acta poloniae pharmaceutica. 1993;**50**(4-5):413-416

[19] Stevenson FJ. Humus Chemistry: Genesis, Composition, Reactions. New York: John Wiley and Sons; 1982

[20] Islam KMS, Schuhmacher A, Gropp JM. Humic acid substances in animal agriculture. Pakistan Journal Nutrition. 2005;**4**:126-134

[21] Vašková J, Veliká B, Pilátová M, Kron I, Vaško L. Effects of humic acids *in vitro*. In Vitro Cellular and Developmental Biology—Animal. 2011;**47**:376-382. DOI: 10.1007/s11626-011-9405-8

[22] Yasar S, Gokcimen A, Altuntas I, Yonden Z, Petekkaya E. Performance and ileal histomorphology of rats treated with humic acid preparations. Journal of Animal Physioly and Animal Nutrition. 2002;**86**:257-264

[23] Kocabagli N, Alp M, Acar N, Kahraman R. The effects of dietary Humate supplementation on broiler growth and carcass yield. Poultry Science. 2002;**81**:227-230

[24] Ceylan N, Ciftci I, Ilhan Z. The effects of some alternative feed additives for antibiotic growth promoters on the performance and gut microflora of broiler chicks. Turkish journal of veterinary and animal. Science. 2003;**27**:727-733

[25] Rath NC, Huff WE, Huff GR. Effects of humic acid on broiler chickens. Poultry Science. 2006;**85**:410-414

[26] Ozturk E, Ocak N, Coskun I, Turhan S, Erener G. Effects of humic substances supplementation provided through drinking water on performance, carcass traits and meat quality of broilers. Journal of Animal Physiology and Animal Nutrition. 2010;**94**(1):78-85

[27] Jin LZ, Ho YW, Abdullah N, Jalaludin S. Growth performance, intestinal microbial populations, and serum cholesterol of broilers fed diets containing Lactobacillus cultures. Poultry Science. 1998;**77**:1259-1265

[28] Yoruk MA, Gul M, Hayirli A, Macit M. The effects of supplementation of Humate and probiotic on egg production and quality parameters during the late laying period in hens. Poultry Science. 2004;**83**:84-88

[29] Van Rensburg CJ, Van Rensburg CE, Van Ryssen JB, Casey NH, Rottinghaus GE. *In vitro* and *in vivo* assessment of humic acid as an aflatoxin binder in broiler chickens. Poultry Science. 2006;**85**:1576-1583

[30] Gahhri H, AHbibian R, Fam MA. Effect of sodium bentonite, mannan oligosaccharide and humate on performance and serum biochemical parameters during aflatoxicosis in broiler chickens. Global Veterinaria. 2010;**5**:129-134

[31] Arafat RY, Khan SH, Saima. Evaluation of humic acid as an aflatoxin binder in broiler chickens. Annals of Animal Science. 2017;**17**:241-255. DOI: 10.1515/aoas-2016-0050

[32] Henderson PA. The growth of tropical fishes. In: Val AL, Vera MF, Randall DJ, editors. The Physiology

of Tropical Fishes. Vol. 21. New York: Academic Press; 2005. pp. 85-100

[33] Zralý Z, Písaříková B, Trčková M, Navrátilová M. Effect of humic acids on lead accumulation in chicken organs and muscles. Acta Veterinaria Brno. 2008;**77**:439-445. DOI: 10.2754/avb200877030439

[34] Cetin E, Guclu BK, Cetin N. Effect of dietary humate and organic acid supplementation on social stress induced by high stocking density in laying hens. Journal of Animal and Veterinary Advances. 2011;**10**:2402-2407. DOI: 10.3923/javaa.2011.2402.2407

[35] Shermer CL, Maciorowski KG, Bailey CA, Byers FM, Ricke S. Caecal metabolites and microbial populations in chickens consuming diets containing amined humate compound. Journal of Science Food and Agriculture. 1998;**77**:479-486

[36] Hayirli A, Esenbuga N, Macit M, Lacin E, Karaoglu M, Karaca H, et al. Nutrition practice to alleviate the adverse effects of stress on laying performance, metabolic profile, and egg quality in peak producing hens: I. The humate supplementation. Asian Australas. Journal of Animal Science. 2005;**18**:1310-1319

[37] Lazcano C, Gómez-Brandón M, Domínguez J. Comparison of the effectiveness of composting and vermicomposting for the biological stabilization of cattle manure. Chemosphere. 2008;**72**(7):1013-1019

[38] Diaz LF, De Bertoldi M, Bidlingmaier W. Microbiology of the composting process. In: Diaz LF, de Bertoldi M, Bidlingmaier W, editors. Compost Science and Technology. Elsevier; 2011. pp. 25-48

[39] Ryckeboer J, Mergaert J, Vaes K, Klammer S, De Clercq D, Coosemans J, et al. A survey of bacteria and fungi occurring during composting and self-heating processes. Annals of Microbiology. 2003;**53**(4): 349-410

[40] Domínguez J, Gómez-Brandón M. The influence of earthworms on nutrient dynamics during the process of vermicomposting. Waste Management & Research. 2015;**31**(8):859-868. Ryckeboer. 2003

[41] Raviv M. Production of high-quality composts for horticultural purposes—A mini-review. HortTechnology. 2005;**15**:52-57

[42] Bernal MP, Alburquerque JA, Moral R. Composting of animal manures and chemical criteria for compost maturity assessment. A review. Bioresource Technology. 2005;**100**(22):5444-5453

[43] Baglieri A, Ioppolo A, Negre M, Gennari M. A method for isolating soil organic matter after the extraction of humic and fulvic acids. Organic Geochemistry. 2007;**38**(1):140-150

[44] De Souza F, Bragança SR. Extraction and characterization of humic acid from coal for the application as dispersant of ceramic powders. Journal of Materials Research and Technology. 2018;7(3):254-260

[45] Dominguez-Negrete A, Gómez-Rosales S, de Angeles ML, López-Hernández LH, Reis-de Souza TC, López-Garc'ia Y, et al. Effect of the addition of humic substances as growth promoter in broiler chickens under two feeding regimens. Animals, Multidisciplinary Digital Publishing Institute. 2019;**9**(12):1101

[46] Lourdes-Angeles M, Gómez-Rosales S, López-García YR, Montayo-Franco A. Growth performance and tibia mineralization of broiler chickens supplemented with a liquid extract of humic substances. Brazilian. Journal of Poultry Science. 2021:1450 (In press)

[47] Annett CB, Viste JR, Chirino-Trejo M, Classen HL, Middleton DM, Simko E. Necrotic enteritis: Effect of barley, wheat and corn diets on proliferation of *Clostridium perfringens* type A. Avian Pathology. 2002;**31**:598-601

[48] Latorre JD, Hernandez-Velasco X, Kuttappan VA, Wolfenden RE, Vicente JL, Wolfenden AD, et al. Selection of *Bacillus* spp. for cellulase and xylanase production as direct-fed microbials to reduce digesta viscosity and *Clostridium perfringens* proliferation using an in vitro digestive model in different poultry diets. Frontiers in Veterinary Science. 2015;**2**(25). DOI: 10.3389/fvets.2015.00025

[49] Tellez G, Latorre JD, Kuttappan VA, Kogut MH, Wolfenden A, Hernandez-Velasco X, et al. Utilization of rye as energy source affects bacterial translocation, intestinal viscosity, microbiota composition, and bone mineralization in broiler chickens. Frontiers in Genetics. 2014;**5**:339. DOI: 10.3389/fgene.2014.00339

[50] Tellez G, Latorre JD, Kuttappan VA, Hargis BM, Hernandez-Velasco X. Rye affects bacterial translocation, intestinal viscosity, microbiota composition and bone mineralization in Turkey poults. PLoS One. 2015;**10**:e0122390. DOI: 10.1371/journal.pone.0122390

[51] Vicuña EA, Kuttappan VA, Galarza-Seeber R, Latorre JD, Faulkner OB, Hargis BM, et al. Effect of dexamethasone in feed on intestinal permeability, differential white blood cell counts, and immune organs in broiler chicks. Poultry Science. 2015;**94**:2075-2080. DOI: 10.3382/ps/pev211

[52] Kuttappan VA, Berghman L, Vicuña E, Latorre JD, Menconi A, Wolchok J, et al. Poultry enteric inflammation model with dextran sodium sulfate mediated chemical induction and feed restriction in broilers. Poultry Science. 2015;**94**:1220-1226. DOI: 10.3382/ps/pev114

[53] Menconi A, Hernandez-Velasco X, Vicuña EA, Kuttappan VA, Faulkner OB, Tellez G, et al. Histopathological and morphometric changes induced by a dextran sodium sulfate (DSS) model in broilers. Poultry Science. 2015;**94**:906-911. DOI: 10.3382/ps/pev054

[54] Kuttappan VA, Vicuña EA, Latorre JD, Wolfenden AD, Téllez GI, Hargis BM, et al. Evaluation of gastrointestinal leakage in multiple enteric inflammation models in chickens. Frontiers in Veterinary Sciemce. 2015;**2**:66. DOI: 10.3389/fvets.2015.00066

[55] Vicuña EA, Kuttappan VA, Tellez G, Hernandez-Velasco X, Seeber-Galarza R, Latorre JD, et al. Dose titration of FITC-D for optimal measurement of enteric inflammation in broiler chicks. Poultry Science. 2015;**94**:1353-1359. DOI: 10.3382/ps/pev111

[56] Maguey-Gonzalez JA, Michel MA, Baxter MF, Solis-Cruz B, Hernandez-Patlan D, Merino-Guzman R, et al. Effects of humic acids on recovery of Salmonella enterica serovar Enteritidis. Annals of Animal Science. 2018;**18**(2):387-399. Sciendo

[57] Maguey-Gonzalez JA, Michel MA, Baxter MF, Tellez G Jr, Moore PA Jr, Solis-Cruz B, et al. Effect of humic

acids on intestinal viscosity, leaky gut and ammonia excretion in a 24 hr feed restriction model to induce intestinal permeability in broiler chickens. Animal Science Journal. 2018;**89**(7):1002-1010

[58] Mudroňová D, Karaffová V, Pešulová T, Koščová J, Maruščáková IC, Bartkovský M, et al. The effect of humic substances on gut microbiota and immune response of broilers. Food and Agricultural Immunology. 2020;**31**(1):137-149

[59] Dominguez-Negrete A, Gómez-Rosales S, de Angeles ML, López-Hernández LH, Reis-de Souza TCY, Latorre-Cárdenas JD, et al. Addition of different levels of humic substances extracted from worm compost in broiler feeds. Animals. 2021;**11**:3199. DOI: 10.3390/ani11113199

[60] López-García YR, Gómez Rosales S, Angeles ML. Efecto de la adición de sustancias húmicas sobre la histología y número de células caliciformes en la mucosa intestinal de pollos de engorda. In: LVI Reunión Nacional de investigación pecuaria (RNIP); 10-12 November 2021; CDMX, México: Instituto Nacional de Investigaciones Forestales, Agrícolas y Pecuarias (INIFAP); 2021. pp. 410-412

[61] Agredo-Pelechor JA. Adición de sustancias húmicas extraídas de lombricomposta como promotor del crecimiento en cerdos al destete. In: LVI Reunión Nacional de investigación pecuaria (RNIP); 10-12 November 2021; CDMX, México: Instituto Nacional de Investigaciones Forestales, Agrícolas y Pecuarias (INIFAP); 2021. pp. 413-415

Chapter 8

Pyoverdine as an Important Virulence Factor in *Pseudomonas aeruginosa* Antibiotic Resistance

Ovidio Durán, Carlos Ramos, Olga Chen, Julio Castillo, Brenda de Mayorga and Magaly de Chial

Abstract

The World Health Organization has designated *P. aeruginosa* as a priority one pathogen due to the rise of multidrug-resistant (MDR) strains. It is a common opportunistic pathogen among humans. Nosocomial pneumonia, hospital-acquired urinary tract infection, and surgical wound infections are all caused by it. *P. aeruginosa* causes significant rates of disease and death in immunocompromised people such as those who have had a bone marrow transplant, have cystic fibrosis, have had burns, or have AIDS. *P. aeruginosa's* ability to cause such a wide range of infections is owing to its arsenal of virulence factors, which includes pyoverdine molecules, which are responsible for MDR strains. Pyoverdines are nonribosomal short peptides that are essential for bacterial pathogenicity because they serve as a signal molecule for the development of other virulence factors and contribute to antibiotic resistance. Because they are formed under iron-limiting conditions in the host environment, siderophores are required for iron uptake in the host.

Keywords: pyoverdine, antibiotic resistance, *Pseudomonas aeruginosa*, *virulent factor*

1. Introduction

1.1 The genus *P. aeruginosa* and its medical importance

The taxonomy is as follows: Kingdom Monera, phylum Proteobacteria, class gamma subdivision, order *Pseudomonadaceae*, genus *Pseudomonas*, and species *P. aeruginosa*, and it was first described by Gessard in 1882 [1]. *Pseudomonas* is derived from two Greek words: Pseudo, which means "false," and monas, which means "single unit," while aeruginosa, which means "greenish-blue," comes from the Latin aerg, which means "rusted copper" [2]. It is a gram-negative, straight, or slightly curved bacillus with a length of 1–5 m and a width of 0.5–1.0 m. The presence of a polar flagellum, which is made up of a complex protein structure that allows for mobility in liquid environments and reaction to chemical stimuli, makes *P. aeruginosa* mobile. It can also bind to cell membranes thanks to this property. It has small filaments called pili, which are located on the outside. These structures are used to

IntechOpen

move in semisolid media and, like the flagellum, adhere to surfaces. Its morphology is heterogeneous, its colonies are generally large, flattened, smooth, or with serrated edges and may show a metallic sheen.

In relation to its metabolism, it is aerobic although it can develop under anaerobic conditions using nitrate as the terminal electron acceptor. It is a ubiquitous organism in the environment and also, it can colonize multiple niches and utilize many environmental compounds as energy sources. It is found mainly in water, soil, swamps, coastal marine habitats, as well as in plant and animal tissues as well as in hospitals. *P. aeruginosa* is characterized by producing a variety of pigments, such as pyocyanin (blue-green in color), pyoverdine (PVD) (yellowish green fluorescent pigment), and pyorubin (red). P. aeruginosa forms biofilms on moist surfaces such as rocks and soils [3–6].

This bacterium is an extremely important pathogen, since it is responsible for a high percentage of nosocomial infections in patients confined in health centers. As an opportunistic human pathogen, it is responsible for infection in immunocompromised patients such as cystic fibrosis, diabetes, cancer, severely burn patients, advanced HIV infections (acquired immunodeficiency syndrome, AIDS), bone marrow transplants, surgical wound infections, and catheterized patients, and this is as a consequence of its resistance to antibiotics and disinfectants that kill other environmental bacteria [7]. A broad range of cell-associated and external factors influence multidrug resistance and thus bacterial pathogenicity. In the colonization, survival, and invasion of tissues of bacteria, virulence factors play a crucial pathogenic function. The pili are responsible for adhesion to the epithelium. Exoenzyme S and other adhesins help epithelial cells stick together. Tissue necrosis is caused by the exotoxin A. Phospholipase C is a hemolysin that is thermolabile. Exoenzyme S's pathogenic involvement is due to its disruption of normal cytoskeletal organization, degradation of immunoglobulin G and A, depolymerization of actin filaments, and contribution to macrophage resistance. At least four proteases produced by *P. aeruginosa* cause bleeding and tissue necrosis. Siderophores (pyoverdin and pyochelin), which allow bacteria to proliferate in the absence of ferrous ions, are involved in chronic infection [4, 8]. The bacteria in strains recovered from cystic fibrosis patients have an alginate pseudocapsule that protects them against phagocytosis, dehydration, and antibiotics. It also enhances biofilm formation by adhering to epithelial cells. Most of these virulence factors are controlled by two different types of regulation systems: the two-component transcriptional regulatory system and the quorum sensing system. These two pathways are required for the microorganism's survival and multiplication in the host (**Figure 1**).

Replace the entirety of this text with the introduction to your chapter. The introduction section should provide a context for your manuscript and should be numbered as first heading. When preparing the introduction, please bear in mind that some readers will not be experts in your field of research.

Prior to 1966, no comprehensive investigation of the aerobic pseudomonads taxonomy had been conducted. It is the work of Stanier and collaborators [10], in which physiological and biochemical features were used to demonstrate the taxonomic basis for the species identification. The genus was amended in 1984 by Palleroni, and five groups were established based on the results of DNA–DNA hybridization and rRNA–DNA hybridization. All five groups were later identified as belonging to the class Proteobacteria, and members of the genus *Pseudomonas* "sensu stricto" were found to belong to the subclass Gammaproteobacteria's RNA–DNA group I. The rRNAI group is also subdivided into fluorescent and nonfluorescent bacterium [11].

The genome size of *P. aeruginosa*, which ranges between 5.5 and 7 megabytes, is the largest in bacteria and has the greatest environmental adaptability. The PAO1 strain is the

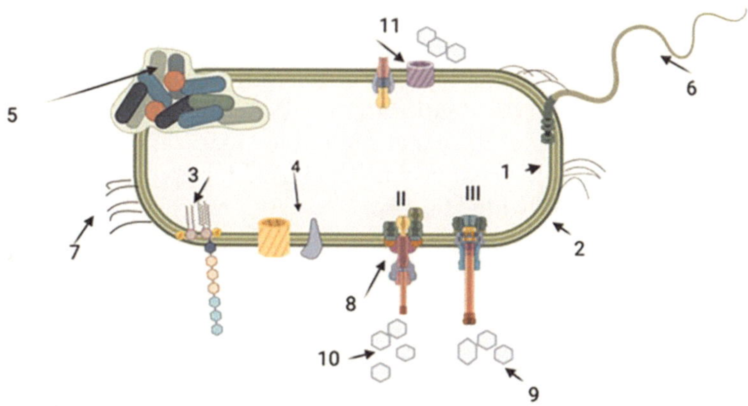

Figure 1.
P. aeruginosa *main virulent factors. Outer membrana(1), inner membrane (2), lipopolysaccharide (3), outermembrane proteins, porins and lipoproteins (4), biofilm (5), flagellum (6), pili (7), secretion systems (8), exotoxins (9), proteases, lipases, elastases, pyocianin (10), pyoverdine siderophore as an iron uptake system [9]. Created with BioRender.com.*

key reference strain for genetic and functional investigations on *P. aeruginosa*. It was first isolated from a wound of an Australian patient in the 1950s. The PAO1 genome is a 6.264-Mbp circular chromosome that encodes 5700 genes, including 5584 projected open read frames (ORFs), and was fully sequenced in 2000. As a result, it has the highest proportion of regulatory genes of any bacterial genome, as well as many genes involved in catabolism, transport, and efflux of organic compounds, as well as four potential chemotaxis systems. The size and complexity of the *P. aeruginosa* genome are thought to be an evolutionary adaptation that allows it to thrive in a variety of habitats while also resisting the effects of antimicrobial drugs. Knowledge of the complete genome sequence and encoded processes provides a wealth of information for the discovery and exploitation of new antibiotic targets, and the hope of developing more effective strategies to treat life-threatening opportunistic infections caused by *P. aeruginosa* in humans [12].

Studies on the P. aeruginosa transcriptome became possible after the genome was completed [13]. Understanding the lifestyle and pathogenicity of *P. aeruginosa* requires gene expression profiling. Following the introduction of next-generation sequencing systems, high-throughput sequencing of cDNA fragments became an alternative to microarray hybridization. The widespread use of this method, dubbed RNA-Seq or RNA-seq, resulted in an exponential rise in the number of whole transcriptome investigations published in the literature [14]. The transcriptomes of *P. aeruginosaclone C* were studied with this technology, and it was discovered that rRNA molecules accounted for 50 to 90 percent of the sense RNAs, followed by mRNA transcripts and noncoding RNA in comparable quantities. Uncharged tRNAs and 29 yet-undescribed antisense tRNAs were found in similar numbers and identified yet-undescribed RNA molecules. The identification of sense-antisense pairs of transfer-messenger RNA (tmRNA), tRNAs, and mRNAs using this RNA seq method implies a new level of gene regulation in bacteria [15].

Although the 16S rRNA gene is the basic tool of the current bacterial classification system, it is known that closely related bacterial species cannot be differentiated based on this gene. Therefore, in the last 10 years, other gene sequences have been used as phylogenetic molecular markers in taxonomic studies, such as atpD, gyrB, rpoB, recA, and rpoD [16]. Mulet and collaborators have shown that analysis of the sequences of four housekeeping genes (16S rRNA, gyrB, rpoB, and rpoD) in all known species of the genus clarified the phylogeny and greatly facilitated the identification of new strains. Multilocus sequence typing (MLST) of the four housekeeping genes is reliable for species delineation and strain identification in Pseudomonas [17]. MLST is enhancing our understanding of the general genome organization of *P. aeruginosa* strains and species, and it is the standard method used for epidemiological surveys on *P. aeruginosa* outbreaks worldwide [18, 19].

2. Iron role in metabolism

Iron is a micronutrient found in almost all living organisms and is an essential component of nearly all of them [20]. It can be present in both reduced (Fe^{2+}) and oxidized (Fe^{3+}) forms in cells, making it simple to insert into an enzyme's catalytic site and serve as an electron carrier in many redox-sensing proteins. Iron forms part of a larger cofactor such as Fe-S clusters and heme, the former is involved in diverse biological processes, including metabolite biosynthesis, DNA replication, RNA modification, gene expression, photosynthesis, and respiration, and the latter is required for cytochrome biogenesis and the transport and storage of oxygen in vertebrates. Iron is associated with oxidative stress. In the presence of oxygen, the ferrous ion is unstable, forming ferric ions and reactive oxygen species (ROS), which can damage biological macromolecules and cause cell death. This process is illustrated by the Fenton Reaction [21].

Fenton Reaction:

$$Fe^{2+} + H_2O_2 \longrightarrow Fe^{3+} + OH^{\cdot} + OH^{-}$$

Even though iron is the fourth most prevalent element in the Earth's crust, only its ferrous form is soluble in water, whereas ferric iron has very low solubility and forms insoluble precipitates hydroxides at neutral pH with solubilities of 10–9 to 10–10 M (i.e. 56 ng/L) [22]. Because the concentration is too low to maintain life, all organisms have evolved unique mechanisms to solubilize iron. After absorption of iron in the ferrous form by the protein ferroportin in the duodenal mucosa, animals absorb it from the meal. The iron is then transported to the glycoprotein transferrin, where it becomes ferric, and is then stored in ferritin as a polymeric ferric complex. This is utilized to feed iron to various apoproteins for them to produce various iron-containing proteins as well as to provide the iron required for erythrocyte development and hemoglobin synthesis [22, 23].

P. aeruginosa can acquire iron from different sources of its host organism, among these sources are (**Figure 2**):

- Transferrin and the related protein lactoferrin:Milk and other extracellutlar fluids contain it (saliva, tears, and nasal mucus). (Transferrin (Tf) is an iron carrier glycoprotein (Fe 3+), synthesized and metabolized mainly in hepatocytes. It is made up of a single polypeptide chain of 679 amino acids with a molar mass of 79,500 g/mol. Each transferrin molecule consists of two lobes with a similar internal structure and is independent for Fe 3+ fixation; the N-terminal lobe contains residues 1–336 and the C-terminal residues 336–679. Each lobe in turn is folded, forming two domains. This conformation of the molecule allows the firm, although reversible, union of Fe.

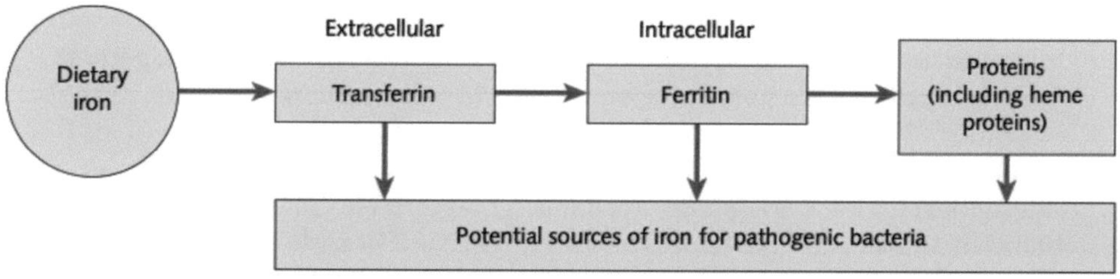

Figure 2.
Outline of the principal iron sources that may be accessed by bacterial pathogens source [23].

- Ferritin: Ferritin is the intracellular protein responsible for the storage and release of iron. Ferritin can store up to 4500 iron atoms as a ferrihydrite mineral in a protein shell and releases these iron atoms when needed by the cell. The ferritin protein coat consists of 24 protein subunits of 2 types, the H subunit and the L subunit.

- Fe-containing proteins such as heme proteins: In these proteins, iron is in its ferrous form and, as such, can be used as an appropriate ligand in which O2 can bind to be transported around the body as oxyhemoglobin.

Pathogens obtain iron from their hosts by three methods that are engaged when the bacterium is in an iron-deficient environment that limits its growth and is not mutually exclusive. First, bacteria get iron by breaking down hemoglobin, such as hemolytic bacteria. FeII does not have enough time to oxidize to insoluble Fe III in this situation. Second, using a particular binding protein, the pathogen can bind to transferrin or lactoferrin. At the bacterial cell surface, the iron is then taken from the molecule. Third, the bacteria create a chelating chemical termed siderophore, which has a stronger affinity for iron than the host organism's iron-containing molecules [22].

3. Microbial siderophores and *P. aeruginosa* siderophores

Bacteria possess specific pathogenicity mechanisms that they exhibit to overcome a host's defenses. A pathogenic microorganism could cause damage, at any level, in a susceptible host organism. Virulence is a quantitative measure of pathogenicity and is measured by the number of microorganisms required to cause disease, that is, it is the degree of pathogenicity.

Throughout evolution, bacteria have acquired characteristics that allow them to invade the host environment, express specialized surface receptors for adhesion, remain in these sites through colonization processes, evade the immune system, and finally cause tissue damage within order to gain access to sources of nutrients necessary for their growth and reproduction [24, 25].

Therefore, the virulence factor or determinant is a microbial component that favors growth or survival during infection; iron being a determining factor of intracellular survival for the growth of most bacteria and especially pathogens, such as *P. aeruginosa*, an opportunistic human pathogen [26, 27].

1. When a microorganism enters a host organism, either in a pathogenic or symbiotic form, it finds a favorable environment with access to practically all the nutrients necessary for its growth except for one, iron. Iron, unlike other elemental sources

for nutrition, such as nitrogen, phosphorus, potassium, and other macro- and micronutrients, is not freely available in host organisms, so it is an important limiting factor for the growth of microorganisms. It is known that one of the responses of host organisms to pathogen attack consists in the reduction of free iron by sequestering this metal in ferritin molecules, structurally known as siderophores. This iron uptake mechanism that operates in bacteria has also been found in animals and plants. In the latter, there is a notable difference, and in the former, the control of ferritin synthesis occurs molecularly at the translational level, while in plants it occurs at the transcriptional level [4, 28–30], Hydroxamates: Siderophores that use a hydroxamate group to bind iron. The most representative siderophore is aerobactin, produced by bacteria of the Salmonella genus and some strains of E. coli, which has a dissociation constant very similar to transferrins, so it competes with other sources such as ferritin.

2. Cathecolates: Enterobactin is the most studied siderophore of this group, produced by strains of E. coli and other enterobacteria.

3. α-Hydroxycarboxylic acids: they are siderophores with a group similar to that of a hydroxamate, in which one of the radicals is replaced by a double bond with oxygen and nitrogen of the skeleton by a carbon. An example is the siderophore achromobactin produced by Erwinia chrysanthemi.

4. Mixed: those are in which two different binding groups are combined in the same molecule.

An example is anguibactin which contains a catechol and a hydroxamate group.

The siderophores, despite the variety in their structures, have similarities between them:

- They contain strongly electron donating atoms (often oxygen and, to a lesser degree, nitrogen or sulfur).
- Their shape is thermodynamically stable.
- They contain high Fe3+ spin species.
- They have a redox potential between −0.33 V (triacetylfusarinine) and − 0.75 V (enterobactin).

More than 500 siderophores, chemically characterized and classified, are currently reported. In addition, some have been shown to have the ability to chelate (subtract) other metals other than iron, such as aluminum, gallium, chromium, copper, zinc, lead, manganese, cadmium, vanadium, indium, plutonium, and uranium. Due to the great variety of siderophores, it is evident that several mechanisms of iron (III) transport exist [31, 32].

3.1 *P. aeruginosa* siderophores

P. aeruginosa synthesizes two types of siderophores, pyoverdine (PVD), and piochelin (PCH). Pyoverdine is the major siderophore of fluorescent pseudomonads

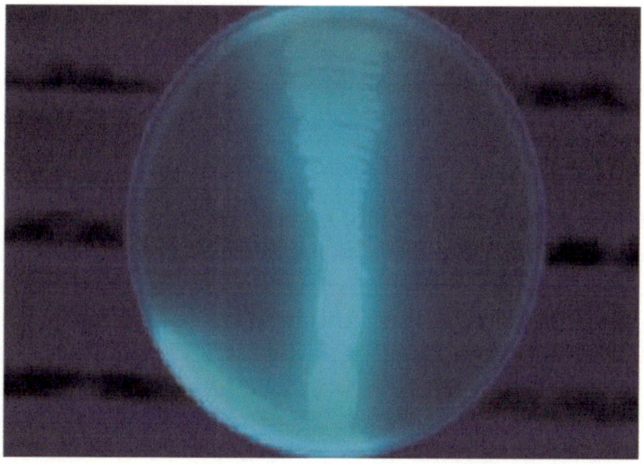

Figure 3.
P. aeruginosa *fluorescent pyoverdine growthing in an iron depleted media.*

(**Figure 3**). Pyoverdines were discovered in 1892, and over the years, they have been given various names: fluorescins, pseudobactins, and finally pyoverdins or pyoverdines. In 1952, J. Totter and F. Moseley observed that the iron levels affected the production of fluorescin by *Pseudomonas aeruginosa*. Today, more than 100 pyoverdines from different strains and species of *Pseudomonas* have been chemically identified and characterized [33]. Pyoverdine are mixed hydroxamate-catecholate siderophores, which are synthesized under iron-limited conditions. Pyoverdine has a high binding affinity for ferric iron. *P. aeruginosa* strains produce four distinct pyoverdines, called PVDI, PVDII, PVDIII, and PVDIV, and PCH as a secondary siderophore [34]. Each is characterized by a different peptide chain and each has a corresponding outer membrane receptor FpvAI, FpvAII, FpvAIII, and FpvAIV [34, 35]. Each outer membrane transporter can recognize and capture back only the ferric form of the produced pyoverdine or one that is structurally related (with a similar peptide sequence) by recognizing the peptide moiety. The methods to identify pyoverdines in different P. aeruginosa strain is referred to as siderotyping [36–40].

P. aeruginosa produces a second siderophore called pyochelin which has a lower binding affinity for the ferric form of iron (**Figure 4**). Pyochelin is the condensation product between salicydic acid and two cysteinyl residues. Pyochelin also chelates Zn (II), Cu(II), Co(II), Mn(II), and Ni(II) [38].

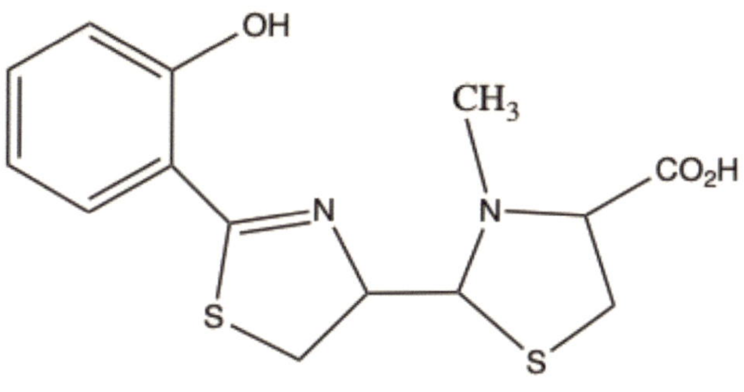

Figure 4.
Pyochelin of P. aeruginosa *PAO1 [41].*

4. Pyoverdine structure

Pyoverdines are a class of fluorescent yellow-green siderophores produced and secreted by many Pseudomonas species. In addition to pyoverdine, other siderophores with lower affinity for ferric ions are also produced such as pyochelin, pseudomonin, corrugatins, yersiniabactin, and thioquinolobactin [42]. Siderophores are small molecules not only produced by many microorganisms but also by plants whose molecular mass range from 200 to 2000 Da. These molecules are used to chelate iron with high affinity and functions in iron acquisition and also as virulence factors in some bacterial. The term siderophores from greek roots "sideros phoros" means iron carrier or transporter. There are different types of siderophores classified according to the ligand used to chelate iron. Catecholates are the more common functional group used to chelate iron in bacterial siderophores (i.e. enterobactin). Hydroxymates (i.e. Ferrioxamine B) are present in bacteria and Ferrichrome in fungi. Carboxylates (i.e. Rhizobactin) are present as functional groups in some bacterial siderophores; however, siderophores such as pyoverdine have a mix of functional groups that form hexadentate coordinates complexes with ferric iron [42]. Plants siderophores are called phytosiderophores, and the mugineic acid is the more common siderophore in plants. Pyoverdine siderophores molecules consist of a hydroxyquinoline chromophore core, a small peptide chain usually contain 6–14 amino acids and acyl side chain (**Figure 5**).

The chromophore is responsible for the color of the molecule and is linked to the peptide chain and acyl group. Both hydroxyl group of the chromophore and side chains oxygens in the peptide chain form interactions with iron. The peptide chain may be partially or completely cyclized and has L and D configuration amino acids. Unusual amino acids such as *N*5-formyl-*N*5-hydroxy ornithine, cyclo-*N*5-hydroxy ornithine, allo-threonine, and others may be present in the peptide chain. The amino acids compositions vary among Pseudomonas species and strains such as *Pseudomonas aeruginosa* strains such as ATCC27853, PAO1, and Pa6 (**Figure 6**). More than 100 different pyoverdines have been identified in Pseudomonas species and strains. Each pyoverdine has a peptide chain with a specific amino acid sequence and length [34, 39].

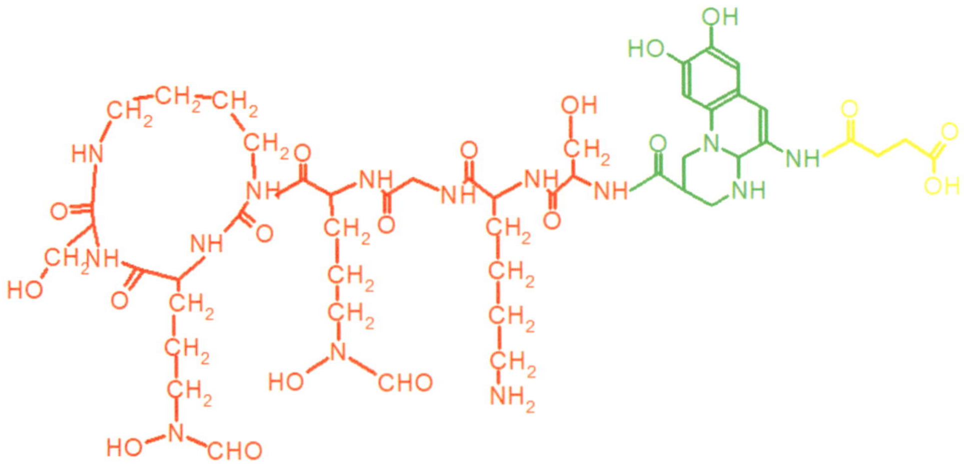

Figure 5.
Pyoverdine structure from P. aeruginosa *PAO1. Red, peptide chain partially cyclized. Green, dihydroxiquinoline chromophore showing the iron hydroxyl interacting groups. Yellow, acyl group [41].*

P. aeruginosa strains	Peptide chain components	PVD
P. aeruginosa ATCC27853	**Ser**-fOH**Orn**-Orn-Gly-a**Thr**-Ser-(OHOrn)	**II**
P. aeruginosa PAO1	Glu-Tyr-**Dab**-Ser-Arg-Ser-fOHOrn-Lys-fOHOrn-Thr-Thr	**I**
P. aeruginosa Pa6	**Ser**-Dab- fOHOrn-Gln-**Gln**-fOH**Orn**-Gly	**III**
P. aeruginosa R'	**Ser**-Dab-fOHOrn-Gln-fOHOrn-Gly	**IV**

Figure 6.
Peptide chain composition of three pseudomonas strains. Amino acids in bold are D configurations. Cyclic structure in the chain is in parenthesis. fOHOrn is N^5-formyl-N^5-hydroxyornithine. aThr correspond to Allo-threonine and dab is L-2,4- diaminobutyrate. The acyl side chain (amide or dicarboxylic acid) is linked to the amino group of the chromophore. The length and type of acyl side chain depend on strain and growth conditions and whose purpose remains unclear [34, 39, 43, 44].

5. Pyoverdines biosynthesis and transport

The siderophores biosynthesis is a complex enzymatic process that requires several specific enzymes whose expression is regulated by iron and different transcriptional factors. The enzymes involved in siderophores biosynthesis are organized into a multi-enzymatic complex, called siderosomes, and are in close vicinity to each other in the cytoplasmic face of the inner membrane. This organization may reduce the diffusion of siderophores precursor. Most of the siderosome enzymes have modular and each module incorporated specific amino acids into a growing peptide chain. Enzymes involved in the biosynthesis of unusual amino acids present in siderophores are also proposed to be part of the siderosome (**Figure 7**) [45].

Pseudomonas aeruginosa strains have four types of pyoverdines designated as PVD1, PVDII, PVDIII, and PVDIV and the biosynthesis of pyoverdine I (PVDI) in *P. aeruginosa* PAO1 involved the action of seven siderosome enzymes. The same number of enzymes is required for pyochelin biosynthesis. A siderosome model for P. fluorescens strain A506 has been included (**Figure 7**).

The initial step in pyoverdines biosynthesis takes place in the cytoplasm where non-ribosomal peptide synthetases (NRPSs) catalyze the formation of the peptide

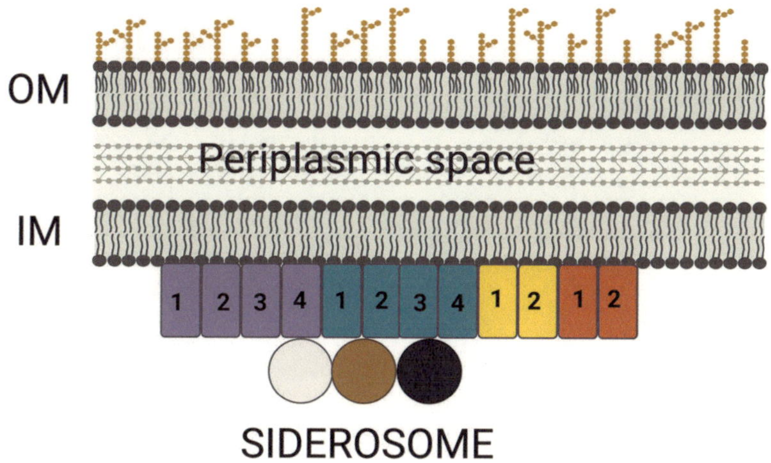

Figure 7.
Model of siderosome of P. aeruginosa *PAOI. The modular enzymes such as PvdL, PvdI, PvdJ, and PvdD are shown in blue, green, yellow, and red color, respectively. The modules in each enzyme are numbered. PvdH, (gray) PvdA (brown), and PvdF (black) are non-modular siderosome enzymes involved in pyoverdine synthesis. OM, outer membrane and IM, inner membrane [45]. Created with BioRender.com.*

precursor for pyoverdines called acylated precursor chain (**Figure 8**) [46, 47]. *P. aeruginosa* PAO1 has four non-ribosomal peptide synthesis (NRPS) enzymes: PvdL, PvdI, PvdJ, and PvdD [48]. Three more enzymes are involved in the formation of unusual amino acids of the peptide chain of PVDI. The unusual amino acids in PVDI are L-2,4 diaminobutyrate (Dab) and L-N5-formyl-N5-hydroxyornithine (fOHOrn). The PvdH enzyme is responsible for the synthesis of Dab, PvDA, and PvDF that catalyze the synthesis of fOHorn through consecutive hydroxylation and formylation reactions. Model of siderosome of *P. fluorescens* strain A506 includes three NRPS enzymes (PvdG, PvdL, and PvdD) and the same *P. aeruginosa* enzymes that catalyze the formation of unusual amino acids.

The NRPS enzymes are modular enzymes with 2–4 modules. PvdL and PVdI have four modules and PvdJ and PvdD are bimodular. The first module (M1) of PvdL catalyzes the incorporation of acyl group (myristic or myristoleic acid) instead of amino acid. This acylation probably links the peptide to the membrane and prevents diffusion during synthesis. The M2 of PvdL catalyzes the activation of L-Glu and its condensation to the acyl group. PvdL, module three (M3), incorporates an L-Tyr that is converted to D-Tyr by domain of this module. M4 adds Dab to generate an acylated tripeptide (Glu-Tyr-Dab). PvdI modules are responsible for adding D-Ser, L-Arg, D-Ser, and fOHOrn to previous acylated tripeptide. L-Lys and fOHorn and two L-Thr are, respectively, added by the bimodular enzymes PvdJ and PvdD. The peptide bound formation is catalyzed by a PCP domain present in the modules. Thioesterase domain of the PvdD module is released by hydrolysis of the 11 amino acid chain from the NRPS [42].

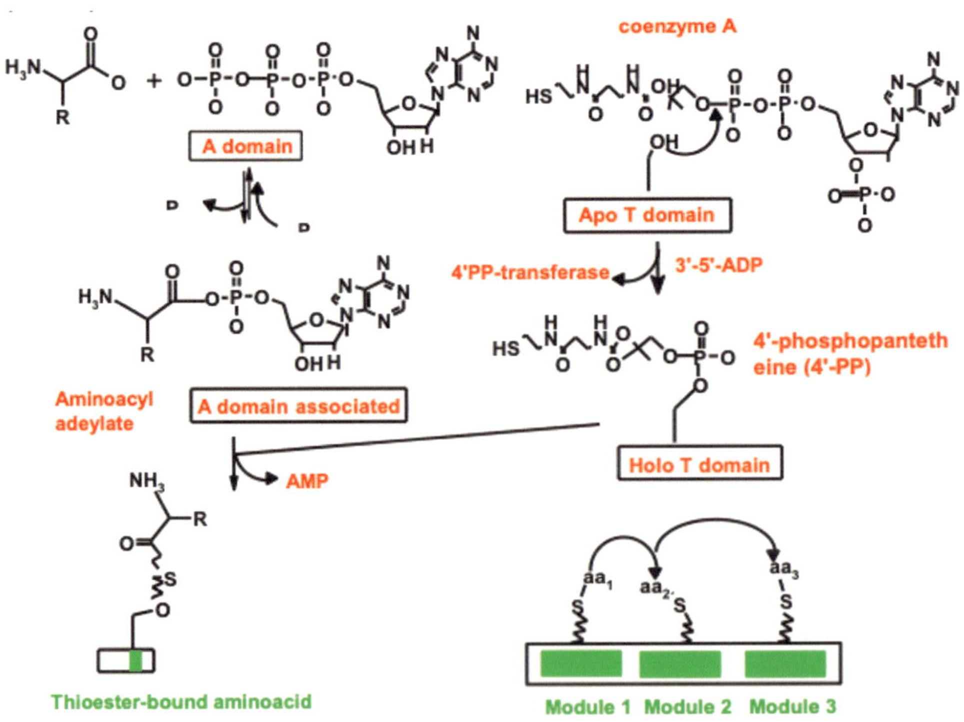

Figure 8.
Mechanism of multiple carrier thiotemplate [47].

The released peptide is transported to the periplasmic space where it is modified. The transport to the periplasmic space involved a class of ABC pumps codified by *pvdE* gene. The modifications that take place in the periplasmic space includes the deacetylation and the removal of myristic or myristoleic acid from the peptide chain and the formation of pyoverdine precursor called ferribactin. The PvdQ enzyme is responsible for this modification. The enzyme PvdP converts ferribactin into dihydropyoverdine. The conversion of dihydropyoverdine to PVDI is catalyzed by the enzyme PvdO.

Finally, the PVDI is secreted from the periplasmic space to the environment via PvdRF-OpmQ ATP pump. The secreted PVDI binds to ferric iron to form PVDI-iron complex (Ferripyoverdine). The Ferripyoverdine is imported via FpvA receptors which interact with TonB-ExbBD complex and the help of transporter FpvB [49]. The Fe^{+3} of the Ferripyoverdine in the periplasmic space is reduced to Fe^{+2} and released from pyoverdine. Liberated Fe^{+2} is transported into the cytoplasm through ABC transporter FpvDE.

6. Regulation of pyoverdine production

The transcriptional control of genes involved in the synthesis of pyoverdine is induced by iron deficiency or depletion (**Figure 9**). The regulation of pyoverdine production involves sensing cytoplasmic levels of iron ions by the regulator protein Fur, which in turn represses regulatory genes involved in iron uptakes, such as FpvR, FpvI, and PvdS [50–54]. PvdS is a sigma factor required for the expression of pyoverdine biosynthesis genes and some virulence-related genes [29, 55–58]. FpvI is a sigma factor required by the genes encoding the outer membrane pyoverdine receptor/importer FpvA, and FpvR is an anti-sigma factor that binds to and inactivates PvdS and FpvI [50, 59]. FpvR autoproteolytic cleaves itself at a periplasmic domain without any further degradation unless it contacts ferripyoverdine-bound FpvA. When FpvR/FpvA contact occurs, which involves the activity of TonB (the transport-energizing inner membrane protein), the protease RseP releases PvdS and FpvI allowing the activation of their regulated genes [50, 60]. The regulation of pyoverdine biosynthesis is more complex because it involves signals other than iron starvation, such as the influence of the regulator protein CysB may imply coordination with sulfur availability or biofilm formation and alginate production [61, 62]. Phosphate starvation has been reported to trigger pyoverdine production in host environments [63]. Additionally, the LexR-type transcriptional regulator AmpR affects the expression of more than 500 genes related to metabolism and virulence in *P. aeruginosa* and has been recently implicated in the regulation of pyoverdine production [64]. Intracellular levels of bis-(3′-5′)-cyclic dimeric guanosine monophosphate (c-di-GMP) may also modulate pyoverdine production [65].

7. Pyoverdine as virulence factor in *Pseudomonas aeruginosa*

The World Health Organization classified *P. aeruginosa* as a priority one pathogen due to the multidrug resistance mechanisms of this bacterium. As we mentioned, P. aeruginosa produces diverse and overwhelming infections due to the wide variety

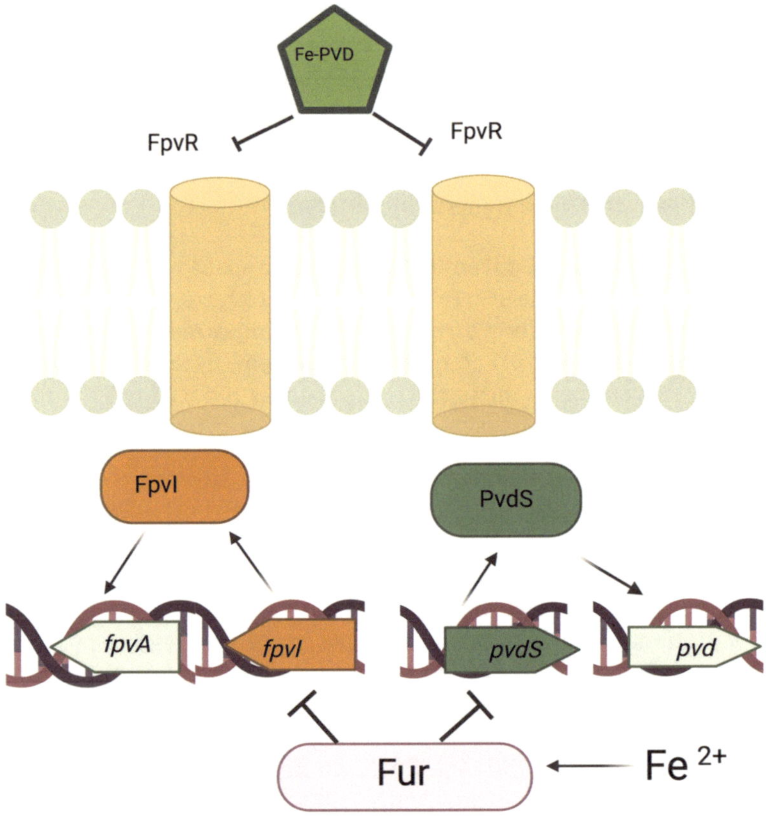

Figure 9.
The ferripyoverdine signaling pathway. The alternative sigma factors σ^{FpvI} and σ^{PvdS} direct expression of the fpvA *gene encoding the cell surface ferripyoverdine receptor protein FpvA, and* pvd *genes encoding enzymes required for synthesis of pyoverdine. σ^{FpvI} and σ^{PvdS} are inhibited by the $FpvR_{20}$ protein, which is degraded in the presence of ferripyoverdine (Fe-Pvd). Expression of the* fpvI *and* pvdS *genes is repressed by the iron (Fe^{2+})-containing form of the fur repressor [66]. Created with BioRender.com.*

of virulence factors that are responsible for multidrug resistance strains (MDRs) [67]. The pyoverdine siderophore is a key virulence factor that provides the bacterium with iron during infection [68].

Regarding the virulence, it has been found that deficient pyoverdine mutants of *P. aeruginosa* were not virulent in the Burn mouse model compared with the wild-type strain PAO1. The mutant strain virulence was restored when pyoverdine purified from the wild-type strain was added during infection. In a similar study, although infected calf muscles of immunocompromised mice were injected with pyoverdine and pyochelin mutants, no lethality was observed, concluding the role of pyoverdine during infection. The pyoverdin-deficient mutant and the double mutant, on the other hand, grew poorly in the lungs when compared to wild-type strain PAO1, and the latter's virulence was significantly reduced [69, 70].

In the model nematode Caenorhabditis elegans, pyoverdine is virulent, even in the absence of the pathogen. A study found that when this siderophore is consumed by C. elegans together with other chemicals in its aqueous environment, pyoverdine gains access to and eliminates ferric iron through an unknown method once within the host. The host mitochondria, which are iron-rich organelles, are a likely target for this abstraction. Mitochondrial function is disrupted, and mitochondria are targeted for turnover when they are removed. In vitro experiments with pyoverdine-treated murine macrophages revealed considerable toxicity, while no pyoverdine production reduced pathogenicity. Furthermore, pyoverdine translocates into cells and impairs host mitochondrial homeostasis, as previously observed in C. elegans [71–73].

Pyoverdine is a multifaceted role in *P. aeruginosa* pathogenesis; in addition, scavenging iron from host proteins also regulates the expression of several virulence factors, including exotoxin A, an endoprotease PrpL, and pyoverdine biosynthesis itself acting as a signaling molecule to control the production of secreted products [74]. A recent study using a double mutation in genes involved in biofilm formation showed that pyoverdine is essential for the development of virulent factors such as exotoxin A and PrpL protease [4, 75].

Exotoxin A is one of *P. aeruginosa's* most potent toxins, capable of inducing apoptosis in host cells and killing model organisms [76]. Under iron-repleted growth conditions, the transcription of exotoxin A was absent, indicating that it is negatively regulated for iron. Interestingly, in the presence of iron ions, pyoverdine was found to activate a signaling pathway for the upregulation of exotoxin A expression [77, 78].

The extracellular protease IV, PrpL, degrades surfactant proteins and interleukin-22 necessary for pulmonary mucosal immunity that made *P. aeruginosa* a major pathogen of ventilator-associated pneumonia and causes considerable lung tissue damage [79].

The sigma factor PvdS is required for the expression of PrpL. The extracellular protein profiles obtained, using PAO1 and a Δ*pvdS*::*Gm* mutant, showed that PrpL ((PvdS-regulated endoprotease, lysyl class) was expressed under the control of PvdS under iron-deficient conditions. In this study, Rnase protection assays confirm that the initiation of transcription is iron-dependent. A study shows that expression of prpL was lower in the biosynthetic gene pvdF mutant than in wild-type bacteria, and expression was increased to wild-type levels by the addition of pyoverdine [80].

The relationship between iron and antibiotic resistance in *P. aeruginosa* has been reviewed [21]. It is required to determine whether iron fluctuations are a critical component for antibiotic resistance in this bacterium, according to this review. A series of studies concluded that increased concentration of iron in the growth medium decreased the resistance of *P. aeruginosa* strains of various origins to antibiotics such as ampicillin, norfloxacin, gentamicin, ofloxacin, and cefsulodin [21]. Conversely, another study showed that increasing iron concentration increased resistance to two antibiotics, tobramycin and tigecycline by using the wild-type strain PAO1 and isolated strains from cystic fibrosis patients in a growth medium with an iron chelator, chelex, and $FeCl_3$ (100 μM) [81]. Inversely, Singh and collaborators found that "decreasing" iron concentration "decreased" resistance to tobramycin [82]. They used an iron chelator, the most used being 2,2 dipyridyl (DIP) or deferoxamine (DFO), a siderophore used by P. aeruginosa to the growth medium, to create conditions of iron limitation. These discrepancies could be attributable to the variability of experimental methods, particularly the iron concentration settings and in addition P. aeruginosa uptake iron by different methods and those are important for antibiotic resistance mechanisms.

Therefore, pyoverdine plays an important role in antibiotic resistance, since it mediates the uptake of iron in *P. aeruginosa*. Oglesby-Sherouse et al. [81] demonstrated that pyoverdine increases the ability of *P. aeruginosa*to resist tigecycline treatment. A recent study with clinical isolates of P. aeruginosa showed the importance of pyoverdine for the bacteria virulence, since most of them were MDR-expressing resistance such as genes of the MexAB-OprM efflux pump system (mexABR) and pyoverdine receptor genes (fpvA) which are induced by iron limitation conditions.

Among the clinical isolates, 22 out 51 were ESBLs (extended-spectrum β-lactamases) producers which represent an important subclass of enzymes that confer resistance to oxyimino-cephalosporins [83]. A similar study related the pyoverdine production, biofilm formation, ESBL, and other virulence genes (OprI, OprL, LasB, PlcH, ExoS, and ToxA) with the antibiotic resistance of 54 clinical isolates, and among them 93% were pyoverdine producers showing its role in antibiotic resistance [83]. A strong relationship between EBSL isolates producers and virulence factor, including pyoverdine, was also found in clinical isolates of P. aeruginosa [84]. Additionally, multidrug resistance, presence of virulence-associated genes, and expression of certain virulence factors, most notably elastase, protease, siderophore, and DNase activity, were strongly related to pyoverdine production. A recent study with clinical isolates of P. aeruginosa related its pathogenicity with pyoverdine production in model's organism in *C. elegans* and an acute murine pneumonia model [85].

8. Concluding remarks

The rise of resistant *P. aeruginosa* strains in hospitals has prompted researchers to look for new therapeutic options. Attenuation of the pathogen's virulence factors is one of these strategies. The pathophysiology of the bacteria is restricted because it does not need to be eliminated or killed. In many cases, attenuation of these products that harm the host is unnecessary for growth or colonization, and their absence causes the pathogen to adopt a commensal lifestyle. Pyoverdine is an important virulence factor, and its role in iron transport as well as its position as a signaling molecule for the synthesis of other virulence factors make it an attractive target for new therapeutics that block its function. A potential therapeutic method involves tagging pyochelin or pyoverdine with an antibiotic or a redox-inactive metal ion such as gallium, which interferes with P. aeruginosa iron metabolism and its synthesis. Some *P. aeruginosa* strains may utilize another iron-chelating pathway using nicotianamine in the absence of siderophores, which may be investigated in the same way.

Author details

Ovidio Durán[1], Carlos Ramos[1,4], Olga Chen[1], Julio Castillo[1,3,4,5], Brenda de Mayorga[2] and Magaly de Chial[1,4*]

1 Departamento de Genética y Biología Molecular, Universidad de Panamá, Panamá

2 Departamento de Microbiología, Universidad de Panamá, Panamá

3 Centro de Biodiversidad y Descubrimiento de Drogas, Instituto de Investigaciones Científicas y Servicios de Alta Tecnología (INDICASAT AIP), Panamá

4 Estación Científica Coiba AIP, Panamá

5 Member of Sistema Nacional de Investigación, SENACYT, Panamá

*Address all correspondence to: mdechial@yahoo.com.mx; magaly.dechial@up.ac.pa

References

[1] Diggle SP, Whiteley M. Microbe profile: *Pseudomonas aeruginosa*: Opportunistic pathogen and lab rat. Microbiology (Reading, England). 2020; **166**(1):30-33. DOI: 10.1099/mic.0.000860

[2] Palleroni, NJ. The Pseudomonas story. *Environmental microbiology*. 2010; **12**(6):1377–1383. DOI: 10.1111/j.1462-2920.2009.02041.x

[3] Hardalo C, Edberg SC. Pseudomonas aeruginosa: Assessment of risk from drinking water. Critical Reviews in Microbiology. 1997;**23**(1):47-75. DOI: 10.3109/10408419709115130

[4] Cornelis P, Dingemans J. *Pseudomonas aeruginosa* adapts its iron uptake strategies in function of the type of infections. Frontiers in Cellular and Infection Microbiology. 2013;**3**:75. DOI: 10.3389/fcimb.2013.00075

[5] Ossa-Giraldo AC, Echeverri-Toro LM, Santos ZM, García MG, Agudelo Y, Ramírez F, et al. Factores de riesgo para infección por Pseudomonas aeruginosa multi-resistente en un hospital de alta complejidad [Risk factors for multidrug-resistant Pseudomonas aeruginosa infection, in a tertiary hospital in Colombia]. Revista chilena de infectologia: organo oficial de la Sociedad Chilena de Infectologia. 2014;**31**(4):393-399. DOI: 10.4067/S0716-10182014000400003

[6] Budzikiewicz H. Siderophores of the Pseudomonadaceae sensu stricto (fluorescent and non-fluorescent pseudomonas spp.). Fortschritte der Chemie organischer Naturstoffe = Progress in the chemistry of organic natural products. Progres dans la chimie des substances organiques naturelles. 2004;**87**:81-237. DOI: 10.1007/978-3-7091-0581-8_2

[7] Bodey GP, Bolivar R, Fainstein V, Jadeja L. Infections caused by Pseudomonas aeruginosa. Reviews of Infectious Diseases. 1983;**5**(2):279-313. DOI: 10.1093/clinids/5.2.279

[8] Strateva T, Mitov I. Contribution of an arsenal of virulence factors to pathogenesis of *Pseudomonas aeruginosa* infections. Annales de Microbiologie. 2011;**61**:717-732. DOI: 10.1007/s13213-011-0273-y

[9] Jurado-Martín I, Sainz-Mejías M, McClean S. *Pseudomonas aeruginosa*: An audacious pathogen with an adaptable arsenal of virulence factors. International Journal of Molecular Sciences. 2021;**22**(6):3128. DOI: 10.3390/ijms22063128

[10] Stanier RY, Palleroni NJ, Doudoroff M. The aerobic pseudomonads: A taxonomic study. Journal of General Microbiology. 1966; **43**(2):159-271. DOI: 10.1099/00221287-43-2-159

[11] Gomila M, Peña A, Mulet M, Lalucat J, García-Valdés E. Phylogenomics and systematics in pseudomonas. Frontiers in Microbiology. 2015;**6**:214. DOI: 10.3389/fmicb.2015.00214

[12] Stover CK, Pham XQ, Erwin AL, Mizoguchi SD, Warrener P, Hickey MJ, et al. Complete genome sequence of Pseudomonas aeruginosa PAO1, an opportunistic pathogen. Nature. 2000; **406**(6799):959-964. DOI: 10.1038/35023079

[13] Palma M, Worgall S, Quadri LE. Transcriptome analysis of the Pseudomonas aeruginosa response to iron. Archives of Microbiology. 2003; **180**(5):374-379. DOI: 10.1007/s00203-003-0602-z

[14] Wang Z, Gerstein M, Snyder M. RNA-Seq: A revolutionary tool for transcriptomics. Nature Reviews Genetics. 2009;**10**(1):57-63. DOI: 10.1038/nrg2484

[15] Pust MM, Davenport CF, Wiehlmann L, Tümmler B. Direct RNA nanopore sequencing of Pseudomonas aeruginosa clone C transcriptomes. Journal of Bacteriology. 2022;**204**(1): e0041821. DOI: 10.1128/JB.00418-21

[16] Hilario E, Buckley TR, Young JM. Improved resolution on the phylogenetic relationships among pseudomonas by the combined analysis of atp D, car a, rec a and 16S rDNA. Antonie Van Leeuwenhoek. 2004;**86**(1):51-64. DOI: 10.1023/B:ANTO.0000024910.57117.16

[17] Curran B, Jonas D, Grundmann H, Pitt T, Dowson CG. Development of a multilocus sequence typing scheme for the opportunistic pathogen Pseudomonas aeruginosa. Journal of Clinical Microbiology. 2004;**42**:5644-5649

[18] Mulet M, Lalucat J, García-Valdés E. DNA sequence-based analysis of the pseudomonas species. Environmental Microbiology. 2010;**12**(6):1513-1530. DOI: 10.1111/j.1462-2920.2010.02181.x

[19] Curran B, Jonas D, Grundmann H, Pitt T, Dowson CG. Development of a multilocus sequence typing scheme for the opportunistic pathogen *Pseudomonas aeruginosa*. Journal of Clinical Microbiology. 2004;**42**:5644-5649

[20] Ganz T. Iron and infection. International Journal of Hematology. 2018;**107**(1):7-15. DOI: 10.1007/s12185-017-2366-2

[21] Ezraty B, Barras F. The' liaisons Dangereuses' between iron and antibiotics. FEMS Microbiology Reviews. 2016;**40**:418-435

[22] Ratledge C. Iron metabolism and infection. Food and Nutrition Bulletin. 2007;**28**(Suppl. 4):S515-S523. DOI: 10.1177/15648265070284S405

[23] Ratledge C, Dover L. Iron metabolism in pathogenic bacteria. Annual Review of Microbiology. 2000; **54**:881-941. DOI: 10.1146/annurev.micro.54.1.881

[24] Arora SK, Ritchings BW, Almira EC, Lory S, Ramphal R. The Pseudomonas aeruginosa flagellar cap protein, FliD, is responsible for mucin adhesion. Infection and Immunity. 1998;**66**(3):1000-1007. DOI: 10.1128/IAI.66.3.1000-1007.1998

[25] Rada B. Neutrophil extracellular trap release driven by bacterial motility: Relevance to cystic fibrosis lung disease. Communicative & Integrative Biology. 2017;**10**(2):e1296610. DOI: 10.1080/19420889.2017.1296610

[26] Khalili Y, Memar MY, Farajnia S, Adibkia K, Kafil HS, Ghotaslou R. Molecular epidemiology and carbapenem resistance of *Pseudomonas aeruginosa* isolated from patients with burns. Journal of Wound Care. 2021;**30**(2):135-141. DOI: 10.12968/jowc.2021.30.2.135

[27] Hilliam Y, Kaye S, Winstanley C. *Pseudomonas aeruginosa* and microbial keratitis. Journal of Medical Microbiology. 2020;**69**(1):3-13. DOI: 10.1099/jmm.0.001110

[28] Visca P, Imperi F, Lamont IL. Pyoverdine siderophores: From biogenesis to biosignificance. Trends in Microbiology. 2007;**15**(1):22-30. DOI: 10.1016/j.tim.2006.11.004

[29] Visca P. Iron regulation and siderophore signalling in virulence by pseudomonas aeruginosa. In: Ramos J-L, editor. Pseudomonas. US, Boston, MA: Springer; 2004. pp. 69-123

[30] Cézard C, Farvacques N, Sonnet P. Chemistry and biology of pyoverdines, pseudomonas primary siderophores. Current Medicinal Chemistry. 2015; **22**(2):165-186. DOI: 10.2174/0929867321666141011194624

[31] Galle M, Carpentier I, Beyaert R. Structure and function of the type III secretion system of Pseudomonas aeruginosa. Current Protein & Peptide Science. 2012;**13**(8):831-842. DOI: 10.2174/138920312804871210

[32] Cornelis P, Matthijs S. Diversity of siderophore-mediated iron uptake systems in fluorescent pseudomonads: Not only pyoverdines. Environmental Microbiology. 2002;**4**(12):787-798. DOI: 10.1046/j.1462-2920.2002.00369.x

[33] Schalk IJ, Guillon L. Pyoverdine biosynthesis and secretion in Pseudomonas aeruginosa: Implications for metal homeostasis. Environmental Microbiology. 2013;**15**(6):1661-1673. DOI: 10.1111/1462-2920.12013

[34] Cornelis P, Hohnadel D, Meyer JM. Evidence for different pyoverdine-mediated iron uptake systems among Pseudomonas aeruginosa strains. Infection and Immunity. 1989;**57**(11): 3491-3497. DOI: 10.1128/iai.57.11.3491-3497.1989

[35] de Chial M, Ghysels B, Beatson SA, Geoffroy V, Meyer JM, Pattery T, et al. Identification of type II and type III pyoverdine receptors from Pseudomonas aeruginosa. Microbiology (Reading, England). 2003;**149**(Pt 4):821-831. DOI: 10.1099/mic.0.26136-0

[36] Meyer JM, Geoffroy VA, Baida N, Gardan L, Izard D, Lemanceau P, et al. Siderophore typing, a powerful tool for the identification of fluorescent and nonfluorescent pseudomonads. Applied and Environmental Microbiology. 2002; **68**(6):2745-2753. DOI: 10.1128/AEM.68.6.2745-2753.2002

[37] Meyer JM, Gruffaz C, Raharinosy V, Bezverbnaya I, Schäfer M, Budzikiewicz H. Siderotyping of fluorescent pseudomonas: Molecular mass determination by mass spectrometry as a powerful pyoverdine siderotyping method. Biometals: an international journal on the role of metal ions in biology, biochemistry, and medicine. 2008;**21**(3):259-271. DOI: 10.1007/s10534-007-9115-6

[38] Visca P, Colotti G, Serino L, Verzili D, Orsi N, Chiancone E. Metal regulation of siderophore synthesis in Pseudomonas aeruginosa and functional effects of siderophore-metal complexes. Applied and Environmental Microbiology. 1992;**58**(9):2886-2893. DOI: 10.1128/aem.58.9.2886-2893.1992

[39] Meyer JM. Pyoverdines: Pigments, siderophores and potential taxonomic markers of fluorescent pseudomonas species. Archives of Microbiology. 2000; **174**(3):135-142. DOI: 10.1007/s002030000188

[40] Regine F, Mathias S, Valerie G, Jean-Marie M. Siderotyping a powerful tool for the characterization of pyoverdines. Current Topics in Medicinal Chemistry. 2001;**1**(1):31-57. DOI: 10.2174/1568026013395542

[41] Poole K, McKay GA. Iron acquisition and its control in Pseudomonas aeruginosa: Many roads lead to Rome. Frontiers in Bioscience: a Journal and Virtual Library. 2003;**8**:d661-d686. DOI: 10.2741/1051

[42] Schalk IJ, Rigouin C, Godet J. An overview of siderophore biosynthesis among fluorescent Pseudomonads and new insights into their complex cellular organization. Environmental

Microbiology. 2020:**22**(4):1447–1466. DOI: 10.1111/1462-2920.14937

[43] Budzikiewicz H, Schäfer M, Fernández DU, Matthijs S, Cornelis P. Characterization of the chromophores of pyoverdins and related siderophores by electrospray tandem mass spectrometry. Biometals: an international journal on the role of metal ions in biology, biochemistry, and medicine. 2007;**20**(2): 135-144. DOI: 10.1007/s10534-006-9021-3

[44] Demange P, Wendenbaum S, Linget C, Mertz C, Cung MT, Dell A, et al. Bacterial siderophores: Structure and NMR assigment of pyoverdins PaA, siderophores of Pseudomonas aeruginosa ATCC 15692. Biology of Metals. 1990;**3**: 155-170

[45] Gasser V, Guillon L, Cunrath O, Schalk IJ. Cellular organization of siderophore biosynthesis in Pseudomonas aeruginosa: Evidence for siderosomes. Journal of Inorganic Biochemistry. 2015;**148**:27-34. DOI: 10.1016/j.jinorgbio.2015.01.017

[46] Konz D, Marahiel MA. How do peptide synthetases generate structural diversity? Chemistry & Biology. 1999; **6**(2):R39-R48. DOI: 10.1016/S1074-5521(99)80002-7

[47] Konz D, Klens A, Schörgendorfer K, Marahiel MA. The bacitracin biosynthesis operon of bacillus licheniformis ATCC 10716: Molecular characterization of three multi-modular peptide synthetases. Chemistry & Biology. 1997;**4**(12):927-937. DOI: 10.1016/s1074-5521(97)90301-

[48] Poppe J, Reichelt J, Blankenfeldt W. Pseudomonas aeruginosapyoverdine maturation enzyme PvdP has a noncanonical domain architecture and affords insight into a new subclass of tyrosinases. The Journal of Biological Chemistry. 2018;**293**(38):14926-14936. DOI: 10.1074/jbc.RA118.002560

[49] Cornelis P, Bodilis J. A survey of TonB-dependent receptors in fluorescent pseudomonads. Environmental Microbiology Reports. 2009;**1**:256-262. DOI: 10.1111/j.1758-2229.2009.00041.x

[50] Rédly GA, Poole K. FpvIR control of fpvA ferric pyoverdine receptor gene expression in Pseudomonas aeruginosa: Demonstration of an interaction between FpvI and FpvR and identification of mutations in each compromising this interaction. Journal of Bacteriology. 2005;**187**(16):5648-5657. DOI: 10.1128/JB.187.16.5648-5657.2005

[51] Cornelis P, Matthijs S, Van Oeffelen L. Iron uptake regulation in pseudomonas aeruginosa. Biometals: An International Journal on the Role of Metal Ions in Biology, Biochemistry, and Medicine. 2009;**22**(1):15-22. DOI: 10.1007/s10534-008-9193-0

[52] Pasqua M, Visaggio D, Lo Sciuto A, Genah S, Banin E, Visca P, et al. Ferric uptake regulator fur is conditionally essential in Pseudomonas aeruginosa. Journal of Bacteriology. 2017;**199**(22): e00472-e00417. DOI: 10.1128/JB.00472-17

[53] Ochsner UA, Wilderman PJ, Vasil AI, Vasil ML. GeneChip expression analysis of the iron starvation response in Pseudomonas aeruginosa: Identification of novel pyoverdine biosynthesis genes. Molecular Microbiology. 2002;**45**(5):1277-1287. DOI: 10.1046/j.1365-2958.2002.03084.x

[54] Ochsner UA, Vasil ML. Gene repression by the ferric uptake regulator in Pseudomonas aeruginosa: Cycle selection of iron-regulated genes.

Proceedings of the National Academy of Sciences of the United States of America. 1996;**93**(9):4409-4414. DOI: 10.1073/pnas.93.9.4409

[55] Vasil ML, Ochsner UA, Johnson Z, Colmer JA, Hamood AN. The fur-regulated gene encoding the alternative sigma factor PvdS is required for iron-dependent expression of the LysR-type regulator ptxR in Pseudomonas aeruginosa. Journal of Bacteriology. 1998;**180**(24):6784-6788. DOI: 10.1128/JB.180.24.6784-6788.1998

[56] Wilderman PJ, Vasil AI, Johnson Z, Wilson MJ, Cunliffe HE, Lamont IL, et al. Characterization of an endoprotease (PrpL) encoded by a PvdS-regulated gene in Pseudomonas aeruginosa. Infection and Immunity. 2001;**69**(9):5385-5394. DOI: 10.1128/IAI.69.9.5385-

[57] Wilderman PJ, Vasil AI, Johnson Z, Wilson MJ, Cunliffe HE, Lamont IL, et al. Characterization of an endoprotease (PrpL) encoded by a PvdS-regulated gene in Pseudomonas aeruginosa. Infection and Immunity. 2001;**69**(9):5385-5394. DOI: 10.1128/IAI.69.9.5385-

[58] Wilson MJ, Lamont IL. Characterization of an ECF sigma factor protein from Pseudomonas aeruginosa. Biochemical and Biophysical Research Communications. 2000;**273**(2):578-583. DOI: 10.1006/bbrc.2000.2996

[59] Rédly GA, Poole K. Pyoverdine-mediated regulation of FpvA synthesis in Pseudomonas aeruginosa: Involvement of a probable extracytoplasmic-function sigma factor. FpvI. Journal of Bacteriology. 2003;**185**(4):1261-1265. DOI: 10.1128/JB.185.4.1261-1265.2003

[60] Wilson MJ, McMorran BJ, Lamont IL. Analysis of promoters recognized by PvdS, an extracytoplasmic-function sigma factor protein from Pseudomonas aeruginosa. Journal of Bacteriology. 2001;**183**(6): 2151-2155. DOI: 10.1128/JB.183.6.2151-2155.2001

[61] Imperi F, Tiburzi F, Fimia GM, Visca P. Transcriptional control of the pvdS iron starvation sigma factor gene by the master regulator of sulfur metabolism CysB in Pseudomonas aeruginosa. Environmental Microbiology. 2010;**12**(6):1630-1642. DOI: 10.1111/j.1462-2920.2010.02210.x

[62] Delic-Attree I, Toussaint B, Garin J, Vignais PM. Cloning, sequence and mutagenesis of the structural gene of Pseudomonas aeruginosa CysB, which can activate algD transcription. Molecular Microbiology. 1997;**24**(6): 1275-1284. DOI: 10.1046/j.1365-2958.1997.4121799.x

[63] Zaborin A, Romanowski K, Gerdes S, Holbrook C, Lepine F, Long J, et al. Red death in Caenorhabditis elegans caused by Pseudomonas aeruginosa PAO1. Proceedings of the National Academy of Sciences of the United States of America. 2009;**106**(15): 6327-6332. DOI: 10.1073/pnas.0813199106

[64] Balasubramanian D, Kumari H, Jaric M, Fernandez M, Turner KH, Dove SL, et al. Deep sequencing analyses expands the Pseudomonas aeruginosa AmpR regulon to include small RNA-mediated regulation of iron acquisition, heat shock and oxidative stress response. Nucleic Acids Research. 2014;**42**(2): 979-998. DOI: 10.1093/nar/gkt942

[65] Chen Y, Yuan M, Mohanty A, Yam JK, Liu Y, Chua SL, et al. Multiple diguanylate cyclase-coordinated regulation of pyoverdine synthesis in Pseudomonas aeruginosa. Environmental Microbiology Reports.

2015;7(3):498-507. DOI: 10.1111/1758-2229.12278

[66] Edgar RJ, Hampton GE, Garcia G, Maher MJ, Perugini MA, Ackerley DF, et al. Integrated activities of two alternative sigma factors coordinate iron acquisition and uptake by Pseudomonas aeruginosa. Molecular Microbiology. 2017;**106**(6):891-904. DOI: 10.1111/mmi.13855

[67] Frèdi LR, Neill DR, Fothergill JL. The building blocks of antimicrobial resistance in Pseudomonas aeruginosa: Implications for current resistance-breaking therapies. Frontiers in Cellular and Infection Microbiology. 2021;**11**: 665759

[68] Meyer JM, Neely A, Stintzi A, Georges C, Holder IA. Pyoverdin is essential for virulence of Pseudomonas aeruginosa. Infection and Immunity. 1996;**64**(2):518-523. DOI: 10.1128/iai.64.2.518-523.1996

[69] Takase H, Nitanai H, Hoshino K, Otani T. Impact of siderophore production on *Pseudomonas aeruginosa* infections in immunosuppressed mice. Infection and Immunity. 2000a;**68**: 1834-1839. DOI: 10.1128/IAI.68.4.1834-1839.2000

[70] Khalil MA, Ibrahim Sonbol F, Mohamed AF, Ali SS. Comparative study of virulence factors among ESbetaL-producing and nonproducing *Pseudomonas aeruginosa* clinical isolates. Turkish Journal of Medical Sciences. 2015;**45**:60-69

[71] Kang D, Kirienko DR, Webster P, Fisher AL, Kirienko NV. Pyoverdine, a siderophore from Pseudomonas aeruginosa, translocates into C. elegans, removes iron, and activates a distinct host response. Virulence. 2018;**9**(1):804-817. DOI: 10.1080/21505594.2018.1449508

[72] Kang D, Kirienko NV. An In vitro cell culture model for pyoverdine-mediated virulence. Pathogens. 2021; **10**(1):9. DOI: 10.3390/pathogens10010009

[73] Kirienko NV, Kirienko DR, Larkins-Ford J, Wahlby C, Ruvkun G, Ausubel FM. Pseudomonas aeruginosa disrupts Caenorhabditis elegans iron homeostasis, causing a hypoxic response and death. Cell Host & Microbe. 2013;**13**: 406-416

[74] Lamont IL, Beare PA, Ochsner U, Vasil AI, Vasil ML. Siderophore-mediated signaling regulates virulence factor production in *Pseudomonas aeruginosa*. Proceedings. National Academy of Sciences. United States of America. 2002;**99**:7072-7077. DOI: 10.1073/pnas.092016999

[75] Vasil ML, Ochsner UA. The response of Pseudomonas aeruginosa to iron: Genetics, biochemistry and virulence. Molecular Microbiology. 1999;**34**(3): 399-413. DOI: 10.1046/j.1365-2958.1999.01586.x

[76] Jenkins CE, Swiatoniowski A, Issekutz AC, Lin TJ. Pseudomonas aeruginosa exotoxin a induces human mast cell apoptosis by a caspase-8 and -3-dependent mechanism. The Journal of Biological Chemistry. 2004;**279**: 37201-37207. DOI: 10.1074/jbc.M405594200

[77] Lory S. Effect of iron on accumulation of exotoxin A-specific mRNA in Pseudomonas aeruginosa. Journal of Bacteriology. 1986;**168**(3): 1451-1456. DOI: 10.1128/jb.168.3.1451-1456.1986.

[78] Beare PA, For RJ, Martin LW, Lamont IL. Siderophore-mediated cell signalling in Pseudomonas aeruginosa: Divergent pathways regulate virulence

factor production and siderophore receptor synthesis. Molecular Microbiology. 2003;**47**(1):195-207. DOI: 10.1046/j.1365-2958.2003.03288.x

[79] Guillon A, Brea D, Morello E, Tang A, Jouan Y, Ramphal R, et al. Pseudomonas aeruginosa proteolytically alters the interleukin 22-dependent lung mucosal defense. Virulence. 2017;**8**(6):810-820. DOI: 10.1080/21505594.2016.1253658

[80] Hunt TA, Peng WT, Loubens I, Storey DG. The *Pseudomonas aeruginosa* alternative sigma factor PvdS controls exotoxin a expression and is expressed in lung infections associated with cystic fibrosis. Microbiology. 2002;**148**: 3183-3193. DOI: 10.1099/00221287-148-10-3183

[81] Oglesby-Sherrouse AG, Djapgne L, Nguyen AT, Vasil AI, Vasil ML. The complex interplay of iron, biofilm formation, and mucoidy affecting antimicrobial resistance of Pseudomonas aeruginosa. Pathogens and Disease. 2014;**70**(3):307-320. DOI: 10.1111/2049-632X.12132

[82] Singh PK, Parsek MR, Greenberg EP, Welsh MJ. A component of innate immunity prevents bacterial biofilm development. Nature. 2002;**417**(6888): 552-555. DOI: 10.1038/417552a

[83] Sambrano H, Castillo JC, Ramos CW, de Mayorga B, Chen O, Durán O, et al. Prevalence of antibiotic resistance and virulent factors in nosocomial clinical isolates of Pseudomonas aeruginosa from Panamá. The Brazilian Journal of Infectious Diseases: An Official Publication of the Brazilian Society of Infectious Diseases. 2021;**25**(1):101038. DOI: 10.1016/j.bjid.2020.11.003

[84] Ullah W, Qasim M, Rahman H, Jie Y, Muhammad N. Beta-lactamase-producing Pseudomonas aeruginosa: Phenotypic characteristics and molecular identification of virulence genes. Journal of the Chinese Medical Association: JCMA. 2017;**80**(3):173-177

[85] Kirienko DR, Kang D, Kirienko NV. Novel pyoverdine inhibitors mitigate *Pseudomonas aeruginosa* pathogenesis. Frontiers in Microbiology. 2019;**9**:3317. DOI: 10.3389/fmicb.2018.03317

Chapter 9

Machine Learning for Antimicrobial Resistance Research and Drug Development

Shamanth A. Shankarnarayan, Joshua D. Guthrie and Daniel A. Charlebois

Abstract

Machine learning is a subfield of artificial intelligence which combines sophisticated algorithms and data to develop predictive models with minimal human interference. This chapter focuses on research that trains machine learning models to study antimicrobial resistance and to discover antimicrobial drugs. An emphasis is placed on applying machine learning models to detect drug resistance among bacterial and fungal pathogens. The role of machine learning in antibacterial and antifungal drug discovery and design is explored. Finally, the challenges and prospects of applying machine learning to advance basic research on and treatment of antimicrobial resistance are discussed. Overall, machine learning promises to advance antimicrobial resistance research and to facilitate the development of antibacterial and antifungal drugs.

Keywords: machine learning, antimicrobial resistance, fungi, bacteria, infection, drug discovery and design

1. Introduction

Antimicrobials are the agents used to prevent and treat the infection caused by bacteria, fungi, viruses, and parasites in plants, animals, and humans. Sir Alexander Fleming in his Nobel Prize lecture emphasized the importance of avoiding resistance to antibiotics [1]. Antimicrobial resistance (AMR) is a phenomenon that occurs when infectious microorganisms do not respond to antimicrobial agents, leading to treatment failure, the spread of the infectious disease, and severe illness and death [2]. Among microorganisms, bacteria and fungi are the most encountered pathogens with resistance in clinical settings. Patients infected with resistant bacteria or fungi have worse clinical outcomes compared to patients with infections caused by the same bacteria or fungi without resistance [3]. It is estimated that by the end of year 2050, if unmitigated, AMR will result in 10 million lives lost per year and cumulative cost of 100 trillion USD [4]. The global burden associated with bacterial AMR alone, considering 204 countries and territories, 23 bacterial pathogens, and 88 drug-pathogen combinations, was 4.95 million deaths during the year 2019 [5]. The majority of these patients succumbed to lower respiratory tract and blood stream infections associated with drug-resistant bacteria, with highest mortality rate of 27.3 per 100,000 patients [5]. Among elderly patients in the USA, the treatment of methicillin resistant *Staphylococcus aureus* (MRSA) infection costs $22,293 more per patient compared to patients infected with non-resistant *Staphylococcus aureus*. Similarly, treating patients infected with resistant carbapenem-resistant *Acinetobacter* species costs $57,390 more per patient compared to patients infected with non-resistant *Acinetobacter* species. These extra costs are attributed to the increased length of hospital stays and health complications, which lead to more medical interventions and higher mortality rates [6].

The most common bacterial pathogens associated with hospital acquired infections and AMR are the ESKAPE pathogens. ESKAPE is an acronym for *Enterococcus faecium*, *Staphylococcus aureus*, *Klebsiella pneumoniae*, *Acinetobacter baumannii*, *Pseudomonas aeruginosa*, and *Enterobacter species* [7]. The priority pathogens recognized by the World Health Organization are extended spectrum beta lactamases (ESBL) producing *Escherichia coli*, MRSA, ESBL-producing *Klebsiella pneumoniae*, *Streptococcus pneumoniae*, carbapenem-resistant *Acinetobacter baumannii*, and multidrug-resistant (MDR; organism resistant to at least one agent in three or more antimicrobial classes) *P. aeruginosa* and vancomycin-resistant *Enterococcus fecalis* [5, 8, 9]. Antimicrobial resistance among fungi is a serious issue because of the limited number of classes of antifungal agents available for treating invasive fungal infections, as compared to antibacterial agents (**Table 1**). Moreover, due to variety of socio-economic reasons it has been over a decade that no new class of antifungal drug has been developed [10]. Global warming and climate change is also predicted to increase the prevalence of fungal infections (as fungi adapt to higher temperatures, humans and animals may lose their

Mechanism of action	Antibacterial class
Inhibitor of cell wall synthesis	β-Lactams, Carbapenems, Cephalosporins, Monobactams, Penicillin, Glycopeptide
Cell membrane depolarizer	Lipopeptides
Inhibitor of protein synthesis	Aminoglycosides, Tetracyclines, Chloramphenicol, Lincosamides, Macrolides, Oxazolidinones, Streptogramins
Inhibitor of nucleic acid synthesis	Quinolones
Inhibitor of metabolic pathways	Sulfonamides, Trimethoprim
Mechanism of action	**Antifungal class**
Inhibitors of ergosterol synthesis	Azoles
Aqueous pores in cell membrane	Polyenes
Inhibitor of glucan synthase	Echinocandins
Inhibitor of squalene epoxidase	Allylamines
Inhibitor of nucleic acid	5-Flurocytosine

Table 1.
Different classes of antibacterial and antifungal drugs and their mechanism of action.

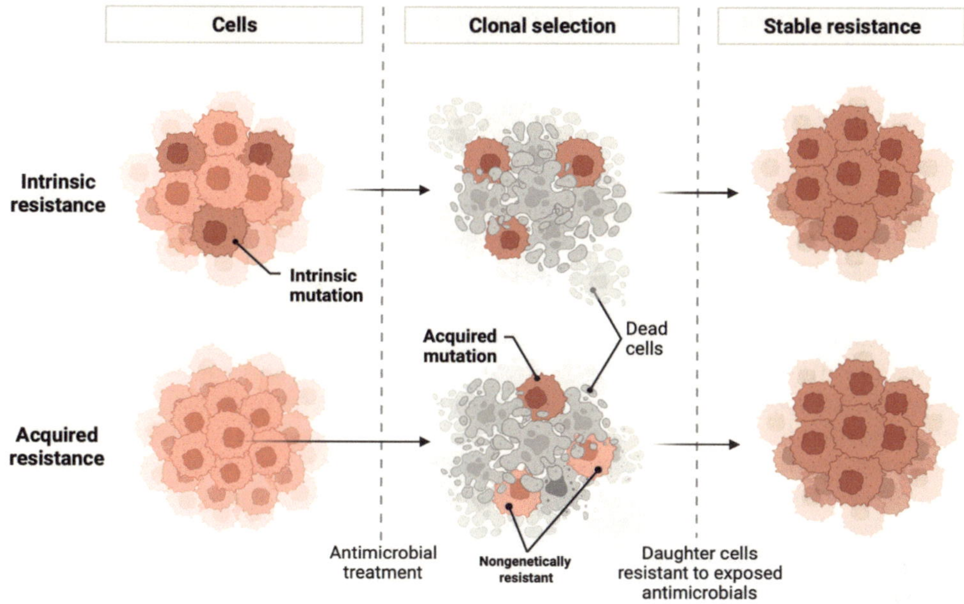

Figure 1.
Depicting the difference between intrinsic and acquired resistance. Microorganisms that are intrinsically resistant can propagate from the moment that they are exposed to the antimicrobial agent. Microorganisms can also acquire resistance during exposure to an antimicrobial agent through genetic and nongenetic mechanisms. Adapted from 'Intrinsic and acquired drug resistance', by BioRender.com (2022). Retrieved from https://app.biorender.com/biorender-templates.

thermal protection provided by their elevated body temperatures) [11]. The majority of the invasive fungal infections are caused by yeasts, especially *Candida albicans*, which can cause mild symptomatic infection to acute sepsis with a mortality rate over 70% in immunocompromised patients [12]. Over the last decade, *Candia auris* has been reported on all continents and in more than 44 countries [13, 14]. The first known appearance of *Candida auris* dates back to 1996 in South Korea, when it was originally misidentified as *Candida hemulonii* (and then later correctly identified as *Candida auris*) [15]. This fungus displays intrinsic resistance and acquired resistant (**Figure 1**) to the major classes of antifungals and hospital disinfectants and has caused several outbreaks [16–19]. The main reason that *Candida auris* attention across globe is due to high mortality rate (45%) among patients with bloodstream infections [20]. Interestingly, *Candida auris* has different resistance profiles based on the genomic sequences identified in different countries; presently, *Candida auris* is classified into four discrete clades, as well as a potential fifth clade [21, 22]. *Candida auris* is less virulent than *C. albicans* because of the 'fitness cost' associated with its MDR nature; as a consequence, *Candida auris* has not been observed to revert back to its susceptible form in the absence of antimicrobial pressure [23]. Recently in the United States the identification of pandrug-resistant (resistant to all agents in all classes of antimicrobial agents) [24] *Candida auris* among skin colonizers has raised alarm [25]. Mycelial fungi, which consisting of network of fine filaments known as hyphae, such as *Aspergillus* species are ubiquitous in nature and commonly cause respiratory disorders. *Aspergillus* species resistant to the azole class of antifungals are a serious threat, as azoles are first line of therapy against *Aspergillus* infection [26]. Another mycelial fungi, *Trichophyton indotinea,* which causes skin infection is spreading across the globe [27, 28].

The emergence of AMR in high-income countries is mainly associated with use, misuse, and overuse of antibiotics in hospitals, agriculture, and communities [29]. Whereas in low- and middle-income countries unhygienic practices, contaminated water supplies, civil conflicts, and an increased number of immunocompromised patients (especially among HIV infections) are the main contributors to AMR [30]. Increased infections, and in turn increased use of antimicrobial agents, has imposed selection pressures that result in the retention of resistant strains. Identifying infectious agents early helps clinicians to promptly choose the appropriate antimicrobial agent to treat the infection based on the intrinsic resistance profiles and local epidemiology data on resistance [31]. Resistance profiling methods, such as culture-based and molecular biology-based methods, currently take up to 72 h from the time of sample collection. During this time, patients often receive broad-spectrum antibiotics, which may lead to acquired resistance (**Figure 1**). Several novel strategies have been developed for rapid detection of AMR. However, most of these methods are based on molecular biology, immunology, biochemistry, and rapid culture techniques [32]. Importantly, the cost and the expertise involved in establishing and maintaining these techniques and related devices is often too high for many hospitals and institutions, especially those in remote and impoverished communities.

Machine learning (ML) has been around for decades, as optical character recognition gained popularity during 1990s with its application as spam filters. A seminal paper by Geoffery Hinton in 2006 on recognizing handwritten digits using 'deep learning' (a ML technique implemented in artificial neural networks) rekindled interest in ML. Recently, during the 14th Critical Assessment of Protein Structure Prediction (CASP14) competition [33], a neural network based model called AlphaFold predicted protein structures with high accuracy (i.e., comparable to the experimental structures), outperforming other protein structural deduction methods [34]. Furthermore, deep learning is increasingly being applied to solve complex multidimensional problems, such as speech recognition [35] and image classification [36].

Machine learning is the application of advanced algorithms that enable a computer to 'learn' and generate predictive mathematical models from data. Arthur Samuel in 1959 described ML as 'the field of study that gives computers the ability to learn without being explicitly programmed' [37]. Tom Mitchell in 1997 provided a more engineer-oriented definition, when he stated that a 'computer program is said to learn from experience E with respect to some task T and some performance measure P, if its performance on T, as measured by P, improves with experience E' [38]. Machine learning can be divided into supervised, unsupervised, and reinforcement learning. In supervised learning, the ML model is trained using labeled datasets, with the resulting model being a function that can take new data and predict an output. To determine the reliability of the trained model, a test set of complete input/output data which was not used during training is employed to determine an unbiased estimate of model performance. Whereas, in unsupervised learning, the training data are supplied without labels. Unsupervised learning algorithms find the similarity among data points and cluster them together. Reinforcement learning (RL) uses algorithms that learn from the accumulation of 'rewards' that a computational agent receives through interactions with its environment. Reinforcement learning, which is often combined with other ML methods such as deep neural networks, has led to some of the most successful artificial intelligence systems ever developed. These range from systems that beat human professionals in the game of Go [39] to systems that help control nuclear fusion reactions [40].

Recent advances in digitizing medical records and data generated in experiments have paved the way for ML applications in the fields of biology and medicine. Many clinical trials are leveraging ML processes to improve the efficiency and quality of clinical research and pre-clinical drug development [41]. Machine learning is also being applied to assess the risk of developing sepsis based on patients' clinical records [42]. Machine learning has also found applications at the cellular level. For instance, convolutional neural networks (CNNs) can predict the interactions of transcription factors and histones within chromosome structures, which in turn aids in analyzing genome architecture as well as gene regulation [43]. Other examples include using neural networks to identify the role of non-coding DNA in humans in regulating gene expression [44] and applying recurrent neural networks (RNNs) to characterize chromatin folding in *Drosophila melanogaster* [45]. Furthermore, the availability of large-scale high-throughput genomic and epigenomic data has led to several studies that have highlighted the potential applications of ML in the field of genomics [46] as well as non-coding RNAs [47]. Machine learning has also been used to assist clinicians treating infectious diseases [48]. However, the use of ML in studying drug-resistant pathogens is less developed.

In this chapter, we first discuss the mechanisms of underlying bacterial and fungal AMR, followed by an overview of ML methods used to detect drug-resistant pathogens. We then highlight the application of ML in the discovery and design of antimicrobial drugs. Finally, we present the challenges and prospects of applying ML to AMR research and drug development.

2. Mechanisms of antibiotic resistance

The major burden of AMR in hospital settings is due to bacteria and fungi. Antimicrobial resistance can be classified into different types, including 'intrinsic resistance' and 'acquired resistance' (**Figure 1**) [49]. Intrinsic resistance occurs when bacteria or fungi are naturally resistant to an AMR drug or to a class of AMR drugs [50]. Bacteria and fungi which were previously susceptible to an antimicrobial drug can acquire resistance, for instance, by modifying the target site of the drug or by gaining a resistance mutation (**Figure 1**). In these scenarios, the microorganism develops resistance post-exposure to the drug. Whereas, if the microorganism does not have a target site for the drug or has a preexisting resistance mutation, then it is classified as intrinsically resistant. Other forms of AMR exist, such as 'clinical resistance', whereby a microorganism is susceptible to a drug *in-vitro,* but the drug is ineffective against the same microorganism in *in-vivo*. Clinical resistance can occur in a patient due to pharmacokinetic and pharmacodynamic factors.

Another aspect of AMR is 'persistence' and 'tolerance', which are phenomena that allow non-growing or slow growing bacterial and yeast pathogens to survive antimicrobial treatment [51, 52]. In the case of genetic resistance to a drug, all the progeny of the resistant microorganism stably inherit resistance to the drug (**Figure 1**). Whereas persistence occurs when a small fraction of a clonal bacterial population is resistant to an antibiotic, but the persistent cells do not harbor resistance mutations or genes. Rather, these persister cells are in a stationary or dormant phase, which reduces the effectiveness of antibiotics that target growth processes [53–55]. Antibiotic persistence is a heterogenous response of a bacterial population to an antibiotic and causes a delay in the clearance of the infection [56]. In contrast, tolerant cells require more time to be affected by an antimicrobial drug compared to susceptible cells [56]. Systemic infections due to persistent and tolerant organisms lead to higher mortality rates compared to infections caused by susceptible microorganisms [57]. Nongenetic drug resistance is another form of AMR. Nongenetically drug-resistant phenotypes can be found in clonal cell populations [58] and results from genetically identical cells differentially expressing genes that confer resistance, along with various epigenetic mechanisms [59, 60].

Bacteria and fungi belong to different kingdoms, have differences in cellular components, and antibacterial and antifungal agents target different sites. Despite this, there are similarities between the AMR agents that are used to treat antifungal and antibacterial infections. For instance, cell wall inhibitors of bacteria target peptidoglycan, an important component of the bacterial cell wall, whereas some antifungal agents inhibit ergosterol, an important component of fungal cell membrane. Antibacterial agents have diverse mechanisms of action, including inhibiting cell wall synthesis, depolarizing cell membranes, as well as inhibiting of protein synthesis, nucleic acid synthesis, and metabolic pathways (**Table 1**) [61]. However, in contrast to many antibacterial agents, antifungal analogues for protein inhibitors, topoisomerase inhibitors, and metabolic pathways inhibitors are not available. Only a limited number of antifungal agents are available that target ergosterol synthesis, cell membrane integrity, glucan synthase, nucleic acid synthesis, and the squalene epoxidase enzyme.

2.1 Antibacterial resistance mechanisms

The main mechanisms of antibiotic resistance among bacteria are (i) limiting uptake of a drug; (ii) modifying a drug target; (iii) inactivating a drug; and (iv) active drug efflux (**Figure 2a**). Limiting uptake due to natural permeability barriers imposed by the cell membrane, drug inactivation by antibiotic inactivating enzymes, and drug efflux resulting non-specific protein efflux pumps are mechanisms of intrinsic resistance. Whereas the transfer of genes between bacteria that encode drug efflux pumps or enzymes that inactivate antibiotics, as well as drug target modifications, are acquired resistance mechanisms. Antibiotic resistance mechanisms differ between gram-negative and gram-positive bacteria due to differences in their cell wall composition. Gram-negative bacteria employ all the drug resistance mechanisms, whereas gram-positive bacteria mainly limit the uptake of a drug [62]. Due to the hydrophobic nature of the cell wall, many of the hydrophilic antibiotic cannot bind to the cell wall and the high lipid content among mycobacteria restricts the entry of hydrophilic antibiotics [63]. However, porin channels found within the cell membrane allow certain hydrophilic antibiotics to enter the cell. Modifications to these porin channels limits drug uptake [64]. Mutations in the gene responsible for porin proteins alter the selectivity of hydrophilic drugs [65]. Drug intake is also restricted by the thickening of cell wall [63]. Another widely observed phenomenon that restricts drug uptake is the formation of bacterial and fungal biofilms. The thick outer layer of a biofilm is composed of extracellular polymeric substances and is impenetrable to many antimicrobial drugs [66].

Antibiotics target multiple cellular components and bacteria can modify these targets leading to AMR. One of the major targets is the cell wall, which is commonly targeted by ß-lactam drugs, specifically among gram positive bacteria. Resistance to ß-lactam antibiotics results from modifications in the cell wall structures as well as a number of penicillin-binding-proteins [67]. Bacteria can alter the precursor of the target by mutating the gene responsible for these precursors, eventually leading to an altered target site. This results in the antibiotic failing to bind to the target site [68]. Ribosomes are also commonly targeted by antibiotics to inhibit protein synthesis.

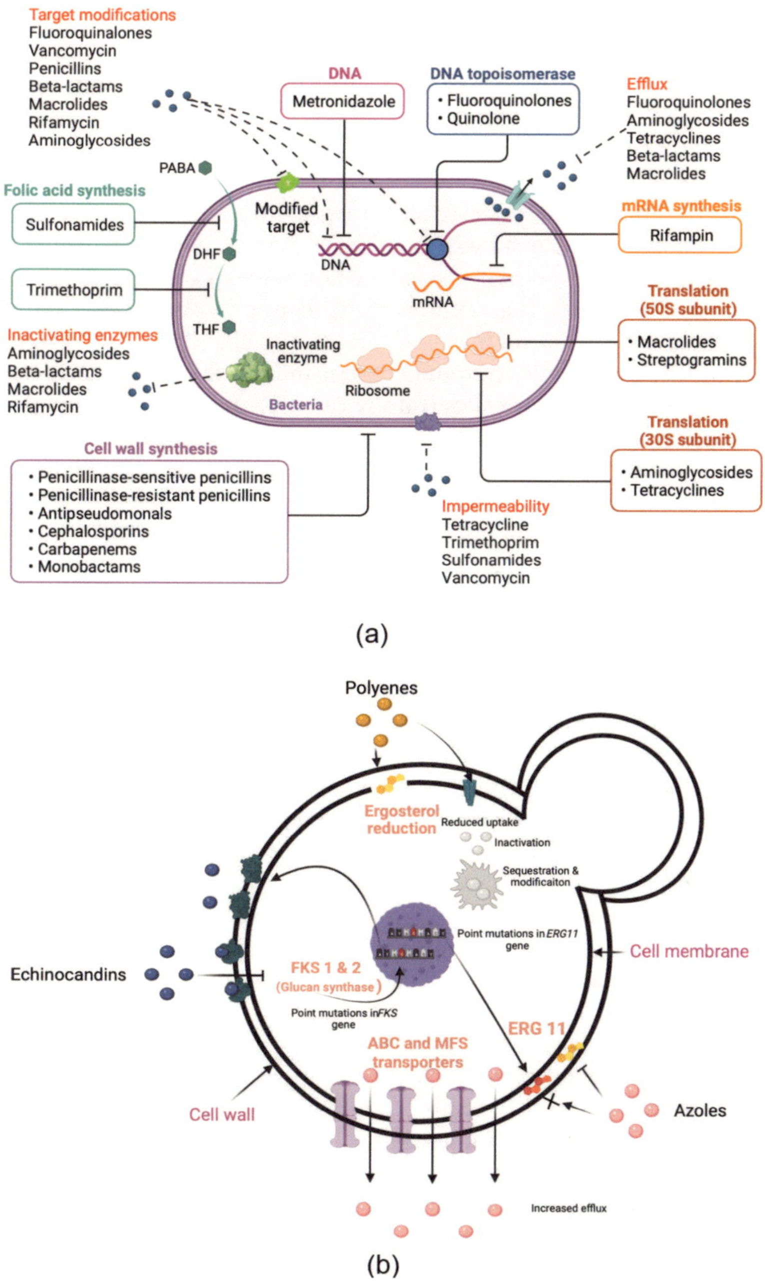

Figure 2.
Mechanisms of action of antimicrobial drugs in bacteria and fungi. (a) Effect of antibacterial drugs on bacterial cellular components and the corresponding resistance mechanism developed by bacteria. Created with Bio-Render. com. (b) Effect of antifungal drugs on fungal cellular components and the resistant mechanisms developed by the fungi. Adapted from "Antimicrobial Therapy Strategies", by BioRender.com (2022). Retrieved from https://app. biorender.com/biorender-templates.

Mutations in the ribosomal gene leading to the protection of the ribosomes and methylation of the ribosomal subunits lower the binding affinity of antibiotics, leading to resistance [69]. Similarly, modifications in the DNA gyrase or topoisomerase enzyme, nucleic acid synthesis inhibitors fail to bind to these enzymes [70]. Drugs that inhibit metabolic pathways inhibit important metabolic byproducts that are essential for bacterial survival. These antibiotics competitively bind to the active sites of enzymes responsible for the synthesis essential metabolites. Mutations in the gene responsible for these enzymes restricts antibiotics from binding [71]. Another mechanism of AMR is the inactivation of the drug by the pathogens. Degrading or transferring a chemical group to the antibiotics modifies its structure and affinity towards the target [72]. Efflux pumps remove toxic substances from the bacterial cell; some efflux pumps are constitutively expressed and others are induced or overexpressed in the presence of antibiotics. There are majorly five families of efflux pumps depending on the energy source they utilize and their structure [64]. Namely, the ATP-binding cassette (ABC) family, the multidrug and toxic compound extrusion family, the small multidrug resistance family, the major facilitator superfamily (MFC), and the resistance-nodulation-cell division family. The majority of the bacteria resistant to antibiotics overexpress efflux pumps from one of these families during antibiotics exposure [73].

2.2 Antifungal resistance mechanisms

Antifungal resistance mechanisms are not as extensively studied as antibacterial resistance mechanisms. Several factors including immunosuppressive treatments, indiscriminate use of broad-spectrum antibiotics, and immune suppressive diseases like HIV led to a surge in fungal infections during 1970s and 1980s [74]. Antifungal drugs including imidazoles and azoles were subsequently approved during late 1980s and 1990. Extensive use, misuse, and overuse of these antifungal drugs since then have led to the emergence of AMR in fungal pathogens. Determining if a fungal isolate is resistant is based on the minimum inhibitory concentration (MIC) of the antifungal drug. The MIC of a fungus isolated from a clinical sample informs the decision on the appropriate course of antifungal therapy.

Currently three major classes of anti-fungal drugs used for treating systemic fungal infections. Namely, azoles (itraconazole, voriconazole, posaconazole, and isavuconazole), polyenes (amphotericin B) and echinocandins (caspofungin, micafungin, and anidulafungin) (**Table 1**). The limited number of classes of antifungal drugs and AMR in fungi restricts treatment options. The emergence of MDR fungal species further hinders treatment options. Azoles target ergosterol biosynthetic pathway, as ergosterol is necessary in the cell membrane to maintain the stability, permeability and the activity of membrane bound enzymes (**Figure 2b**) [75]. The substitution of an amino acid in the binding site of the enzyme is a common mechanism of azole resistance among *Candida* species. Overexpression of *ERG11* gene is also common among azole-resistant strains [76]. Furthermore, the overexpression of drug targets decreases the effectiveness of a drug, as more drug is required for inhibition [77]. Like bacterial efflux pumps, fungi have two main membrane associated efflux pumps superfamilies, the ABC superfamily and the MFC superfamily. Overexpression of *Candida* drug resistance (CDR) genes such as *CDR1* and *CDR2* of the ABC superfamily lead to the efflux of azoles and decreased drug accumulation [78, 79]. Gain-of-function mutation in the gene responsible for a transcription factor UPC2 leads to upregulation of many ergosterol biosynthesis genes, conferring azole resistance [80]. Another transcription factor TAC1 regulates the activity of efflux pumps in *Candida* species. TAC1 is responsible for upregulation of *CDR1* and *CDR2* in the presence of azoles [81]. Chromosomal abnormalities and mitochondrial defects also contribute to azole resistance [82, 83]. Stress response pathways related to the heat shock protein Hsp90 provide critical strategies for the survival in the presence azoles leading to resistance [84]. Echinocandin resistance is mainly due to mutations in the *FKS* gene. *FKS* gene is responsible for the synthesis of glucan synthase enzyme involved in the synthesis of ß-glucan in the fungal cell wall [85, 86]. In certain cases, echinocandin induces chitin synthesis via protein kinase-C, high osmolarity glycerol, and calcineurin pathways [87] by activating two chitin synthases (Chs2 and Chs8) [88], leading to masked target sites. Polyene resistance in fungal pathogens is less understood because of its various mechanisms of action on the fungal cell. Polyenes act on the fungal cell membrane by interacting with ergosterol and impairs the membrane barrier function [89]. Polyene resistance is mainly attributed to the alterations in the sterol content of the cell membrane, a defense mechanism developed against oxidative stress created by the drug and reorientation of ergosterol structures within the cell membrane [90]. Furthermore, *Candida* species harboring mutations in the *ERG3* and *ERG6* genes exhibited polyene resistance [91]. However, increased catalase activity by the fungal cell also reduces the oxidative stress imparted by the amphotericin leading to resistance [92]. Polyene and azole resistance in combination has been reported among *Candida* species as well as *Cryptococcus neoformans*, and has mostly been attributed to the reduction of ergosterol in the cell membrane and accumulation of its intermediates [93].

Current methods for detecting AMR among the infecting pathogens take up to 72 h from the time of sample collection. All the isolated bacterial and fungal pathogens must undergo standard antimicrobial susceptibility testing (AST) as recommended by the European Committee on Antimicrobial Susceptibility Testing and the Clinical Laboratory Standards Institute [94, 95]. Early detection of the infecting pathogen along with its drug resistance profile are critical for initiating prompt antimicrobial therapy. However, several challenges are faced during this process, such identifying the pathogen, differentiating between commensal and pathogenic microorganisms in a clinical sample [96]. After successful isolation of the pathogen, a round of subculture must be performed so that contamination can be excluded before commencing AST. Microbroth dilution and disk diffusion AST methods can get delayed due to contamination, leading to delays in initiating the appropriate antimicrobial therapy. Several new technologies and methods are being used for early and rapid detection of AMR. For example, technologies based on nucleic acid amplification, hybridization, microscopy, electrochemical, mass spectroscopy, and nanotechnology [97, 98]. However, these methods require sophisticated instruments, expertise, and expensive consumables restricts their deployment in low-income countries. Point-of-care tests (POCTs) used at patient bedsides are now being used to determine AMR; POCTs can be also used among outpatients. Some types of POCTs like microscopy stations, single molecule biosensors, and microfluidic platforms are being tested [99, 100]. The drawbacks of POCTs, including small sample size, lack of internal standards, and their inability to detect nongenetic forms of AMR resistance still need to be resolved. More advanced methods such as ML approaches to detect AMR could further reduce turn-around times and could be deployed across diagnostic laboratories. Machine learning methods can be also applied to detect certain features that are present in resistant bacteria and fungi, but absent in sensitive isolates, which the human eye or other diagnostic technologies may fail to recognize [101]. For instance, real-time high-throughput screening of modified proteins within the resistant isolates [102] has been less explored and is an ideal application for ML methods. The application of ML methods (Section 3) may lead to a deeper understanding of AMR mechanisms, which in turn could lead to rapidly detecting AMR pathogens in patients (Section 4) and to developing new drugs (Section 5).

3. Machine learning basics

Machine learning enables us to investigate and draw conclusions from information contained in data that would otherwise be inaccessible to humans. Problems that benefit from the application of ML are endless, but they have a few defining features [103]. First, the problem may have a known solution, but converting it into a computer program is not feasible or requires extensive resources. For example, humans can easily identify a dog within a group of other four-legged animals but writing a computer program to explicitly describe all possible aspects of a dog and its differences to other similar animals would be error prone and practically infeasible. On the other hand, training a ML algorithm to identify a dog may only take a few lines of code, given modern ML software tools. Second, complex problems where traditional methods have failed to identify a solution may benefit from the use of ML algorithms (**Figures 3** and **4**), such as the use of deep learning systems to master the game of Go [104] or to make highly accurate predictions of protein structure [34]. Not only does this enable the use of the resulting ML model in practical applications, but it can also guide researchers towards a deeper understanding of the system they are studying. For instance, ML can guide mathematicians by finding patterns and relations between mathematical objects that can lead to the formation of new conjectures and theorems [105].

Although the defining feature of all ML approaches is to learn from a given dataset, ML techniques can be separated into three broad categories based on the amount of human input: Supervised learning, unsupervised learning, and reinforcement learning [103, 106–108]. Each of these approaches have their own concepts, techniques, and areas of applicability, with the differences between them not always clear. Nonetheless, these categories are useful to provide a means to determine the best approach for a particular problem at hand. Understanding the available tools is

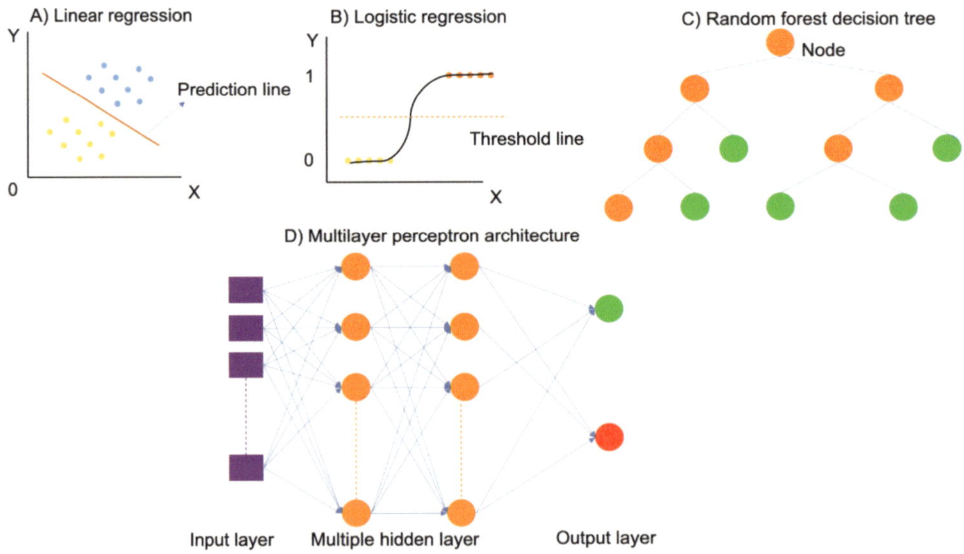

Figure 3.
A selection of common machine learning methods. (A) Linear regression model using a prediction line to distinguish the test dataset. (B) Logistic regression model using a threshold to distinguish the test dataset into two groups. (C) Random forest model using a visually generated decision tree for datapoints to estimate each samples outcome by voting. (D) Multilayer perceptron architecture consisting of an input layer, multiple hidden layers, and an output layer.

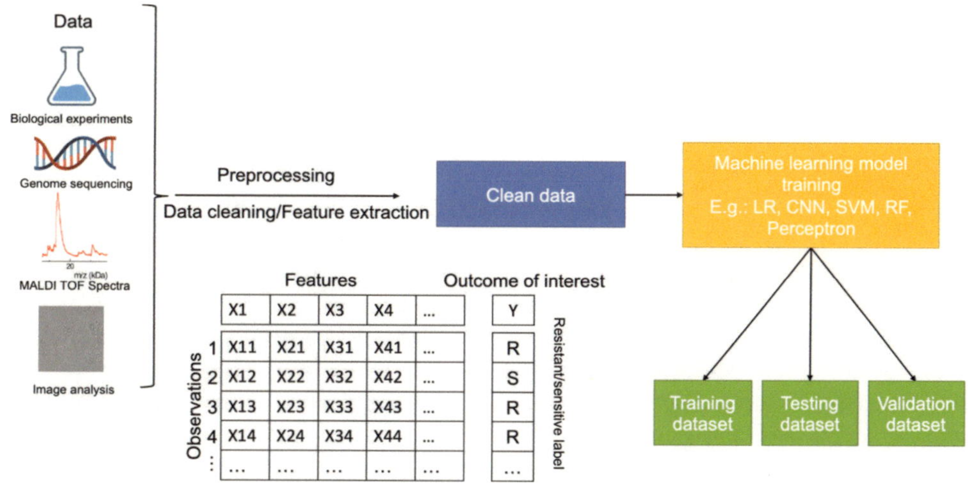

Figure 4.
The machine learning pipeline. This pipeline consists of data originating from different biological experiments, preprocessing steps for cleaning the data, along with the feature extraction process. Machine learning methods are then applied to the clean data by dividing this data into training, testing, and validation sets. 'MALDI TOF' stands for 'matrix assisted laser desorption ionization time of flight', 'LR' for 'logistic regression, 'CNN' for 'convoluted neural network, 'SVM' for 'support vector machine', and 'RF' for 'random forest'.

crucial for choosing the best ML technique to solve a particular problem. Although an extensive overview of each ML category is outside the scope of this chapter, we provide an overview of some of the common ML methods below.

3.1 Supervised learning

Supervised learning consists of algorithms that learn using a training set consisting of labeled data [106]. The goal of supervised learning is to find a model for the relationship between the inputs (called 'features') and known outputs, which can then be used to predict outputs for future inputs, where the actual outputs are unknown. Supervised learning techniques can be separated into two categories, 'classification' and 'regression' [109, 110].

Classification problems generally aim to classify future inputs into predefined categories through training on examples, where the inputs are labeled with their corresponding category [107]. Given enough quality training data, models created with classification techniques can provide accurate classification of future data, without requiring the details of the input data to be explicitly programed [103, 106–108]. For instance, a researcher may desire to have a computer take a microscopy image of a cell and return the name of the species, without requiring a human to identify the species. Using a training set of microscopy images for a variety of different species labeled with the name of the species, a classification model can be trained to learn the relationships between the visual aspects of the species and their labels. The model produced can then be used on unlabeled microscopy images to determine the species, saving researchers time and effort, along with producing a model that can be shared in the scientific community. Classification learning algorithms are not restricted to images; any form of data that can be separated into predefined categories can be fed into a classification learning algorithm for training to produce a classifier model [107, 108].

While classification methods aim to predict discrete class labels for inputs, regression methods aim to predict continuous numerical values for given numerical inputs [107, 108]. Regression techniques also learn from training data containing inputs and outputs, but in this case the data consists of numerical inputs and their corresponding numerical outputs, with the resulting model being a continuous mathematical relationship between inputs (independent variables) and outputs (dependent variables) [107]. The resulting model can then be provided with future inputs to make numerical predictions. For example, a researcher may be interested in finding a mathematical relationship between the inputs of an experiment (e.g., preset voltages) and the corresponding outputs they detect (e.g., electrical currents), for systems where theory is unable to make accurate predictions. By training a regression model on a large amount of set inputs and detected outputs, the researcher may be able to find a model that accurately predicts numerical outputs when given future inputs. Not only is this useful in a practical sense, but the resulting model can also be used to guide fundamental research by providing an accurate mathematical and physical relationships that can be further analyzed and understood in terms of theoretical ideas [105, 111].

Through extensive research on supervised learning, many different learning algorithms for classification and regression have been developed and programmed into readily available software packages. Linear regression, logistic regression [107, 108], support vector machines (SVMs) [112], decision trees and random forests [113] and most artificial neural networks [114] are some examples of supervised learning systems, each having their own advantages and disadvantages.

3.2 Unsupervised learning

Unsupervised learning methods, unlike supervised learning, attempt to learn from unlabeled data [115]. This often takes the form of data clustering, but other methods such as anomaly detection and dimensionality reduction also fall under this category [107, 108]. Clustering algorithms attempt to separate unlabeled data into groups with similar components, which can be useful for extracting information from high-dimensional data, which is often infeasible for a human to do. Anomaly detection involves finding anomalous outliers in large datasets by comparing data points to learned patterns, which can be helpful when working with noisy experimental data [116, 117]. Dimensionality reduction methods attempt to simplify high-dimensional data without losing important information, making the analysis and use of such data easier [118, 119]. Unsupervised learning methods can also be combined with supervised learning, referred to as 'semi-supervised' learning, to learn from data that is partially labeled [120, 121]. This is useful when working with large amounts of data, where labeling every data point is infeasible. Some examples of unsupervised learning methods include k-means clustering [122, 123], hierarchical clustering [124, 125], DBSCAN [126], isolation forests [127], principal component analysis [128], autoencoders [107, 108], locally linear embedding [129], and expectation-maximization algorithms [130].

3.3 Reinforcement learning

Reinforcement learning approaches rely on the idea of learning from 'rewards' obtained through interactions with an environment [131]. Reinforcement learning problems are formulated as a discrete-time stochastic control processes known as

'Markov decision processes', with the goal of training a computational system (or 'agent') to determine the best strategy (or 'policy') for reaching a defined goal [132]. The environment is defined by 'states' that the agent can be in, while the agent is able to perform certain 'actions' to interact with the environment. As the agent interacts with its environment, numerical values called rewards that model performance are collected for performing certain actions [132]. The goal of the agent is then to maximize these rewards (using sophisticated statistical methods) by learning the best policy for making decisions in particular situations through repeated interactions with its environment [132]. For example, a reinforcement learning system may be programmed into a cleaning robot to maximize the amount of cleaning it can do while still being able to return to its charging station. In this case, a positive reward would be given for picking up trash, while a negative reward would be given for letting its battery die without reaching the charging station. Using reinforcement learning methods, the robot can learn to optimize its own behavior through repeated experience with its environment.

3.4 Validating machine learning models

To ensure the model created using ML is accurate it must be validated on data independent of the training set [103, 106–108, 133]. Applying the trained model directly to a certain problem is one method of testing, but this is often impractical for real-world applications where model performance matters. The usual method of validation is to split the initial dataset into training and testing sets, where the model is trained on the training set and its accuracy is determined by comparing its predictions using the testing set inputs to the true outputs from the test set [107, 108]. This analysis provides the 'generalization error' estimate of the model, which is used to determine whether the model is accurate, and the errors associated with using the model on new data [107]. Many different metrics are used to determine the generalization error, such as the root mean square error or false-positive/false-negative rates [103, 107, 108], and the choice of method depends on the problem and the learning algorithm. Through iterative training and testing cycles, model performance is improved until a satisfactory accuracy is achieved.

A major issue when using ML is overfitting the model to the training set [103, 106–108, 133]. This corresponds to the case where the 'training error' (i.e., how well the model matches the training data) is low, but the generalization error (i.e., how well the model can predict outcome values for previously unseen data) is high [107, 108]. This is a common occurrence, especially when using models that are more complex than the actual relationships contained in the data. For example, if the actual relationship between inputs and outputs is linear but we attempt to fit a third-degree polynomial to the data, we may produce a model that passes through each of the training set data points exactly (low training error) but cannot generalize to data outside of the training set (high generalization error). Avoiding overfitting (as well as underfitting) requires the use of appropriate training and validation methods to determine model performance before deploying a trained ML model. The quantity of training data is also important. A lack of training data can lead to inaccurate or biased predictions. The amount of data required to create accurate models ultimately depends on the problem and ML method being used [103, 106–108, 133].

During the testing stage, it is important to tune the 'hyperparameters' of the model to improve training accuracy [103, 106–108, 133–135]. Hyperparameters refer to the parameters that are not being learned, such as gradient time steps or data batch size. Many cross-validation techniques for hyperparameter tuning are available, such as k-fold cross validation [135], and can be implemented directly in ML software packages. It is also often necessary for datasets to be pre-processed before applying ML techniques [136]. Pre-processing is application/software dependent and involves converting the collected data into data structures that can be read by the ML algorithm/software package being used.

3.5 Machine learning software

The extensive and increasing use of ML in industry and scientific research has led to the development of many tools for applying ML techniques quickly and accurately. With almost every well-established ML algorithm being implemented in free dedicated software packages, deploying a ML solution has in some cases become as simple as writing a few lines of code. Although the researcher must determine whether their problem may benefit from the application of ML, the availability of extensively tested and optimized tools to apply ML has made doing so much easier once the relevant data has been collected and organized.

Python is currently the most used programming language for ML, as it contains well-developed and optimized ML libraries. However, other languages such as Julia are also becoming popular with ML researchers. Below is a list of some of the free software packages used for ML applications, along with the programming languages they can be used with.

- TensorFlow (https://www.tensorflow.org/) [137]. Developed by Google, TensorFlow can be used with a variety of programming languages, including Python, C++, Julia, and Java.

- Keras (https://keras.io/) [138]. Keras is a widely used, user-friendly Python interface for the TensorFlow library.

- Scikit-learn (https://scikit-learn.org/) [139]. Scikit-learn is a Python library that contains many ML algorithms, optimized for Python data structures. Wrappers to use Scikit-learn with other programming languages, such as Julia, are also available.

- PyTorch (https://pytorch.org/) [140]. Developed by Facebook, PyTorch is a ML framework primarily for Python, but it also has a C++ version.

4. Machine learning for detecting drug resistance

Over the last decade, an increase in AMR has occurred across the world. At the same time, ML methods have been successfully applied in numerous scientific fields. The availability of large datasets from whole genome sequencing (WGS), matrix assisted laser desorption ionization time of flight mass spectroscopy (MALDI TOF MS), transcriptional response to antibiotics and proteome profiles have facilitated the application of ML algorithms to detect AMR. Specifically, ML methods have been used to detect AMR in bacterial and fungal pathogens based on the data obtained from WGS and MALDI TOF MS (**Figure 4**) [102, 141–143]. Reduced genomic sequencing cost and high-throughput data from WGS has enabled application of

ML methods to sequence data. A few studies have utilized genome sequencing data to predict resistance phenotypes among bacterial pathogens using ML methods [144–149]. A ML method called 'adaptive boosting' was employed to detect carbapenem resistance in *A. baumannii*, MRSA, and beta-lactam and co-trimoxazole resistance in *S. pneumoniae* with accuracies ranging from 88 to 99% [145]. Similarly, another ML method called 'gradient-boosting' was able to detect MIC in *K. pneumoniae* against 20 antibiotics [146]. A software package called 'Mykrobe predictor' detected resistance in *S. aureus* and *Mycobacterium tuberculosis* against 12 antibiotics [147]. These models were able to classify the pathogens as either resistant or sensitive, however, the features used by the algorithm to classify them are not known. In this regard, classification and regression trees (CART) and set covering machines (SCM) models were employed to detect resistance among 12 bacterial species against 56 antibiotic combinations. Both CART and SET are rule-based learning algorithms, which helped to interpret the resistance mechanisms by identifying the presence or absence of 'k-mers' (all of a gene sequence's subsequences of length *k*). These type of methods help to interpret the model's results based on the features it has used, thus overcoming the 'interpretability problem' (i.e., non-availability of data or features used to reach the conclusion by the ML method) [150]. MALDI TOF MS is being extensively used for identifying bacteria and fungi in diagnostic laboratory across the world. The fluconazole resistance in *C. albicans* was detected using three ML methods (Random Forest, Logistic regression and Linear discriminant analysis (LDA)) using spectral data. Of these three models, authors found that LDA was most robust method in detecting AMR with the accuracy, sensitivity, and specificity of 85.7%, 88.9%, and 83.3% respectively. Furthermore, another study employed the MALDI TOF spectral data from *S. aureus*, *E. coli*, and *K. pneumoniae* to predict the resistance phenotype. They used multilayer perceptron and gradient boost methods to get an area under receiver operator curve (AUROC) of 0.80, 0.74, and 0.74 [102]. AUROC is the metric used to measure the accuracy of the ML model in predicting the label (in this case, sensitive or resistant). A few studies have utilized patient data to predict if patients could develop resistant infections along with suitable therapies based on the local epidemiology of the pathogens. Microsoft's Azure ML algorithm determined the appropriate therapy based on patient demographic data and the resistance profiles of previously isolated microorganisms [151]. Another study applied ML methods to patients' medical records to predict antibiotic resistance against five antibiotics [152]. Patient demographic data and previous clinical and antibiotic history was used to predict AMR in pathogens isolated from urinary tract infection, such that the appropriate antibiotic could be prescribed [153].

5. Machine learning in drug design and drug discovery

The success rate of a potential therapeutic drug is extremely very low. Between 2000 and 2015, the success rate of drug development in oncology alone was as low as 3.4% [154]. Drug discovery involves various steps from target identification, optimization, validation, and hit discovery [155]. Machine learning is being implemented in the drug discovery process, from identifying the potential molecules or compounds against a particular disease to clinical trials [156]. A new drug, from its discovery through to clinical trials, involves huge cost (approximately 2.5 billion USD) and may take up to 10–15 years to come to market [157, 158]. The advent of high-throughput screening methods and the associated 'omics' data, along with the computer-assisted drug design (CADD) technologies, encouraged pharmaceutical companies to focus on leveraging ML methods to identify potential drug targets as well as new drugs. These *in-silico* methods not only provide the molecular properties of the potential drug molecules, but they also have an impact on the attrition rate in the drug discovery pipeline, especially in pre-clinical experiments.

The first step in the drug discovery is to associate the target with the disease of interest. Here, it is hypothesized that inhibiting or modifying the target results in the alleviation of the disease. Machine learning has been applied to find the target using protein-protein, transcriptional, and metabolic interactions within cells and tissues. In this regard, semi-supervised learning models based on drug-protein interaction network information, chemical structures and genomic sequence data were able to predicted drug-protein interactions on enzyme, ion channel, GPCR (G protein coupled receptor), and nuclear receptor datasets [159]. A decision tree-based meta-classifier was employed to predict genes based on the aforementioned interactions that are associated with morbidity and that can be used as targets [160]. Similarly, a SVM model was able to classify proteins as drug targets and non-drug targets, for breast, pancreatic, and ovarian cancers [156]. In this study, after predicting multiple targets, two of the predicted targets were validated using peptide inhibitors, which had antiproliferative activity on cell culture models. Other studies have utilized ML methods for identifying drug targets, including for Huntington's disease [161]. The drug-protein interaction (DPI) databases consist of drugs that interact with therapeutic protein targets. However, these drugs might interact with the non-target proteins *in-vivo*, leading to side-effects or toxicity. Furthermore, knowledge on the drug and non-target interaction is limited. To address this knowledge gap, a study used a pool of 35 ML methods to predict DPIs based on the similarities between drugs and protein targets [162].

Support vector machines have been extensively used in drug development. The SVM method has been applied to raw data to predict the radiation protection function and toxicity for radioprotectors targeting p53 [163]. A regression-SVM model was used to assess target-ligand interactions [164]. Support vector machines were also able to predict the 'druggability' based on the structure of target [165] and have been used for other applications such as identifying drug-target interaction [109], cancer cell properties, drug resistance [110], selection of therapeutic compounds from public database [166], predicting properties of organic compound [167], designing new ligands [168], and virtual screening [169]. Random forest algorithms have been used to improve scoring function performance in ligand-protein binding affinity [169]. Random forest approaches have also been used to select molecular descriptors to achieve better accuracy for the compounds designed for drugs used in immune network technology [170]. Multilayer perceptron (MLP) algorithm is another ML approach that has been mainly used to generate compounds automatically for *de novo* drug design [171]. Yavuz et al. used MLP approach to predict the secondary structure of the proteins, which are used in drug design [133]. Deep learning approaches such as deep neural networks (DNNs), CNNs, RNNs, and autoencoders have been exploited in the drug discovery process. Deep learning algorithms increase the prediction performance on quantitative structure-activity relationship by retrieving feature extractions and capabilities in chemical characters automatically. 'DeepChem' is a multi-task neural network platform that helps in performing drug development process [172]. Convolutional neural networks have been utilized to predict affinities in protein-ligand binding [114, 173, 174]. Additionally, RNNs have been employed to virtually screen of molecular libraries to find anti-cancer agents via molecular fingerprints [175]. Finally, autoencoders have been used to generate molecules in *de novo* drug design [176, 177].

Machine learning approaches have been used to discover antibiotics. Stokes et al. discovered an antibiotic from the 'Drug Repurposing Hub' called halicin. This drug is effective against *E. coli*, *Clostridioides difficile*, and pan-resistant *Acinetobacter bahumanii* [178]. Machine learning methods can mine large databases of genes and metabolites to identify molecule types that may include novel antibiotics [179, 180]. Machine learning methods are also being applied to the databases such as 'ChEMBL', which contains 1.9 million compounds with biological activity against 12,500 targets [181], 'BindingDB', which consists of 805,000 compounds with their binding affinities and 7500 protein targets [182], and 'AnitbioticDB', which consists of 1100 compounds that are in different stage of development for therapeutic use [183]. Antimicrobial peptides (AMPs) are found in all classes of life and are an important component of the innate immune response. Xiao et al. used fuzzy k-nearest neighbor algorithm to identify and define the functions of AMPs [184]. Another study used a semi-supervised density-based clustering algorithm model on linear AMPs that are active against gram-negative strains. Wang et al., applied four ML methods to discover new agents against MRSA. In this study, the authors derived *in-silico* models from 5451 cell-based anti-MRSA assay data using Bayesian, SVM, recursive partitioning, and k-nearest neighbor methods. By applying a ML approach to the 'Guangdong Small molecule Tangible Library' (which contains over 7500 small molecules), 56 hits were found, of which 12 novel anti-MRSA compounds were reported [185]. Targeting components in bacteria that are absent in humans can lead to new treatments against infections. DNA gyrase present in bacteria was targeted by Li et al. to discover anti-DNA gyrase compound using a ML approach [186]. In the same study, the authors also used *in-vitro* models to verify the virtual hits to check the hit activities against *E. coli,* MRSA, and other bacteria. Machine learning approaches have also been applied to discover antifungal drugs. For instance, a ML approach was employed to generate genome-wide gene essentiality predictions for *C. albicans* using a functional genomics resource named 'Gene Replacement and Conditional Expression' to identify three primary targets out of 866 genes. These three genes were involved in kinetochore function, mitochondrial integrity, and translation; glutaminyl-tRNA synthetase Gln4 was then identified as the target of N-pyrimidinyl-β-thiophenylacrylamide, which is an antifungal compound [187]. Temporal convolutional networks (TCNs) have been developed and deployed for antifungal peptide (AFP) prediction using deep learning models [188]. Similarly, Mousavizadegan et al. used pseudo amino acid composition to predict AFPs using a SVM algorithm [138]. Three peptides with highest prediction score were subsequently used in *in-vitro* assays. Sharma et al. proposed 'Deep-AFPpred', a deep learning classifier that predicts AFPs from protein sequence data [189].

6. Challenges and prospects

Antimicrobial resistance is an emerging global health crisis. As infectious microorganisms are evolving resistance through genetic and nongenetic mechanisms, new methods are required to rapidly diagnose and treat drug-resistant infections. The recent discovery of novel forms of AMR, including tolerance, persistence, and nongenetic resistance highlights the ingenuity of pathogenic microorganisms as well as the multifaceted nature of this problem. Digitization of clinical records presents opportunities for leveraging ML methods for fast and accurate identification of resistant microorganisms. However, applying ML methods to detect AMR is still in the nascent stage. Importantly, the quantity and quality of the data required to detect resistance among bacteria and fungi are still limited. Furthermore, ML models currently used elsewhere require optimization to successfully detect AMR. Advancement in the areas of laboratory diagnosis of infectious agents and sharing of data across different centers could pave the way forward for using ML methods identify and detecting drug-resistant microorganisms.

Machine learning has played an important role in the discovery of drugs by identifying novel drug targets and drug molecules. Several new drugs discovered using ML methods have been successful in clinical trials after spending comparatively less time in the drug discovery pipeline. Though ML methods are proving to useful in drug design and drug discovery, several challenges still exist. For instance, the absence of sufficient training data as well as biased, faulty, or noisy training data results in poor ML model predictions. To address this, methods to remove outliers, and filter out unwanted features are being developed to increase the predictive power of ML models.

Another issue is that ML algorithms employ a 'black box' approach to train ML models. Specifically, how the features are being interpreted during each stage of the training to come to an accurate prediction is largely still not understood. An area of research called explainable artificial intelligence (XAI) has emerged to address this issue. XAI consists of processes and methods that help the human users to comprehend the results generated by ML algorithms. Also, XAI helps to characterize the model accuracy, transparency, and outcomes [190]. Applying XAI in the field of AMR research may lead to the discovery of novel resistance mechanisms. Finally, the heterogeneity of many databases restricts the incorporation of ML algorithms to these databases. However, the data on disease, drug compounds, and AMR mechanisms are growing day-by-day, leading to the continuous curation of ML models. Other challenges for deploying ML algorithms include cross-platform normalization, statistical issues, and the division of testing datasets. Many of these issues may be resolved through sophisticated data preprocessing methods. Importantly, these data and interpretability issues will need to be resolved before ML methods are more widely adopted in scientific research and trusted in clinical settings.

Acknowledgements

DC was supported by a seed grant from AI4Society and funding from University of Alberta.

Author details

Shamanth A. Shankarnarayan[1], Joshua D. Guthrie[1] and Daniel A. Charlebois[1,2*]

1 Department of Physics, University of Alberta, Edmonton, Canada

2 Department of Biological Sciences, University of Alberta, Edmonton, Canada

*Address all correspondence to: dcharleb@ualberta.ca

References

[1] Fleming A. Sir Alexander Fleming—Nobel Lecture: Penicillin. Nobel Lect; 1945

[2] WHO. Antimicrobial Resistance [Internet]. 2021. Available from: https://www.who.int/news-room/fact-sheets/detail/antimicrobial-resistance

[3] MacGowan AP. Clinical implications of antimicrobial resistance for therapy. The Journal of Antimicrobial Chemotherapy. 2008;**62**(SUPPL. 2):105-114

[4] O'Neill J. Review on Antimicrobial Resistance: Tackling Drug-Resistant Infections Globally: Final Report and Recommendations. London: Wellcome Trust; 2016. p. 80

[5] Murray CJ, Ikuta KS, Sharara F, Swetschinski L, Robles Aguilar G, Gray A, et al. Global burden of bacterial antimicrobial resistance in 2019: A systematic analysis. Lancet. 2022;**6736**(21):629-655

[6] Nelson RE, Hatfield KM, Wolford H, Samore MH, Scott RD, Reddy SC, et al. National estimates of healthcare costs associated with multidrug-resistant bacterial infections among hospitalized patients in the United States. Clinical Infectious Diseases. 2007;**2021**(72):S17-S26

[7] Mulani MS, Kamble EE, Kumkar SN, Tawre MS, Pardesi KR. Emerging strategies to combat ESKAPE pathogens in the era of antimicrobial resistance: A review. Frontiers in Microbiology. 2019;**10**(APR):539

[8] Jernigan JA, Hatfield KM, Wolford H, Nelson RE, Olubajo B, Reddy SC, et al. Multidrug-resistant bacterial infections in U.S. hospitalized patients, 2012-2017. The New England Journal of Medicine. 2020;**382**(14):1309-1319

[9] Tacconelli E, Carrara E, Savoldi A, Harbarth S, Mendelson M, Monnet DL, et al. Discovery, research, and development of new antibiotics: The WHO priority list of antibiotic-resistant bacteria and tuberculosis. The Lancet Infectious Diseases. 2018;**18**(3):318-327

[10] Wall G, Lopez-Ribot JL. Current antimycotics, new prospects, and future approaches to antifungal therapy. Antibiotics. 2020;**9**(8):1-10

[11] Nnadi NE, Carter DA. Climate change and the emergence of fungal pathogens. PLoS Pathogens. 2021;**17**(4):1-6

[12] Pappas PG, Lionakis MS, Arendrup MC, Ostrosky-Zeichner L, Kullberg BJ. Invasive candidiasis. Nature Reviews Disease Primers. 2018;**4**:18026

[13] Tracking Candida auris | Candida auris | Fungal Diseases | CDC [Internet]. 2022. Available from: https://www.cdc.gov/fungal/candida-auris/tracking-c-auris.html#historical

[14] Centers for Disease Control and Prevention. Tracking Candida auris: Candida auris Fungal Diseases CDC [Internet]. Centers for Disease Control and Prevention. 2019. Available from: https://www.cdc.gov/fungal/candida-auris/tracking-c-auris.html

[15] Oh BJ, Shin JH, Kim MN, Sung H, Lee K, Joo MY, et al. Biofilm formation and genotyping of *Candida haemulonii*, *Candida pseudohaemulonii*, and a proposed new species (*Candida auris*) isolates from Korea. Medical Mycology. 2010;**49**(1):98-102

[16] Rhodes J, Fisher MC. Global epidemiology of emerging *Candida auris*. Current Opinion in Microbiology. 2019;**52**:84-89

[17] Biswal M, Rudramurthy SM, Jain N, Shamanth AS, Sharma D, Jain K, et al. Controlling a possible outbreak of *Candida auris* infection: Lessons learnt from multiple interventions. The Journal of Hospital Infection. 2017;**97**(4):363-370

[18] European Centre for Disease Prevention and Control. Candida Auris Outbreak in Healthcare Facilities in Northern Italy, 2019-2021. ECDC: Stockholm; 2022

[19] Schelenz S, Hagen F, Rhodes JL, Abdolrasouli A, Chowdhary A, Hall A, et al. First hospital outbreak of the globally emerging *Candida auris* in a European hospital. Antimicrobial Resistance and Infection Control. 2016;**5**(1):35

[20] Chen J, Tian S, Han X, Chu Y, Wang Q, Zhou B, et al. Is the superbug fungus really so scary? A systematic review and meta-analysis of global epidemiology and mortality of *Candida auris*. BMC Infectious Diseases. 2020;**20**(1):1-10

[21] Du H, Bing J, Hu T, Ennis CL, Nobile CJ, Huang G. Candida auris: Epidemiology, biology, antifungal resistance, and virulence. PLoS Pathogens. 2020;**16**(10):1-18

[22] Chow NA, de Groot T, Badali H, Abastabar M, Chiller TM, Meis JF. Potential fifth clade of *Candida auris*, Iran, 2018. Emerging Infectious Diseases. 2019;**25**(9):1780-1781

[23] Osei SJ. *Candida auris*: A systematic review and meta-analysis of current updates on an emerging multidrug-resistant pathogen. Microbiology. 2018;**7**(4):1-29

[24] Magiorakos A-P, Srinivasan A, Carey RB, Carmeli Y, Falagas ME, Giske CG, et al. Multidrug-resistant, extensively drug-resistant and pandrug-resistant bacteria: An international expert proposal for interim standard definitions for acquired resistance. Clinical Microbiology and Infection. 2012;**18**(3):268-281

[25] Lyman M, Forsberg K, Reuben J, Dang T, Free R, Seagle EE, et al. Notes from the field: Transmission of pan-resistant and Echinocandin-resistant *Candida auris* in health care facilities—Texas and the District of Columbia, January–April 2021. MMWR. Morbidity and Mortality Weekly Report. 2021;**70**(29):1022-1023

[26] Verweij PE, Lucas JA, Arendrup MC, Bowyer P, Brinkmann AJF, Denning DW, et al. The one health problem of azole resistance in *Aspergillus fumigatus*: Current insights and future research agenda. Fungal Biology Reviews. 2020;**34**(4):202-214

[27] Rudramurthy SM, Shankarnarayan SA, Dogra S, Shaw D, Mushtaq K, Paul RA, et al. Mutation in the squalene epoxidase gene of *Trichophyton interdigitale* and *Trichophyton rubrum* associated with Allylamine resistance. Antimicrobial Agents and Chemotherapy. May 2018;**62**(5):1-9

[28] Kano R, Kimura U, Kakurai M, Hiruma J, Kamata H, Suga Y, et al. *Trichophyton indotineae* sp. nov.: A new highly terbinafine-resistant anthropophilic dermatophyte species. Mycopathologia. 2020;**185**(6):947-958

[29] Laxminarayan R, Heymann DL. Challenges of drug resistance in the developing world. BMJ. 2012;**344**(7852):3-6

[30] Laxminarayan R, Duse A, Wattal C, Zaidi AKM, Wertheim HFL,

Sumpradit N, et al. Antibiotic resistance-the need for global solutions. The Lancet Infectious Diseases. 2013;**13**(12):1057-1098

[31] Huang AM, Newton D, Kunapuli A, Gandhi TN, Washer LL, Isip J, et al. Impact of rapid organism identification via matrix-assisted laser desorption/ionization time-of-flight combined with antimicrobial stewardship team intervention in adult patients with bacteremia and candidemia. Clinical Infectious Diseases. 2013;**57**(9):1237-1245

[32] Burnham CAD, Leeds J, Nordmann P, O'Grady J, Patel J. Diagnosing antimicrobial resistance. Nature Reviews. Microbiology. 2017;**15**(11):697-703

[33] Moult J, Fidelis K, Kryshtafovych A, Schwede T, Topf M. Critical Assessment of Techniques for Protein Structure Prediction, Fourteenth Round. 2020. pp. 1-344

[34] Jumper J, Evans R, Pritzel A, Green T, Figurnov M, Ronneberger O, et al. Highly accurate protein structure prediction with AlphaFold. Nature. 2021;**596**(7873):583-589

[35] Mikolov T, Deoras A, Povey D, Burget L, Černocký J. Strategies for training large scale neural network language models. In: 2011 IEEE Work Autom Speech Recognit Understanding, ASRU 2011, Proc. 2011. pp. 196-201

[36] Krizhevsky A, Sutskever I, Hinton GE. ImageNet classification with deep convolutional neural networks. Communications of the ACM. 2017;**60**(6):84-90

[37] Samuel AL. Some studies in machine learning using the game of checkers. IBM Journal of Research and Development. 1959;**3**:210-229

[38] Awad M, Khanna R. Machine learning. In: Efficient Learning Machines: Theories, Concepts, and Applications for Engineers and System Designers. Berkeley, CA: Apress; 2015. pp. 1-18

[39] Silver D, Huang A, Maddison CJ, Guez A, Sifre L, van den Driessche G, et al. Mastering the game of go with deep neural networks and tree search. Nature. 2016;**529**(7587):484-489

[40] Degrave J, Felici F, Buchli J, Neunert M, Tracey B, Carpanese F, et al. Magnetic control of tokamak plasmas through deep reinforcement learning. Nature. 2022;**602**(7897):414-419

[41] Weissler EH, Naumann T, Andersson T, Ranganath R, Elemento O, Luo Y, et al. The role of machine learning in clinical research: Transforming the future of evidence generation. Trials. 2021;**22**(1):537

[42] Ripoli A, Sozio E, Sbrana F, Bertolino G, Pallotto C, Cardinali G, et al. Personalized machine learning approach to predict candidemia in medical wards. Infection. 2020;**48**(5):749-759

[43] Jaroszewisz A, Ernst J. An integrative approach for fine-mapping chromatin interactions. Bioinformatics. 2020;**36**(6):1704-1711

[44] Movva R, Greenside P, Marinov GK, NairS,ShrikumarA,KundajeA.Deciphering regulatory DNA sequences and noncoding genetic variants using neural network models of massively parallel reporter assays. PLoS One. 2019;**14**(6):1-20

[45] Rozenwald MB, Galitsyna AA, Sapunov GV, Khrameeva EE, Gelfand MS. A machine learning framework for the prediction of chromatin folding in Drosophila using epigenetic features. PeerJ Computer Science. 2020;**6**:2-21

[46] Talukder A, Barham C, Li X, Hu H. Interpretation of deep learning in genomics and epigenomics. Briefings in Bioinformatics. 2021;**22**(3):1-16

[47] Chen X, Clarence Yan C, Luo C, Ji W, Zhang Y, Dai Q. Constructing lncRNA functional similarity network based on lncRNA-disease associations and disease semantic similarity. Scientific Reports. 2015;**5**(June):1-12

[48] Peiffer-Smadja N, Rawson TM, Ahmad R, Buchard A, Pantelis G, Lescure FX, et al. Machine learning for clinical decision support in infectious diseases: A narrative review of current applications. Clinical Microbiology and Infection. 2020;**26**(5):584-595

[49] Martinez JL. General principles of antibiotic resistance in bacteria. Drug Discovery Today: Technologies. 2014;**11**:33-39

[50] Zhang G, Feng J. The intrinsic resistance of bacteria. Yi chuan = Hered. 2016;**38**(10):872-880

[51] Brauner A, Fridman O, Gefen O, Balaban NQ. Distinguishing between resistance, tolerance and persistence to antibiotic treatment. Nature Reviews. Microbiology. 2016;**14**(5):320-330

[52] Berman J, Krysan DJ. Drug resistance and tolerance in fungi. Nature Reviews. Microbiology. 2020;**18**(6):319-331

[53] Wood TK, Knabel SJ, Kwan BW. Bacterial persister cell formation and dormancy. Applied and Environmental Microbiology. 2013;**79**:7116-7121

[54] Balaban NQ, Merrin J, Chait R, Kowalik L, Leibler S. Bacterial persistence as a phenotypic switch. Science (80-). 2004;**305**(5690):1622-1625

[55] Moyed HS, Bertrand KP. hipA, a newly recognized gene of Escherichia coli K-12 that affects frequency of persistence after inhibition of murein synthesis. Journal of Bacteriology. 1983;**155**(2):768-775

[56] Balaban NQ, Helaine S, Lewis K, Ackermann M, Aldridge B, Andersson DI, et al. Definitions and guidelines for research on antibiotic persistence. Nature Reviews. Microbiology. 2019;**17**(7):441-448

[57] Hammoud MS, Al-Taiar A, Fouad M, Raina A, Khan Z. Persistent candidemia in neonatal care units: Risk factors and clinical significance. International Journal of Infectious Diseases. 2013;**17**(8):e624-e628

[58] Elowitz MB, Levine AJ, Siggia ED, Swain PS. Stochastic gene expression in a single cell. Science. 2002;**297**(5584):1183-1186

[59] Adam M, Murali B, Glenn NO, Potter SS. Epigenetic inheritance based evolution of antibiotic resistance in bacteria. BMC Evolutionary Biology. 2008;(8):52

[60] Farquhar KS, Rasouli Koohi S, Charlebois DA. Does transcriptional heterogeneity facilitate the development of genetic drug resistance? BioEssays. 2021;**43**(8):1-7

[61] Reygaert WC. An overview of the antimicrobial resistance mechanisms of bacteria. AIMS Microbiology. 2018;**4**(3):482-501

[62] Chancey ST, Zähner D, Stephens DS. Acquired inducible antimicrobial resistance in Gram-positive bacteria. Future Microbiology. 2012;7(8):959-978

[63] Lambert PA. Cellular impermeability and uptake of biocides and antibiotics in

gram-positive bacteria and mycobacteria. Symposium Series (Society for Applied Microbiology). 2002;**31**:46S-54S

[64] Blair JMA, Richmond GE, Piddock LJV. Multidrug efflux pumps in Gram-negative bacteria and their role in antibiotic resistance. Future Microbiology. 2014;**9**(10):1165-1177

[65] Gill MJ, Simjee S, Al-Hattawi K, Robertson BD, Easmon CS, Ison CA. Gonococcal resistance to beta-lactams and tetracycline involves mutation in loop 3 of the porin encoded at the penB locus. Antimicrobial Agents and Chemotherapy. 1998;**42**(11):2799-2803

[66] Mah T-F. Biofilm-specific antibiotic resistance. Future Microbiology. 2012;**7**(9):1061-1072

[67] Reygaert W. Methicillin-resistant *Staphylococcus aureus* (MRSA): Molecular aspects of antimicrobial resistance and virulence. Clinical Laboratory Science. 2009;**22**(2):115-119

[68] Cox G, Wright GD. Intrinsic antibiotic resistance: Mechanisms, origins, challenges and solutions. International Journal of Medical Microbiology. 2013;**303**:287-292

[69] Roberts MC. Resistance to macrolide, lincosamide, streptogramin, ketolide, and oxazolidinone antibiotics. Applied Biochemistry and Biotechnology—Part B Molecular Biotechnology. 2004;**28**:47-62

[70] Redgrave LS, Sutton SB, Webber MA, Piddock LJV. Fluoroquinolone resistance: Mechanisms, impact on bacteria, and role in evolutionary success. Trends in Microbiology. 2014;**22**(8):438-445

[71] Huovinen P, Sundström L, Swedberg G, Sköld O. Trimethoprim and sulfonamide resistance. Antimicrobial Agents and Chemotherapy. 1995;**39**(2):279-289

[72] Blair JMA, Webber MA, Baylay AJ, Ogbolu DO, Piddock LJV. Molecular mechanisms of antibiotic resistance. Nature Reviews. Microbiology. 2015;**13**(1):42-51

[73] Kumar A, Schweizer HP. Bacterial resistance to antibiotics: Active efflux and reduced uptake. Advanced Drug Delivery Reviews. 2005;**57**(10):1486-1513

[74] Beck-Sagué C, Jarvis WR. Secular trends in the epidemiology of nosocomial fungal infections in the United States, 1980-1990. National Nosocomial Infections Surveillance System. The Journal of Infectious Diseases. 1993;**167**(5):1247-1251

[75] White TC, Holleman S, Dy F, Mirels LF, Stevens DA. Resistance mechanisms in clinical isolates of *Candida albicans*. Antimicrobial Agents and Chemotherapy. 2002;**46**(6):1704-1713

[76] White TC. Increased mRNA levels of ERG16, CDR, and MDR1 correlate with increases in azole resistance in Candida albicans isolates from a patient infected with human immunodeficiency virus. Antimicrobial Agents and Chemotherapy. 1997;**41**(7):1482-1487

[77] Franz R, Kelly SL, Lamb DC, Kelly DE, Ruhnke M, Morschhäuser J. Multiple molecular mechanisms contribute to a stepwise development of fluconazole resistance in clinical *Candida albicans* strains. Antimicrobial Agents and Chemotherapy. 1998;**42**(12):3065-3072

[78] Braun BR, van het Hoog M, d'Enfert C, Martchenko M, Dungan J, Kuo A, et al. A human-curated annotation of the Candida albicans genome. PLoS Genetics. 2005;**1**:0036-0057

[79] Sanglard D, Coste A, Ferrari S. Antifungal drug resistance mechanisms in fungal pathogens from the perspective of transcriptional gene regulation. FEMS Yeast Research. 2009;**9**(7):1029-1050

[80] Flowers SA, Barker KS, Berkow EL, Toner G, Chadwick SG, Gygax SE, et al. Gain-of-function mutations in UPC2 are a frequent cause of ERG11 upregulation in azole-resistant clinical isolates of *Candida albicans*. Eukaryotic Cell. 2012;**11**(10):1289-1299

[81] Sanglard D. Diagnosis of antifungal drug resistance mechanisms in fungal pathogens: Transcriptional gene regulation. Current Fungal Infection Reports. 2011;**5**(3):157-167

[82] Selmecki A, Forche A, Berman J. Genomic plasticity of the human fungal pathogen *Candida albicans*. Eukaryotic Cell. 2010;**9**(7):991-1008

[83] Gulshan K, Moye-Rowley WS. Multidrug resistance in fungi. Eukaryotic Cell. 2007;**6**(11):1933-1942

[84] Cowen LE, Steinbach WJ. Stress, drugs, and evolution: The role of cellular signaling in fungal drug resistance. Eukaryotic Cell. 2008;**7**(5):747-764

[85] Perlin DS. Current perspectives on echinocandin class drugs. Future Microbiology. 2011;**6**(4):441-457

[86] Katiyar S, Pfaller M, Edlind T. *Candida albicans* and *Candida glabrata* clinical isolates exhibiting reduced Echinocandin susceptibility. Antimicrobial Agents and Chemotherapy. 2006;**50**(8):2892-2894

[87] Munro CA, Selvaggini S, de Bruijn I, Walker L, Lenardon MD, Gerssen B, et al. The PKC, HOG and Ca2+ signalling pathways co-ordinately regulate chitin synthesis in *Candida albicans*. Molecular Microbiology. 2007;**63**(5):1399-1413

[88] Walker LA, Munro CA, de Bruijn I, Lenardon MD, McKinnon A, Gow NAR. Stimulation of chitin synthesis rescues *Candida albicans* from Echinocandins. Cormack BP, editor. PLoS Pathogens. 2008;**4**(4):e1000040

[89] Loo AS, Muhsin SA, Walsh TJ. Toxicokinetic and mechanistic basis for the safety and tolerability of liposomal amphotericin B. Expert Opinion on Drug Safety. 2013;**12**(6):881-895

[90] Vanden Bossche H, Marichal P, Odds FC. Molecular mechanisms of drug resistance in fungi. Trends in Microbiology. 1994;**2**(10):393-400

[91] Nolte FS, Parkinson T, Falconer DJ, Dix S, Williams J, Gilmore C, et al. Isolation and characterization of fluconazole- and amphotericin B-resistant *Candida albicans* from blood of two patients with leukemia. Antimicrobial Agents and Chemotherapy. 1997;**41**(1):196-199

[92] Blum G, Hörtnagl C, Jukic E, Erbeznik T, Pümpel T, Dietrich H, et al. New insight into amphotericin B resistance in *Aspergillus terreus*. Antimicrobial Agents and Chemotherapy. 2013;**57**(4):1583-1588

[93] Eddouzi J, Parker JE, Vale-Silva LA, Coste A, Ischer F, Kelly S, et al. Molecular mechanisms of drug resistance in clinical Candida species isolated from Tunisian hospitals. Antimicrobial Agents and Chemotherapy. 2013;**57**(7):3182-3193

[94] CLSI. Reference Method for Broth Dilution Antifungal Susceptibility Testing of Filamentous Fungi; Approved Standard—CLSI Document M38-A2. Vol. 28. Clinical and Laboratory Standards Institute (CLSI); 2008. p. 52

[95] European Committee on Antimicrobial Susceptibility Testing—EUCAST. EUCAST reading guide for broth microdilution. Read Guid broth microdilution. 2020;**1.0**(March):17

[96] McEwen SA, Collignon PJ. Antimicrobial resistance: A one health colloquium. Microbiology Spectrum. 2018;**6**(2):1-26

[97] Pulido MR, García-Quintanilla M, Martín-Peña R, Cisneros JM, McConnell MJ. Progress on the development of rapid methods for antimicrobial susceptibility testing. The Journal of Antimicrobial Chemotherapy. 2013;**68**(12):2710-2717

[98] Vasala A, Hytönen VP, Laitinen OH. Modern tools for rapid diagnostics of antimicrobial resistance. Frontiers in Cellular and Infection Microbiology. 2020;**10**:308

[99] Boyle D. Unitaid TB Diagnostics—NAAT for Microscopy Stations [Internet]. 2017. Available from: http://unitaid.org/assets/2017-Unitaid-TB-Diagnostics-Technology-Landscape.pdf

[100] Peytavi R, Raymond FR, Gagné D, Picard FJ, Jia G, Zoval J, et al. Microfluidic device for rapid (<15 min) automated microarray hybridization. Clinical Chemistry. 2005;**51**(10):1836-1844

[101] Dougherty K, Smith BA, Moore AF, Maitland S, Fanger C, Murillo R, et al. Multiple phenotypic changes associated with large-scale horizontal gene transfer. PLoS One. 2014;**9**(7):e102170

[102] Weis C, Cuénod A, Rieck B, Dubuis O, Graf S, Lang C, et al. Direct antimicrobial resistance prediction from clinical MALDI-TOF mass spectra using machine learning. Nature Medicine. 2022;**28**(1):164-174

[103] Mitchell TM. Machine Learning. McGraw Hill; 1997. p. 414

[104] Silver D, Schrittwieser J, Simonyan K, Antonoglou I, Huang A, Guez A, et al. Mastering the game of go without human knowledge. Nature. 2017;**550**(7676):354-359

[105] Davies A, Veličković P, Buesing L, Blackwell S, Zheng D, Tomašev N, et al. Advancing mathematics by guiding human intuition with AI. Nature. 2021;**600**(7887):70-74

[106] Russell S, Norvig P. Artificial Intelligence: A Modern Approach. New Jersey: Pearson; 2010

[107] Trevor H, Jerome F, Robert T. The elements of statistical learning data mining, inference, and prediction. The Elements of Statistical Learning. 2009;**27**:83-85

[108] Gareth J, Daniela W, Hastie T, Robert T. An Introduction to Statistical Learning with Applications in R. 2nd ed. New York: Springer Text in Statistics; 2011. 110p

[109] Wang Q, Feng Y, Huang J, Wang T, Cheng G. A novel framework for the identification of drug target proteins: Combining stacked auto-encoders with a biased support vector machine. PLoS One. 2017;**12**(4):e0176486

[110] Gupta S, Chaudhary K, Kumar R, Gautam A, Nanda JS, Dhanda SK, et al. Prioritization of anticancer drugs against a cancer using genomic features of cancer cells: A step towards personalized medicine. Scientific Reports. 2016;**6**(1):23857

[111] Lemos P, Jeffrey N, Cranmer M, Ho S, Battaglia P. Rediscovering orbital mechanics with machine learning. arXiv. 2022

[112] Cortes C, Vapnik V, Saitta L. Support-vector networks. Machine Learning. 1995;**20**(3):273-297

[113] Breiman L. Random forests. Machine Learning. 2001;**45**(1):5-32

[114] Lecun Y, Bengio Y, Hinton G. Deep learning. Nature. 2015;**521**:436-444

[115] Hinton G, Sejnowski T. Unsupervised learning: Foundations of neural computation. Computers & Mathematics with Applications. 1999;**38**(5-6):256

[116] Steinwart I, Gov D, Gov J. A classification framework for anomaly detection Don hush Clint Scovel. Journal of Machine Learning Research. 2005;**6**:211-232

[117] Shon T, Moon J. A hybrid machine learning approach to network anomaly detection. Information Sciences. 2007;**177**(18):3799-3821

[118] Tenenbaum JB, De Silva V, Langford JC. A global geometric framework for nonlinear dimensionality reduction. Science (80-). 2000;**290**(5500):2319-2323

[119] Van Der Maaten L, Postma E, Van den Herik J. Dimensionality reduction: A comparative review. Journal of Machine Learning Research. 2009;**10**:66-71

[120] Chapelle O, Schölkopf B, Zien A. Semi-supervised learning. 2010;**508**:373-440

[121] van Engelen JE, Hoos HH. A survey on semi-supervised learning. Machine Learning. 2020;**109**(2):373-440

[122] Hartigan JA, Wong MA. Algorithm AS 136: A K-means clustering algorithm. Applied Statistics. 1979;**28**(1):100

[123] Likas A, Vlassis N, J. Verbeek J. The global k-means clustering algorithm. Pattern Recognition. 2003;**36**(2):451-461

[124] Johnson SC. Hierarchical clustering schemes. Psychom. 1967;**32**(3):241-254

[125] Murtagh F, Contreras P. Algorithms for hierarchical clustering: An overview, II. Wiley Interdisciplinary Reviews: Data Mining and Knowledge Discovery. 2017;**7**(6):e1219

[126] Birant D, Kut A. ST-DBSCAN: An algorithm for clustering spatial–temporal data. Data & Knowledge Engineering. 2007;**60**(1):208-221

[127] Liu FT, Ting KM, Zhou ZH. Isolation forest. In: Proc—IEEE Int Conf Data Mining. ICDM; 2008. pp. 413-422

[128] Wold S, Esbensen K, Geladi P. Principal component analysis. Chemometrics and Intelligent Laboratory Systems. 1987;**2**(1-3):37-52

[129] Roweis ST, Saul LK. Nonlinear dimensionality reduction by locally linear embedding. Science (80-). 2000;**290**(5500):2323-2326

[130] Moon TK. The expectation-maximization algorithm. IEEE Signal Processing Magazine. 1996;**13**(6):47-60

[131] Kaelbling LP, Littman ML, Moore AW. Reinforcement learning: A survey. Journal of Artificial Intelligence Research. 1996;**4**:237-285

[132] Sutton RS, Barto AG. Reinforcement learning. In: An Introduction. 2nd ed. United States: MIT Press; 2018. pp. 1-3

[133] Carkli Yavuz B, Yurtay N, Ozkan O. Prediction of protein secondary structure with clonal selection algorithm and multilayer perceptron. IEEE Access. 2018;**6**:45256-45261

[134] Bergstra J, Bardenet R, Bengio Y, Kégl B. Algorithms for hyper-parameter optimization. In: Shawe-Taylor J,

Zemel R, Bartlett P, Pereira F, Weinberger KQ, editors. Advances in Neural Information Processing Systems. Curran Associates, Inc.; 2011

[135] Refaeilzadeh P, Tang L, Liu H. Cross-validation. In: Encyclopedia of Database Systems. 2016. pp. 1-7

[136] García S, Luengo J, Herrera F. Dealing with missing values. IntelligentSystems Reference Library. 2015;**72**:59-105

[137] Abadi M, Barham P, Chen J, Chen Z, Davis A, Dean J, et al. {TensorFlow}: A system for {large-scale} machine learning. In: 12th USENIX Symposium on Operating Systems Design and Implementation (OSDI 16). Savannah, GA: USENIX Association; 2016. pp. 265-283

[138] Mousavizadegan M, Mohabatkar H. Computational prediction of antifungal peptides via Chou's PseAAC and SVM. Journal of Bioinformatics and Computational Biology. 2018;**16**(4):1850016

[139] Fabian P, Michel V, Varoquaux G, Thirion B, Dubourg V, Passos A, et al. Scikit-learn: Machine learning in python. Journal of Machine Learning Research. 2011;**12**:2825-2830

[140] Paszke A, Gross S, Massa F, Lerer A, Bradbury Google J, Chanan G, et al. PyTorch: An Imperative Style, High-Performance Deep Learning Library. Conference proceedings: Advances in Neural Information Processing Systems: 2019

[141] Hicks AL, Wheeler N, Sánchez-Busó L, Rakeman JL, Harris SR, Grad YH. Evaluation of parameters affecting performance and reliability of machine learning-based antibiotic susceptibility testing from whole genome sequencing data. PLoS Computational Biology. 2019;**15**(9):e1007349

[142] Li D, Wang Y, Hu W, Chen F, Zhao J, Chen X, et al. Application of machine learning classifier to *Candida auris* drug resistance analysis. Frontiers in Cellular and Infection Microbiology. 2021;**11**:742062

[143] Delavy M, Cerutti L, Croxatto A, Prod'hom G, Sanglard D, Greub G, et al. Machine learning approach for *Candida albicans* fluconazole resistance detection using matrix-assisted laser desorption/ ionization time-of-flight mass spectrometry. Frontiers in Microbiology. 2020;**10**(January):3000

[144] Liu Z, Deng D, Lu H, Sun J, Lv L, Li S, et al. Evaluation of machine learning models for predicting antimicrobial resistance of *Actinobacillus pleuropneumoniae* from whole genome sequences. Frontiers in Microbiology. 2020;**11**(February):1-7

[145] Davis JJ, Boisvert S, Brettin T, Kenyon RW, Mao C, Olson R, et al. Antimicrobial resistance prediction in PATRIC and RAST. Scientific Reports. 2016;**6**:27930

[146] Nguyen M, Brettin T, Long SW, Musser JM, Olsen RJ, Olson R, et al. Developing an in silico minimum inhibitory concentration panel test for *Klebsiella pneumoniae*. Scientific Reports. 2018;**8**(1):421

[147] Bradley P, Gordon NC, Walker TM, Dunn L, Heys S, Huang B, et al. Rapid antibiotic-resistance predictions from genome sequence data for *Staphylococcus aureus* and *Mycobacterium tuberculosis*. Nature Communications. 2015;**6**:10063

[148] Her HL, Wu YW. A pan-genome-based machine learning approach for

predicting antimicrobial resistance activities of the *Escherichia coli* strains. Bioinformatics. 2018;**34**(13):i89-i95

[149] Gordon NC, Price JR, Cole K, Everitt R, Morgan M, Finney J, et al. Prediction of staphylococcus aureus antimicrobial resistance by whole-genome sequencing. Journal of Clinical Microbiology. 2014;**52**(4):1182-1191

[150] Drouin A, Letarte G, Raymond F, Marchand M, Corbeil J, Laviolette F. Interpretable genotype-to-phenotype classifiers with performance guarantees. Scientific Reports. 2019;**9**(1):4071

[151] Feretzakis G, Sakagianni A, Loupelis E, Kalles D, Skarmoutsou N, Martsoukou M, et al. Machine learning for antibiotic resistance prediction: A prototype using off-the-shelf techniques and entry-level data to guide empiric antimicrobial therapy. Healthcare Informatics Research. 2021;**27**(3): 214-221

[152] Lewin-Epstein O, Baruch S, Hadany L, Stein GY, Obolski U. Predicting antibiotic resistance in hospitalized patients by applying machine learning to electronic medical records. Clinical Infectious Diseases. 2021;**72**(11):e848-e855

[153] Didelot X, Pouwels KB. Machine-learning-assisted selection of antibiotic prescription. Nature Medicine. 2019;**25**(7):1033-1034

[154] Wong CH, Siah KW, Lo AW. Estimation of clinical trial success rates and related parameters. Biostatistics. 2019;**20**(2):273-286

[155] Vohora D, Singh G. Pharmaceutical Medicine and Translational Clinical Research. Elsevier; 2018. pp. 1-497

[156] Jeon J, Nim S, Teyra J, Datti A, Wrana JL, Sidhu SS, et al. A systematic approach to identify novel cancer drug targets using machine learning, inhibitor design and high-throughput screening. Genome Medicine. 2014;**6**(7):57

[157] DiMasi JA, Grabowski HG, Hansen RW. Innovation in the pharmaceutical industry: New estimates of R&D costs. Journal of Health Economics. 2016;**47**:20-33

[158] Turner JR. New Drug Development. New York, NY: Springer New York; 2010

[159] Xia Z, Wu L-Y, Zhou X, Wong STC. Semi-supervised drug-protein interaction prediction from heterogeneous biological spaces. BMC Systems Biology. 2010;**4**(Suppl 2):S6

[160] Costa PR, Acencio ML, Lemke N. A machine learning approach for genome-wide prediction of morbid and druggable human genes based on systems-level data. BMC Genomics. 2010;**11**(Suppl. 5):S9

[161] Ament SA, Pearl JR, Cantle JP, Bragg RM, Skene PJ, Coffey SR, et al. Transcriptional regulatory networks underlying gene expression changes in Huntington's disease. Molecular Systems Biology. 2018;**14**(3):e7435

[162] Wang C, Kurgan L. Survey of similarity-based prediction of drug-protein interactions. Current Medicinal Chemistry. 2020;**27**(35):5856-5886

[163] Matsumoto A, Aoki S, Ohwada H. Comparison of random forest and SVM for raw data in drug discovery: Prediction of radiation protection and toxicity case study. International Journal of Machine Learning and Computing. 2016;**6**(2):145-148

[164] Li L, Wang B, Meroueh SO. Support vector regression scoring of receptor-ligand complexes for rank-ordering and virtual screening of chemical libraries.

Journal of Chemical Information and Modeling. 2011;**51**(9):2132-2138

[165] Volkamer A, Kuhn D, Grombacher T, Rippmann F, Rarey M. Combining global and local measures for structure-based druggability predictions. Journal of Chemical Information and Modeling. 2012;**52**(2):360-372

[166] Bundela S, Sharma A, Bisen PS. Potential compounds for oral cancer treatment: Resveratrol, nimbolide, lovastatin, bortezomib, vorinostat, berberine, pterostilbene, deguelin, andrographolide, and colchicine. PLoS One. 2015;**10**(11):e0141719

[167] Maltarollo VG, Kronenberger T, Espinoza GZ, Oliveira PR, Honorio KM. Advances with support vector machines for novel drug discovery. Expert Opinion on Drug Discovery. 2019;**14**:23-33

[168] Schneider G, Hartenfeller M, Proschak E. De novo drug design. Lead Generation Approaches in Drug Discovery. 2010. pp. 165-185

[169] Kinnings SL, Liu N, Tonge PJ, Jackson RM, Xie L, Bourne PE. A machine learning-based method to improve docking scoring functions and its application to drug repurposing. Journal of Chemical Information and Modeling. 2011;**51**(2):408-419

[170] Samigulina G, Zarina S. Immune network technology on the basis of random forest algorithm for computer-aided drug design. In: Lecture Notes in Computer Science (Including Subseries Lecture Notes in Artificial Intelligence and Lecture Notes in Bioinformatics). Cham: Springer; 2017. pp. 50-61

[171] Gómez-Bombarelli R, Wei JN, Duvenaud D, Hernández-Lobato JM, Sánchez-Lengeling B, Sheberla D, et al. Automatic chemical design using a data-driven continuous representation of molecules. ACS Central Science. 2018;**4**(2):268-276

[172] Ramsundar B, Liu B, Wu Z, Verras A, Tudor M, Sheridan RP, et al. Is multitask deep learning practical for pharma? Journal of Chemical Information and Modeling. 2017;**57**(8):2068-2076

[173] Leelananda SP, Lindert S. Computational methods in drug discovery. Beilstein Journal of Organic Chemistry. 2016;**12**:2694-2718

[174] Jiménez J, Škalič M, Martínez-Rosell G, De Fabritiis G. KDEEP: Protein-ligand absolute binding affinity prediction via 3D-convolutional neural networks. Journal of Chemical Information and Modeling. 2018;**58**(2):287-296

[175] Kadurin A, Aliper A, Kazennov A, Mamoshina P, Vanhaelen Q, Khrabrov K, et al. The cornucopia of meaningful leads: Applying deep adversarial autoencoders for new molecule development in oncology. Oncotarget. 2017;**8**(7):10883-10890

[176] Zhavoronkov A, Ivanenkov YA, Aliper A, Veselov MS, Aladinskiy VA, Aladinskaya AV, et al. Deep learning enables rapid identification of potent DDR1 kinase inhibitors. Nature Biotechnology. 2019;**37**(9):1038-1040

[177] Kingma DP, Welling M. An introduction to variational autoencoders. Foundations and Trends in Machine Learning. 2019;**12**:307-392

[178] Stokes JM, Yang K, Swanson K, Jin W, Cubillos-Ruiz A, Donghia NM, et al. A deep learning approach to antibiotic discovery. Cell. 2020;**180**(4):688-702.e13

[179] Mohimani H, Kersten RD, Liu WT, Wang M, Purvine SO, Wu S, et al.

Automated genome mining of ribosomal peptide natural products. ACS Chemical Biology. 2014;**9**(7):1545-1551

[180] Cao L, Gurevich A, Alexander KL, Naman CB, Leão T, Glukhov E, et al. MetaMiner: A scalable peptidogenomics approach for discovery of ribosomal peptide natural products with blind modifications from microbial communities. Cell Systems. 2019;**9**(6):600-608.e4

[181] Gaulton A, Hersey A, Nowotka M, Bento AP, Chambers J, Mendez D, et al. The ChEMBL database in 2017. Nucleic Acids Research. 2017;**45**(D1):D945-D954

[182] Gilson MK, Liu T, Baitaluk M, Nicola G, Hwang L, Chong J. BindingDB in 2015: A public database for medicinal chemistry, computational chemistry and systems pharmacology. Nucleic Acids Research. 2016;**44**(D1):D1045-D1053

[183] Farrell LJ, Lo R, Wanford JJ, Jenkins A, Maxwell A, Piddock LJV. Revitalizing the drug pipeline: AntibioticDB, an open access database to aid antibacterial research and development. The Journal of Antimicrobial Chemotherapy. 2018;**73**(9):2284-2297

[184] Xiao X, Wang P, Lin W-Z, Jia J-H, Chou K-C. iAMP-2L: A two-level multi-label classifier for identifying antimicrobial peptides and their functional types. Analytical Biochemistry. 2013;**436**(2):168-177

[185] Wang L, Le X, Li L, Ju Y, Lin Z, Gu Q, et al. Discovering new agents active against methicillin-resistant *Staphylococcus aureus* with ligand-based approaches. Journal of Chemical Information and Modeling. 2014;**54**(11):3186-3197

[186] Li L, Le X, Wang L, Gu Q, Zhou H, Xu J. Discovering new DNA gyrase inhibitors using machine learning approaches. RSC Advances. 2015;**5**(128):105600-105608

[187] Fu C, Zhang X, Veri AO, Iyer KR, Lash E, Xue A, et al. Leveraging machine learning essentiality predictions and chemogenomic interactions to identify antifungal targets. Nature Communications. 2021;**12**(1):6497

[188] Singh V, Shrivastava S, Kumar Singh S, Kumar A, Saxena S. Accelerating the discovery of antifungal peptides using deep temporal convolutional networks. Briefings in Bioinformatics. 2022;**23**(2):bbac008

[189] Sharma R, Shrivastava S, Kumar Singh S, Kumar A, Saxena S, Kumar SR. Deep-AFPpred: Identifying novel antifungal peptides using pretrained embeddings from seq2vec with 1DCNN-BiLSTM. Briefings in Bioinformatics. 2022;**23**(1):1-16

[190] Linardatos P, Papastefanopoulos V, Kotsiantis S. Explainable ai: A review of machine learning interpretability methods. Entropy. 2021;**23**(1):1-45

Chapter 10

Carriage of Beta-Lactamase and Antibiotic Resistance in *Staphylococcus aureus*

Eghe Izevbizua

Abstract

Studies have shown that *Staphylococcus aureus* is one of the leading causes of bacteria infections in human and it has developed different resistance patterns to different antibiotics, making it a bit difficult to eradicate completely. This study focuses on the resistance of *S. aureus* producing beta-lactamase.

Keywords: *Staphylococcus aureus*, beta-lactamase

1. Introduction

Staphylococcus aureus is an organism that causes various forms of diseases ranging from simple diseases with little or no cause for alarm to severe life threatening diseases [1, 2]. Antibiotic resistant *S. aureus* have been involved in high death rates across the world [3]. For many years till now, antibiotics have been employed in the treatment of Staphylococcus infections but resistance to antibiotics commonly used against them has reduced the effectiveness of the antibiotics [4, 5]. Mec A gene which is located on the staphylococcal cassette chromosome enables the organism to develop resistance against antibiotics used against it [6–8]. An altered 75 kb penicillin binding protein 2A (PBP2a) is produced inside the mec A gene which also gives it its resistance properties to antibiotics especially the beta-lactam [9]. Beta lactam antibiotics are the most common antibiotics used in the treatment of *S. aureus* infections. They act on the synthesis of peptidoglycan during the cell wall formation and stops it by acting on transpeptidases and carboxypeptidases [10]. Beta-lactamase is also known as penicillinase. Beta-lactam antibiotics contain a beta lactam ring and they act by stopping cell wall synthesis of the bacteria. Exampes of beta-lactam antibiotics include penicillin, cephalosporin, carbapenems and carbacephems. As at 2003, beta-lactam antibiotics were the world most used antibacterial drugs. These antibiotics stop the cell wall synthesis by inhibiting penicillin binding protein involved in the cross linking of the peptidoglycan which then causes the bacteria cell to succumb to osmotic pressure [11].

2. Beta-lactamases in *S. aureus*

Beta-lactamases are enzymes produced by *S. aureus* in order to breakdown the beta lactam ring of beta lactam antibiotics which causes resistance to beta lactam antibiotics [12]. Beta-lactamase is an enzyme produced outside the cell of *S.aureus* after it is exposed to beta-lactam antibiotics [13]. Beta lactamase is constitutive or inductive plasmid mediated. The ability of beta-lactamase to resist antibiotics used against it is largely dependent on its chemical reaction, location and physiochemical conditions (see **Figure 1**) [14].

Beta lactamase is currently classified into two groups which are the molecular classification and Class B metalloenzymes [16]. The molecular classification classifies beta lactamase into Class A, C and D. Class A, C and D enzymes form acyl enzyme by making use of serine for the breakdown of beta-lactam bonds. Class B metalloenzymes makes use of divalent zinc ions for the breakdown of substrate [16].

2.1 Test for Beta-lactamase

After antibiotic susceptibility testing, the test for beta-lactamase is performed. There are different methods of testing for the presence of beta-lactamae which includes:

1. **Nitrocefin disks**: after acquiring nitrocefin commercial discs, stock them at −10 degree Celsius until the time of use. Soak the discs in a saline solution and use a sterilized wire loop to take some colonies from the plate and streak on the surface of the disc. Observe the disc within 60 minutes. When the color changes from yellow to pink, it indicates a positive reaction which shows the presence of beta-lactamase [17].

2. **Clover leaf test (Hodge test):** the presence of an irregular inhibition zone is considered as a positive result [18].

3. **Polymerase Chain Reaction detection of blaZ gene:** According to a study performed by Lynette *et al.* [19] to detect blaZ genes in *S.aureus* using real time PCR assay, after the colonies of *S.aureus* on Mueller hinton agar were tested with nitrocefin impregnated disc, presence of beta-lactamase was detected with a sharp cliff edge and negative with a tapered beach edge. Phenotypic tests were performed using two independent observers. Two to five loop-full of *S.aureus* were inoculated in 1 ml demineralized water and placed in a centrifuge for 2 mins. 8 micro liter of the supernatant solution was used with 12 microliter of real time PCR amplification mixes. The mixes contained buffer of PCR, 4 mM MgC12, 0.2 mM deoxynucleoside triphosphates, 0.75 micro AmpliTaq Gold DNA polymerase, forward and reverse primer [19].

4. **Filter paper iodometric method:** 1 g of starch is dissolved in 100 ml of distilled water by boiling. 1 g of penicillin was added to the solution. Whatman filter paper cut into small pieces are soaked in the solution and later dried for about 2 hours. The filter papers are inoculated with a loopful of freshly isolated *Staphylococcus aureus*. The filter papers are then covered with a petri dish

and incubated for 30mins at 30 degree Celsius. Iodine solution is poured over the filter paper. Beta-lactamase is positive when the site of inoculation turns colorless within 10mins and negative when it retains the blue-black color of the iodine solution [20].

5. **Acidometric test:s** Mueller hinton agar is prepared at a pH of 7.4, a small colony of *S. aureus* is inoculated on the plate. 5 μg methicillin disc is added on the surface of the prepared agar and incubate. After incubation, the plate is opened and dried for some time 1.5 ml of 0.25% stock solution of PNCB (k &k laboratories, Inc., Plainview, N.Y.) in N,N-dimethyl formamide (Fischer Scientific Co., Pittsburgh, Pa) is flooded on the closest area to the zone of inhibition. 6% NaOH is added and air dried for 45 mins. 1.5 ml of 10% aqueous benzyl-penicillinase is flooded on the 6 area with the PNCB indicators. Hydrolysis of benzyl penicillin to penicilloic acid shows the presence of penicillinase. Absence of penicillinase occurs when benzyl penicillin is not changed and when there is no color change in the indicator used.

6. **Chromogenic cephalosporin detection methods:** Beta-lactamase producers produce a color change with cephalosporin. The color change can be measured by measuring the absorption changes that takes place in the 380 to 500 nm region.

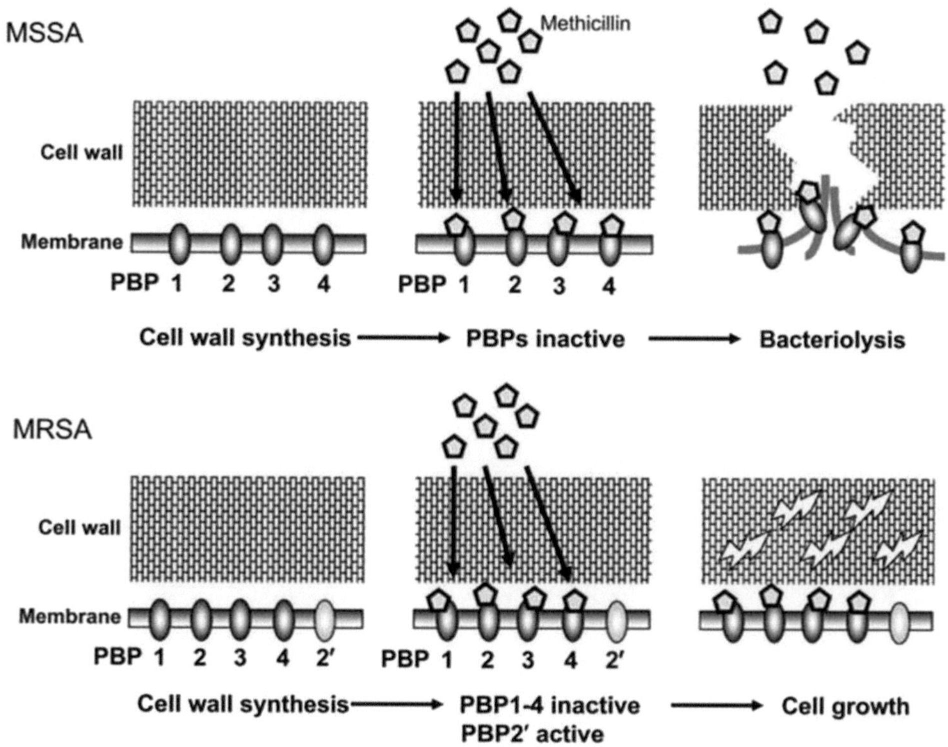

Figure 1.
Resistance mechanism of S .aureus to beta-lactam antibiotics. Source: [15]. Keys: MSSA, Sethicillin susceptible S.aureus; *MRSA, methicillin Resistant* S. aureus.

3. Conclusion

This study shows that *S. aureus* is increasing in its resistance to the beta-lactam antibiotics. *S. aureus* has developed resistance patterns to beta-lactam antibiotics such as the production of beta-lactamase and production of an altered penicillin binding protein 2A. Different tests have been used to detect the presence of beta-lacatmases in *S. aureus*. Examples of such tests include nitrocefin disks, clover leaf tests, polymerase chain reaction detection of blaZ gene, chromogenic cephalosporin tests, acidometric tests and filter paper iodometric tests. The iodometric filter paper method is simple, rapid and can be performed in any bacteriological laboratory and the materials can be stocked and kept under optimal conditions.

Acknowledgements

I want to acknowledge My Lord Jesus for his wisdom and understanding in writing this paper.

Conflict of interest

The author declare no conflict of interest.

Thanks

I thank Intechopen for the opportunity given to me to contribute in this journal.

Author details

Eghe Izevbizua
Cross River University of Technology, Benin City, Edo, Nigeria

*Address all correspondence to: eghevictory123@gmail.com

References

[1] Lowry FD. *S. aureus* infections. New England Journal of Medicine. 1998;**339**:520-532

[2] Diekema DJ, Pfaller MA, Schmitz FF, Schmitz J, Smayevsky J, Bell R, et al. Survey of infections due to staphylococcus species: Frequency of occurrence and antimicrobial susceptibility of isolates collected in the United States, Canada, Latin America, Europe and the Western Pacific region for the sentry antimicrobial surveillance program. Clinical Infectious Diseases. 2001;**32**:114-132

[3] Can HY, Celik TK. Detection of enterogenic and antimicrobial resistant *Staphhylococcus aureus* in Turkish cheese. Food Control. 2012;**24**:100-103

[4] Grundman HN, Aires-de Sousa J, Boyce E, Tiemers M. Emergence and resurgence of methicillin resitant *Staphylococcus aureus* as a public health threat. Lancet. 2006;**368**:874-885

[5] Drooling KL, Kandeal A, Hicks LA, El-Shoubary W, Fawzi K, Kandeel Y, et al. Understanding antibiotic use in Minya district, Egypt: Physician and pharmacist prescribing and the factors influencing their practices. Antibiotics. 2014;**3**(2):233-243

[6] Dashan H, Abd-Elal LAMM, Meghad AM, Abd-Elkader MA, Nabawy EE. Veterinary antibiotic resistance residues and ecological risks in environmental samples obtained from poultry farms, Egypt. Environmental Monitoring and Assessment. 2015;**187**:1007-1014

[7] Ito T, Katayama YL, Asada K, Mori N, Tsutsumimoto K, Tiensitorn C, et al. Structural comparison of three types of staphylococcal cassette chromosome integrated in the chromosome of methiciliin resistant *S.aureus*. Antimicrobial Agents. 2001;**45**:1323-1336

[8] Weese JS, Van Duijkeren E. Methicillin resistant *S.aureus and S. pseudintemiedius* in veterinary medicine. Veterinary Microbiology. 2010;**140**:418-429

[9] Kwon NH, Park KT, Jung WK, Youn HY, Lee Y, Kim SH, et al. Characteristics of methicillin resistant *S.aureus* isolated from chicken meat and hospitalized pigs in Korea and their epidemiological relatedness. Veterinary Microbiology. 2006;**117**:304-312

[10] Lowy FD. Antimicrobial resistance, the example of S. aureus. The Journal of Clinical Investigation. 2003;**3**(9):1265-1273

[11] Miyachiro MM, Contreras MC, Dessen A. Penicillin binding proteins (PBPS) and bacterial cell wall elongation complexes. Sub-Cellular Biochemistry. 2019;**93**:273-289

[12] Kok FK, Schnefer L, Mathee. Beta-lactam antibiotics: From antibiosis to resistance and bacteriology. APMIS. 2010;**118**:1-36

[13] McDougal LK, Thornsberry C. The role of beta-lactamse in staphylococcal resistance to penicillinase resistant penicillins and cephalosporins. Journal of Clinical Microbiology. 1986;**23**(5):832-839

[14] Livermore DM. Beta-lactamase in laboratory and clinical resistances. Clinical Microbiology Reviews. 1995;**8**(4):557-584

[15] Kawada-Matsuo M, Komatsuzawa H. Factors affecting susceptibility of

Staphylococcus aureus to antibacterial agents. Journal of Oral Biosciences. 2012;**54**:86-91

[16] Karen B, George AJ. Updated functional classification of Beta lactamases. Antimicrobial Agents Chemotheraphy. 2010;**54**(3):969-976

[17] Bruno FR, Diego B, Nobrega, Felipe FG, Guido GW, Langoni H. Beta-lactamase detection in *S.aureus* and coagulase negative staphylococcus isolated from bovine mastisis. Research Gate. 2014;**4**:325-328

[18] Bergan T, Bruun JN, Digranes A, Lingaas E, Molby KK, Sander J. Susceptibility testing of bacteria and fungi. Report from the Norwegian working group on antibiotics. Scandinavian Journal of Infectious Diseases (Suppl). 1997;**103**:1-36

[19] Lynette AD, Gerald BH, Meredith MH, Julia AC, David JS. Real time PCR assay for detection of blaz genes in *S.aureus* clinical isolates. Journal of Clinical Microbiology. 2014;**52**(4)

[20] Ekrem K, Meltem C. Comparison of Staphylococcus beta-lactamase detection method. Journal of Pharmaceutical Sciences. 2006;**31**:79-84

Chapter 11

Alternatives to Antibiotics in Semen Extenders Used in Artificial Insemination

Jane M. Morrell, Pongpreecha Malaluang, Aleksandar Cojkic and Ingrid Hansson

Abstract

Antimicrobial resistance is a serious global threat requiring a widespread response. Both veterinarians and medical doctors should restrict antibiotic usage to therapeutic use only, after determining the sensitivity of the causal organism. However, the addition of antibiotics to semen extenders for animal artificial insemination represents a hidden, non-therapeutic use of antimicrobial substances. Artificial insemination for livestock breeding is a huge global enterprise with hundreds of million sperm doses prepared annually. However, reporting of antimicrobial resistance in semen is increasing. This review discusses the consequences of bacteria in semen samples, as well as the effect of antimicrobial substances in semen extenders on bacteria in the environment and even on personnel. Alternatives to antibiotics have been reported in the scientific literature and are reviewed here. The most promising of these, removal of the majority of bacteria by colloid centrifugation, is considered in detail, especially results from an artificial insemination study in pigs. In conclusion, colloid centrifugation is a practical method of physically removing bacteria from semen, which does not induce antibiotic resistance. Sperm quality in stored semen samples may be improved at the same time.

Keywords: antimicrobial resistance, assisted reproduction technologies, semen preservation, environmental bacteria, food of animal origin, livestock production

1. Introduction

Increasing use of antimicrobials is driving antimicrobial resistance (AMR), which is amongst this era's defining global health challenges [1]. By 2030, global antimicrobial use in both humans and food producing animals is projected to increase to 236,757 tons annually [2]. Since the production of milk, meat and eggs requires healthy animals, the ability to treat bacterial infections effectively is of paramount importance to the human population, and therefore the spectre of AMR is an increasing threat in animal husbandry [3]. The latter authors report that a dramatic increase in antimicrobial use in animal production is primarily a consequence of intensive animal production systems introduced to meet growing global food demand [3]. Thus, the situation is complex, involving the interaction between humans, animals and the environment, i.e. One Health [4]. Therefore, an integrated approach to tackling AMR is required, coordinating efforts by the World Health Organisation (WHO), the Food and Agriculture Organisation (FAO) and the World Organisation for Animal Health (OIE) [5].

One area where significant quantities of antibiotics are used is in artificial insemination (AI), which is the method of choice for breeding food-producing animals in most parts of the world. The technique was developed originally to reduce disease transmission since it allows animals to be bred without coming into contact with each other or being transported to different farms for mating [6]. Additional benefits include safety and allowing access to more males, thus permitting more rapid genetic improvement than is possible with natural mating [7]. Success with the technique is dependent on a number of factors, especially a readily available supply of good quality semen. However, regulations stipulate that antibiotics should be added to semen doses for international trade, which contradict current recommendations on prudent use of antimicrobial substances. The latter state that antibiotics should be used only for therapeutic purposes, and only after determining the susceptibility of the causative agent to the relevant antibiotic [8]. Therefore, surprisingly, the subject of widespread use of antibiotics in semen extenders for all animal species has received little attention [9].

The questions surrounding this issue that are relevant to animal breeding are: do we need antibiotics in semen extenders? Is this use of antibiotics in line with recommendations for prudency? Are their alternatives to antibiotics for semen extenders? The purpose of this review is to provide some clarity for these questions.

2. Addition of antibiotics to semen extenders

2.1 Origin of bacteria in semen samples

Almost all semen samples contain some bacteria. The mucosa of the reproductive tract becomes colonised with microbes from the environment and from the animal itself [10]; these bacteria are transferred to semen during ejaculation [11]. The cleanliness of the environment in which the animals are kept plays an important role in the extent of the contamination. Thus, in stables where the bedding was changed every day, there was less microbial contamination of stallion semen than where the bedding was changed less frequently [10].

Post-collection contamination of semen by bacteria from the environment or from personnel can occur during processing; therefore, strict attention to hygiene is required during the whole process. Exposure of sterile semen extender to the air in a semen processing laboratory for a short period followed by culture resulted in the growth of 2000 colony forming units/mL (cfu/mL) *Escherichia coli*, compared to 30,000 cfu/mL of the same bacterium in extended stallion semen samples [12]. Therefore, it is vital that the laboratory should be kept clean and semen processing should preferably be done in a laminar airflow bench to prevent post-collection contamination. In addition, seminal plasma may contain aminopeptidases that promote bacterial proliferation [13], thus compounding the problem of contamination.

2.2 Effect of bacterial contamination

Bacteria not only compete with spermatozoa for nutrients; they could produce metabolic byproducts and toxins, thus reducing the "shelf-life" of semen doses [9]. Therefore, sperm quality may decline in the presence of bacteria. Indeed, bacteria in semen were shown to be negatively associated with sperm quality and fertility [14]. Bacteria may even cause infection and infertility in the female after insemination, or result in the spread of infection [15].

Since the semen extender used to maintain sperm viability also acts as a nutrient medium for bacteria, the contaminating bacteria are in a favourable environment for multiplication. Although bacterial growth is temperature-dependent, some bacteria continue to grow to some extent during cooling. Most commensal bacteria are mesophiles, with an optimum temperature range between 20 and 40°C. They stop growing at temperatures below 15°C, although some can continue to multiply to some extent at lower temperatures [16]. Therefore, one method for inhibiting bacterial growth in semen is to reduce the temperature. However, spermatozoa are sensitive to cooling below 20°C, with species-specific differences in their susceptibility to "cold shock" [17]. Although protocols have been developed to maintain bull sperm survival during cryopreservation with good results, efforts to achieve a successful outcome for other species have met with more variable success. Thus, although it is possible to freeze semen from at least some boars, pregnancy rate and litter size are lower than with liquid semen, providing little incentive for pig breeders to use frozen semen [11]. Post-thaw sperm motility and fertility is acceptable in some stallion ejaculates, whereas others have very poor post-thaw motility and fertilising capacity. On the other hand, artificial insemination with cooled stallion semen the day after semen collection results in a per cycle pregnancy rate of approximately 65% [18]. Therefore, the majority of artificial inseminations in horses currently use cooled semen rather than cryopreserved semen.

It should be noted that bacteria can survive to some extent in frozen semen samples, resuming growth after thawing. Therefore, the time between thawing semen and insemination should be kept to a minimum and strict hygiene measures observed at all times.

To avoid these negative effects of bacterial contamination, regulations governing international trade in semen for artificial insemination specify that antibiotics must be added to extended semen, e.g. in Appendix C of the Council of Europe [19]. These regulations state "Where antibiotics or a mixture of antibiotics are added, their bactericidal activity must be at least equivalent to that of the following mixtures in each ml of semen: gentamicin (250 µg), tylosin (50 µg), lincomycin-spectinomycin (150/300 µg); penicillin (500 IU), streptomycin (500 µg), or amikacin (75 µg), divekacin (25 µg)". Therefore, all semen doses for international trade should contain antibiotics. National regulations governing insemination doses destined for the domestic market may be less rigid; thus, antibiotics might not be added to semen that is to be used for AI on the same premises soon after collection (e.g. 30 minutes). However, it is customary to add antibiotics to all semen doses that are to be cooled and transported to other premises for AI.

2.3 Resistance to antibiotics in semen extenders

Bacteria have been cultured from bull [20, 21] and boar semen [22, 23] despite the presence of antibiotics. Gentamicin, tylosin, spectinomycin and lincomycin did not inhibit the growth of bacteria in bull semen [24] although no growth occurred in semen samples containing ceftiofur/tylosin or ofloxacin. There are anecdotal reports that additional antibiotics are added to boar semen extenders if resistance is suspected, leading to even faster development of antibiotic resistance. Growth of bacteria was reported from stallion semen frozen in extender containing amikacin in a Portugese study [25]. Hernández-Avilés et al. [26] observed that several antibiotics were effective against low levels of *P. aeruginosa* or *K. pneumoniae* inoculated into semen, although they were ineffective against high numbers of these bacteria. All bacteria (n = 55) present in bull ejaculates in a study in Brazil were reported to be resistant to penicillin, and most of them (n = 54) were also resistant to tylosin and lincomycin [27].

Resistant bacteria have been isolated from the reproductive tract of mares in several studies over the last 20 years (reviewed by Malaluang et al. [28]), although semen extenders may not be the only source of antibiotics in the reproductive tract if the mares had been treated for fertility problems or other diseases. A change in the AMR patterns of vaginal bacteria after exposure to antibiotic in semen extenders was reported in horses [29]. Apart from exposure of the vaginal flora to antibiotics, bacteria in the environment are also exposed since the bulk of the inseminated fluid (semen extender and seminal plasma) is expelled from the reproductive tract via backflow. It is not known how much exposure to antibiotics is needed for AMR to develop: some authors consider that even a small exposure to antibiotics is sufficient to induce AMR [30] whereas others are of the opinion that only therapeutic concentrations are needed. However, the duration and length of exposure are likely to contribute to the resistance-inducing potential of the substance. Since environmental pollution events, such as incorrect disposal of antimicrobials, can result in AMR [31], it is essential that any antibiotic-containing substances are destroyed correctly, i.e. by boiling or incineration [32]. Thus, it is particularly important that unused semen extender and semen doses are not poured down the drain, since they could affect environmental bacteria and thereafter be transferred to humans or animals.

2.4 Potential spread of antimicrobial resistance to personnel

Resistance to antimicrobial substances is passed between bacteria, regardless of where these bacteria are present. Therefore, resistance genes can be transmitted between bacteria in different host species, such as between animals and people, with or without involvement of environmental bacteria [33]. If AMR develops in bacteria in the host animal, these resistant bacteria can be spread in the environment and to the human population. Some of the organisms that WHO is most concerned about in the human population i.e. *Staphylococcus aureus, K. pneumoniae* and non-typhoidal *Salmonella* [1], are commonly associated with horses, and *Mycobacterium tuberculosis* [34] can be transmitted from infected cattle.

Transmission of diseases (and AMR) from animals to humans occurs through a variety of routes, although the food-borne route is probably the most important for enteric organisms, such as Salmonella and *Campylobacter coli/jejuni* [35, 36], and *Yersinia enterocolitica* [35]. Contaminated water may also be a source of Campylobacter spp. [37, 38]. However, resistance genes may be transferred from livestock to environmental bacteria and thence to people. Furthermore, animal-associated methicillin-resistant *S. aureus* (MRSA), can be transmitted to, and cause infections in, humans [39].

This transfer of AMR is not only from animals to humans: tourists were thought to be the origin of antibiotic-resistant bacteria in the faeces of reptiles on the Galapagos Islands [40]. A recent review on MRSA concluded that continued resistance in farm animals was likely to be due to contact with human carriers [39]. Therefore, it is advisable to try to restrict the development of AMR in livestock and horses, to protect both animal and human populations.

3. Prudent use of antimicrobial substances

As previously mentioned, the addition of antibiotics to semen extenders does not fit with the current recommendations for prudent use of antimicrobial substances, in the light of increasing incidence of AMR [41]. Although some of the bacteria isolated from semen have also been isolated from some cases of endometritis, they are not present in all cases. Therefore, there is no clear evidence of a *therapeutic* need for the inclusion of antibiotics in semen extenders, if the semen is collected and processed with strict attention to hygiene protocols. However, to comply with "Prudent use", the addition of antimicrobial substances to semen extenders should be avoided, but are there credible alternatives? Possible options to antibiotics are considered in the next section.

4. Alternative to antibiotics in semen extenders

4.1 Reducing the temperature

Low temperature extenders, although recently available for boar semen (e.g. [42, 43]), have not proved popular with the pig breeding industry. Cooling of the semen has to take place over several hours to avoid cold shock, thus interfering with established routines at the semen collection station. It will take time to replace the infrastructure currently in place to support storage at 16–18°C with refrigerators for storage at 4–6°C, and to replace room temperature transport with refrigerated transport. However, even if bacteria do not multiply during refrigeration, some of them will continue to grow and produce toxic substances during the period before refrigeration temperature is reached, as well as after removing the insemination dose from the refrigerator while preparing it for insemination [44]. Thus, the bacteria still have the potential to affect sperm quality and the health of the inseminated sow [15].

4.2 Unconventional antimicrobial substances

Other methods to inhibit bacteria in semen are based on the addition of unconventional antimicrobial substances, such as plant extracts [45, 46] or antimicrobial peptides [47]. However, addition of these novel antimicrobial agents does not preclude the emergence of bacterial resistance, as bacteria adopt new survival mechanisms to evade their effect.

4.2.1 Plant extracts

The effects of extracts from 45 plant species added to semen extenders are reported in the scientific literature, as reviewed by Ros-Santaella et al. [46]. Most of the beneficial effects were due to antioxidant activity but a few extracts also had antibacterial activity. Thus, rosmarinic acid was reported to have an antibacterial effect in boar semen [45, 46, 48, 49] but, according to another report, lacked such an effect in bull semen [50]. Whether these contrasting results reflect the species of animal, type of bacteria present or the source of the rosmarinic acid, or could be attributable to the development of resistance, is not known. Moringa had an antibacterial effect in bull and ram semen [51, 52]. Furthermore, it removed all bacterial contamination from banana shrimp spermatophores [53]; ginger had a similar effect. Tea tree oil exerted an antibacterial effect in boar semen [45, 54]. Omaji was reported to be effective against Gram positive bacteria in bull semen [55], although the minimum inhibitory concentration values shown were for cultured bacterial strains rather than on bacteria in semen.

Antimicrobial peptides are substances produced by the immune system of some mammals and are active against a range of microorganisms [56]. They have a cationic charge, exerting a selective action on negatively charged lipids in bacterial membranes [57]. A cationic peptide derived from human semenogelin was found to have antimicrobial activity [58], although the mechanism of action was not described. Another peptide, GL13K, was found to be active against *Pseudomonas aeruginosa* in biofilms [59], which are notoriously difficult to inhibit. The peptides act by destabilisation of the bacterial membrane. A cyclic hexapeptide was proposed as a potential antimicrobial agent for boar semen, as it apparently did not affect pregnancy rates in AI when used in combination with a low dose of gentamicin, in contrast to other peptides that negatively affected sperm membrane integrity [47]. Recently, two studies investigated a combination of antibiotics and antimicrobial peptides in a low temperature extender for boar semen [60, 61]; the former used semen to which cultures of bacteria were added while the latter used conventional boar semen. Their theory was that bacteria will be exposed to a low concentration of antibiotics together with the antimicrobial peptides while being cooled over several hours for low temperature storage of several days. The authors considered that the antimicrobial activity of the treated samples was similar to the controls with the usual levels of antibiotics. However, since bacteria are still present and viable, there is no guarantee that the low concentration of antibiotics would not be conducive to the development of antimicrobial resistance [62].

4.2.2 Nanoparticles

Nanoparticles were reported to have antimicrobial activity against certain bacteria [63]. The addition of iron oxide (Fe_3O_4) nanoparticles during boar semen processing was reported to produce a slight antibiotic effect with no adverse effects on sperm characteristics [64]. In contrast, although iron oxide nanoparticles were not toxic to ram spermatozoa, they did not have the desired antibacterial effect [65]. A combination of silver and iron oxide nanoparticles produced a greater antibacterial effect than iron oxide but showed higher spermatotoxicity. Therefore, more research is needed before these nanoparticles can be a contender to inhibit the presence of bacteria in semen. Interestingly, selenium nanoparticles [66] and zinc nanoparticles [67] were thought to improve membrane integrity of bull spermatozoa; although no microbiological analysis of the samples was reported, it is conceivable that the improved membrane integrity observed could have been due to an antimicrobial effect. A reduction in microbial activity would theoretically result in a reduction in the production of reactive oxygen species and hence a decrease in sperm membrane damage [9].

Other nanoparticles that have been investigated for their potential antibacterial effect in semen extenders include chitosan, a glycosaminoglycans that interacts with bacterial cell membranes causing lysis, together with ethylene diaminetetraacetic acid (EDTA), which increases the permeability of the cell wall of gram negative bacteria, and bestatin [68]. The combination was considered to have a bacteriostatic effect without causing a decrease in sperm quality.

4.3 Colloid centrifugation

Apart from potentially inducing AMR, one of the problems with antibiotics is that the killed bacteria remain in the sperm suspension after death. Therefore, intracellular substances and reactive oxygen species, or endotoxins from the lipopolysaccharides of the outer membrane of the cell wall of Gram-negative bacteria, are released into the extender and can have a negative effect on sperm quality. Intuitively, it might be better to remove the bacteria from the sperm sample rather than inhibiting their growth or killing them and leaving them in situ. One such method for separating spermatozoa from bacteria is colloid centrifugation [12, 69, 70]. The method used in these latter studies was a modified density gradient in which only one layer of colloid was used, hence its name "Single Layer Centrifugation" (SLC). This simplified method enables large volumes of semen to be processed easily [71]; even voluminous ejaculates can be processed, provided that the centrifuge rotor can accommodate large tubes. A detailed description of the methodology for 50 mL tubes can be found in Morrell and Nunes [72] and for 500 mL tubes in Morrell et al. [71].

4.3.1 Single layer centrifugation

Initially, SLC was shown to be effective in separating aliquots of boar sperm samples from bacteria in 12 ml tubes [69]. A scaled-up version of this technique was used with stallion semen samples that had been inoculated with various bacterial suspensions [12]. In this case, the usual SLC was modified by inclusion of an inner tube - a 5 mL plastic semen straw or similar tubing - inserted through a hole in the cap before loading the semen on the colloid, to facilitate retrieval of the sperm pellet after centrifugation. The sperm sample, inoculated with known amounts of *E.coli*, was added through another small hole near the edge of the cap. The resulting sperm suspensions were not entirely free of bacteria; removal varied from 68–100% and appeared to depend on the bacterial load. Similarly, Varela et al. [73] reported removal of 93% of the microbial load from stallion semen samples using the original SLC method.

In the study by Al-Kass et al. [70, 74], split ejaculates were used, with antibiotic added to half of each ejaculate. This protocol enabled comparisons to be made of control samples with and without antibiotics, as well as SLC samples with and without antibiotics. Although there were clear differences in sperm quality between the SLC and corresponding control samples, in favour of SLC, there were few obvious differences between the samples with and without antibiotics within a treatment. In other words, the presence of antibiotics had little effect on sperm quality. The only exception was for the DNA fragmentation index, where the control samples with antibiotics showed more chromatin damage than control samples without antibiotics. Whether this increased damage was due to the effects of the antibiotics *per se* or whether it was due to release of intracellular contents or LPS following the death of the bacteria, which subsequently damaged the sperm chromatin, is unknown. Since increased levels of chromatin fragmentation are often associated with decreased fertility [75], it would be advisable to avoid semen handling protocols that promote DNA damage.

The experiments by Al-Kass et al. [70, 74] show several important points. The first is that it is possible to remove most of the bacteria from stallion semen relatively easily using SLC. Second, any bacteria remaining in the sperm samples did not have an adverse effect on sperm quality during cooled storage for 96 h. Third, the addition of antibiotics did not enhance sperm quality during storage and, in the case of sperm DNA fragmentation, actually increased DNA damage in the control samples by 96 h. Thus, **from the point of view of sperm quality**, there is no justification for adding antibiotics to sperm samples if the semen could be processed by SLC instead. Since the number of bacteria in the SLC sample is considerably reduced compared to the raw ejaculate, the uterine immune response should be able to deal with any remaining environmental bacteria. However, since no inseminations were carried out with the processed samples, this supposition is still speculative for equine AI.

Experiments with boar semen have progressed further. The economics of pig production are such that the cost of the high-density colloid used for sperm selection would be prohibitive for the industry. Thus, instead of using a high-density colloid to select robust spermatozoa from the rest of the ejaculate, a low-density colloid was tested to determine if the majority of the spermatozoa could be separated from the seminal plasma and its bacterial load [76]. Since the price of the colloid formulations is determined by the cost of the silane-coated silica base, using a formulation with a lower content of this material will be cheaper to manufacture than formulations with a higher density. Preparing the semen using the low-density colloid formulation for boar spermatozoa, Porcicoll, it was possible to remove most of the bacteria from boar semen, and sperm quality did not decline during subsequent storage of the processed samples [76]. In a small AI trial, pregnancy rates and litter sizes were not adversely affected by the low-density SLC-preparation of the sperm samples, and there were fewer mummified piglets in the litters derived from SLC sperm samples than in the controls (**Table 1**; [77]). At current prices (i.e. not adjusting for economies of scale) the cost of processing the whole ejaculate by low-density colloid would add approximately $2 to the price of each insemination dose. Economies of scale in the manufacture of the colloid could further reduce its cost. These results are very encouraging and a larger breeding trial in pigs is planned; the possibility of using this method with stallion semen should be investigated.

4.3.2 Advantages and disadvantages of single layer centrifugation

At this point, it is worth considering both a risk–benefit analysis of the use of antibiotics in semen extenders (**Table 2**; [44]) and a cost–benefit analysis of SLC.

A cost–benefit analysis of performing SLC will include the advantages and disadvantages of not using antibiotics, as summarised in **Table 2**, but will also include the cost of purchasing a centrifuge (unless one is already available), extra time spent by personnel in processing semen, and the cost of the colloid. In many cases, the stud will already possess a centrifuge, for example, if semen is being frozen on the premises or removal of some seminal plasma by sperm washing is done routinely in the preparation of cooled stallion semen samples [78]. The centrifugation time is longer for SLC than for sperm washing (20 minutes versus 10 minutes), but personnel are able to perform other duties while the centrifugation is operating. At present, the cost of Equicoll would add approximately $10–$15 to the cost of the semen dose for equine AI, which is insignificant compared to the price of most stallion semen doses.

No. of sows	No. Farrowing	Total born/ litter	Live born/ litter	Born dead/ litter	Mummified/ litter
Control (17)	13	14.5 ± 5.1	13.6 ± 5.1	0.9 ± 1.0	1.1 ± 1.8*
SLC (12)	9	16.0 ± 3.4	15.3 ± 2.9	0.8 ± 0.9	0.7 ± 0.71*

*$p < 0.022$.

Table 1.
Reproductive efficiency following insemination with spermatozoa after centrifugation through low-density Porcicoll [77].

	Risk	Benefit
No antibiotics	Bacteria compete with sperm for nutrients, produce toxic byproducts and LPS that cause decreased sperm quality; can cause disease in inseminated females.	No antibiotic toxicity to spermatozoa; no risk of spreading antibiotic resistance.
Antibiotics	Can be toxic to spermatozoa; sperm quality may be affected by dead bacteria; bacteria in semen may become resistant; contamination of the environment with antibiotics from inseminated mares; can contribute to the spread of antibiotic resistance in animals and personnel.	Kill most contaminating bacteria before they have a chance to grow; no competition for nutrients in semen extender; bacteria are not transferred to the female via AI unless resistant.

Table 2.
Risk–benefit analysis of antimicrobial substances in semen extenders (modified from [44]).

However, if it were possible to use the low-density colloid instead of the usual colloid formulation for stallion semen, as discussed previously for pig semen, the cost of the colloid would be approximately halved. Perhaps the question should be re-phrased: can we afford *not* to use SLC instead of antibiotics?

It should be noted that the efficiency of the technique will be affected by increasing bacterial load. Therefore, SLC should not be a substitute for attention to hygiene in animal husbandry, or in semen collection and processing protocols. Strict attention to hygiene should occur at all times when dealing with breeding animals and semen handling [79].

5. Conclusion

Bacteria appear in ejaculates during semen collection and processing. Some types of bacteria have a detrimental effect on sperm quality during storage and some may cause disease in inseminated females. The addition of antibiotics to semen extenders reduces the bacterial load in semen doses for artificial insemination but can facilitate the development of antimicrobial resistance. Furthermore, this application is non-therapeutic and does not fit with current guidance on the prudent use of these substances. Bacterial growth is reduced or prevented by cooling semen below 15°C but resumes when the temperature rises again in preparation for artificial insemination. Alternatives to antibiotics, such as plant-based extracts and nanoparticles, are available but may not be effective in all situations or may be spermatotoxic, which is counter-productive. It remains to be seen whether bacteria can develop resistance to such plant-based extracts. Colloid centrifugation of semen separates spermatozoa from most bacteria, thus avoiding the possibilities of AMR development. Since the process is purely physical, bacteria cannot mutate as an avoidance mechanism. The technique can be used to prepare whole stallion and boar ejaculates at the semen collection center and is cost-effective compared to the price of AMR. It could be used in conjunction with reducing the temperature of the semen. However, strict attention to hygiene should still occur at all stages of semen collection and processing, and in the routine husbandry of breeding males.

Acknowledgements

Funded by the Faculty of Veterinary Medicine, Swedish University of Agricultural Sciences, FORMAS Stockholm (Research council for sustainable development; project number 2017-00957), and Carl Tryggers Foundation (project number CTS 21: 1163), both awarded to JMM. The funders were not involved in the writing of this article or in the decision to publish.

Conflict of interest

The author, Prof. Jane M. Morrell, is the inventor and one of the patent holders of the colloids mentioned in this article.

Author details

Jane M. Morrell[1*], Pongpreecha Malaluang[1], Aleksandar Cojkic[1] and Ingrid Hansson[2]

1 Clinical Sciences, Swedish University of Agricultural Sciences, Uppsala, Sweden

2 Biomedical Science and Veterinary Public Health, Swedish University of Agricultural Sciences, Uppsala, Sweden

*Address all correspondence to: jane.morrell@slu.se

References

[1] Prestinaci F, Pezzotti P, Pantosi A. Antimicrobial resistance: A global multifaceted phenomenon. Pathogens and Global Health. 2015;**109**:309-138

[2] Schar D, Klein EY, Laxminarayan R, Gilbert M, van Boeckel TP. Global trends in antimicrobial use in aquaculture. Scientific Reports. 2020;**10**:21878. DOI: 0.1038/s41598-020-78849-3

[3] Tiseo K, Huber L, Gilbert M, Robinson TP, van Boeckel TP. Global trends in antimicrobial use in food animals from 2017 to 2030. Antibiotics. 2020;**9**:918-931. DOI: 10.3390/antibiotics 9120918

[4] Eriksen J, Björkman I, Röing M, Essack SY, Stålsby LC. Exploring the one health perspective in Sweden's policies for containing antibiotic resistance. Antibiotics. 2021;**10**:526-538. DOI: 10.3390/antibiotics10050526

[5] Aslam B, Khurshid M, Arshad MI, Muzammil S, Rasool M, Yasmeen N, et al. Antibiotic resistance: One health one world outlook. Frontiers in Cellular and Infection Microbiology. 2021;**11**:771510 (20 pages). DOI: 10.3389/fcimb.2021.771510

[6] Givens MD. Review: Risks of disease transmission through semen in cattle. Animal. 2018;**12**(s1):s165-s171. DOI: 10.1017/S1751731118000708

[7] Morrell JM. Artificial insemination: Current and future trends. In: Manafi M, editor. Artificial Insemination in Farm Animals [Internet]. London: IntechOpen; 2011. Available from: https://www.intechopen.com/chapters/16096. DOI: 10.5772/17943

[8] Teale CJ, Moulin G. Prudent use guidelines: A review of existing veterinary guidelines. Revue Scientifique et Technique. 2012;**31**:343-354. DOI: 10.20506/rst.31.1.2119

[9] Morrell JM, Wallgren M. Alternatives to antibiotics in semen extenders: A review. Pathogens. 2014;**3**:934-946. DOI: 10.3390/pathogens3040934

[10] Picket BW, Voss JL, Jones RL. Control of bacteria in stallions and their semen. Journal of Equine Veterinary Science. 1999;**19**:424-441. DOI: 10.1016/S0737-0806(99)80254-8

[11] Nitsche-Melkus E, Bortfeldt R, Jung M, Schulze M. Impact of hygiene on bacterial contamination in extended boar semen: An eight-year retrospective study of 28 European AI centers. Theriogenology. 2020;**146**:133-139. DOI: 10.1016/j.theriogenology.2019.11.031

[12] Morrell JM, Klein C, Lundeheim N, Erol E, Troedsson MHT. Removal of bacteria from stallion semen by colloid centrifugation. Animal Reproduction Science. 2014;**145**:47-53. DOI: 10.1016/j.anireprosci.2014.01.005

[13] Viudes-de-Castro MP, Mocé E, Lavara R, Marco-Jiménez F, Vicente JS. Aminopeptidase activity in seminal plasma and effect of dilution rate on rabbit reproductive performance after insemination with an extender supplemented with buserelin acetate. Theriogenology. 2014;**81**:1223-1228. DOI: 10.1016/j.theriogenology.2014.02.003

[14] Bennett DG. Therapy of endometritis in mares. Journal of the American Veterinary Medical Association. 1986;**188**:1390-1392

[15] Kuster CE, Althouse GC. The impact of bacteriospermia on boar sperm storage and reproductive performance. Theriogenology. 2016;**85**:21-26. DOI: 10.1016/j.theriogenology.2015.09.049

[16] Nedwell DB. Effect of low temperature on microbial growth: Lowered affinity for substrates limits growth at low temperature. FEMS Microbial Ecology. 1999;**30**:101-111. DOI: 10.1111/j.1574-6941.1999.tb00639.x

[17] Pickett BW, Komarek RJ. Effect of cold shock and freezing on loss of lipid from spermatozoa. Journal of Dairy Science. 1967;**59**:753-757

[18] Aurich J, Aurich C. Developments in European horse breeding and consequences for veterinarians in equine reproduction. Reproduction in Domestic Animals. 2006;**41**:275-279. DOI: 10.1111/j.1439-0531.2006.00719

[19] Council of Europe. Directive 92/65/EEC of 13 July 1992, laying down animal health requirements governing trade in and imports into the community of animals, semen, ova and embryos not subject to animal health requirements laid down in specific community rules referred to in annex a to directive 90/425/EEC. Official Journal of the European Communities. 1992;**268**:54-72. Updated 2017

[20] Zampieri D, Santos VG, Braga PA, Ferreira CR, Ballottin D, Tasic L, et al. Microorganisms in cryopreserved semen and culture media used in the in vitro production (IVP) of bovine embryos identified by matrix-assisted laser desorption ionization mass spectrometry (MALDI-MS). Theriogenology. 2013;**80**:337-345. DOI: 10.1016/j.theriogenology.2013.04.020

[21] Kilburn C, Rooks DJ, McCarthy AJ, Murray RD. Antimicrobial resistance in some gram-negative bacteria isolated from the bovine ejaculate. Reproduction in Domestic Animals. 2013;**48**:525-528. DOI: 10.1111/rda.12127

[22] Bresciani C, Cabassi CS, Morini G, Taddei S, Bettini R, Bigliardi E, et al. Boar semen bacterial contamination in Italy and antibiotic efficacy in a modified extender. Italian Journal of Animal Science. 2014;**13**:3082. DOI: 10.4081/ijas.2014.3082

[23] Gòdia YR-C, Zingaretti L, Darwich L, López S, et al. A pilot RNA-seq study in 40 pietrain ejaculates to characterize the porcine sperm microbiome. Theriogenology. 2020;**157**:525-533. DOI: 10.1016/j.theriogenology.2020.08.00

[24] Gloria A, Contri A, Wegher L, Vignola G, Dellamaria D, Carluccio A. The effects of antibiotic additions to extenders on fresh and frozen-thawed bull semen. Animal Reproduction Science. 2014;**150**:15-23. DOI: 10.1016/j.anireprosci.2014.08.012

[25] Guimaraes T, Lopes G, Pinto M, Silva E, Miranda C, Correia MJ, et al. Colloid centrifugation of fresh stallion semen before cryopreservation decreased microorganism load of frozen-thawed semen without affecting seminal kinetics. Theriogenology. 2015;**83**:186-191. DOI: 10.1016/j.theriogenology.2014.09.003

[26] Hernández-Avilés C, Serafini R, Love CC, Teague SR, LaCaze KA, Lawhon SD, et al. The effects of antibiotic type and extender storage method on sperm quality and antibacterial effectiveness in fresh and cooled-stored stallion semen. Theriogenology. 2018;**122**:23-29. DOI: 10.1016/j.theriogenology.2018.08.022

[27] Goularte KL, Voloski FLS, Redú JFM, Ferreira CER, Vieira AD,

Duval EH, et al. Antibiotic resistance in microorganisms isolated in a bull semen stud. Reproduction in Domestic Animals. 2020;**55**:318-324. DOI: 10.1111/rda.13621

[28] Malaluang P, Wilén E, Lindahl J, Hansson I, Morrell JM. Antimicrobial resistance in equine reproduction. Animals. 2021;**11**:3035-3047. DOI: 10.3390/ani11113035

[29] Malaluang P, Wilén E, Hansson I, Lindahl J, Morrell JM. Antibiotic resistance in vaginal flora of inseminated mares. Reproduction in Domestic Animals. 2021;**57, S2**:YSC7

[30] Johansson A, Greko C, Engström BE, Karlsson M. Antimicrobial susceptibility of Swedish, Norwegian and Danish isolates of Clostridium perfringens from poultry, and distribution of tetracycline resistance genes. Veterinary Microbiology. 2004;**99**:251-257. DOI: 10.1016/j.vetmic.2004.01.009

[31] Martinez JL, Rojo F. Metabolic regulation of antibiotic resistance. FEMS Microbiology Reviews. 2011;**35**:768-789. DOI: 10.1111/j.1574-6976.2011.00282

[32] Anwar M, Iqbal Q, Saleem F. Improper disposal of unused antibiotics: An often overlooked driver of antimicrobial resistance. Expert Review of Antibiotic Therapy. 2020;**18**:697-699. DOI: 10.1080/14787210.2020.1754797

[33] Young C, Kharmacharya D, Bista M, Sharma A, Goldstein T, Anthony SJ, et al. Sharing of antimicrobial resistance genes among animals, humans, and the environment in Nepal: A one health case study. International Journal of Infectious Diseases. 2019;**79**:20

[34] Lyashchenko KP, Greenwald R, Esfandiari J, Lecu A, Waters WR, Posthaus H, et al. Pulmonary disease due to mycobacterium tuberculosis in a horse: Zoonotic concerns and limitations of antemortem testing. Veterinary Medicine International. 2012:642145. 6 pages

[35] EFSA and ECDC (European Food Safety Authority and European Centre for Disease Prevention and Control). The European Union one health 2020 Zoonoses report. EFSA Journal. 2021;**19**(12):6971, 324. DOI: 10.2903/j.efsa.2021.6971

[36] WHO Estimates of the Global Burden of Foodborne Diseases: Foodborne Disease Burden Epidemiology Reference Group 2007-2015. World Health Organization. ISBN: 978 92 4 156516 5

[37] Kuhn KG, Falkenhorst G, Emborg HD, Ceper T, Torpdahl M, Krogfelt KA, et al. Epidemiological and serological investigation of a waterborne campylobacter jejuni outbreak in a Danish town. Epidemiology and Infection. 2017;**2017**(145):701-709. DOI: 10.1017/S0950268816002788

[38] Hyllestad S, Iversen A, MacDonald E, Amato E, Borge BÅS, Bøe A, et al. Large waterborne campylobacter outbreak: Use of multiple approaches to investigate contamination of the drinking water supply system, Norway, June 2019. Euro Surveillance. 2020;**25**:2000011. DOI: 10.2807/1560-7917.ES.2020.25.35.2000011. Erratum in: Euro Surveill. 2020;25(43)

[39] Crespo-Piazuelo D, Lawlor PG. Livestock-associated methicillin-resistant *Staphylococcus aureus* (LA-MRSA) prevalence in humans in close contact with animals and measures to reduce on-farm colonisation. Irish Veterinary Journal. 2021;**74**:21. DOI: 10.1186/s13620-021-00200-7

[40] Wheeler E, Hong PY, Bedon LC, Mackie RI. Carriage of

antibiotic-resistant enteric bacteria varies among sites in Galapagos reptiles. Journal of Wildlife Diseases. 2012;**48**:56-67. DOI: 10.7589/0090-3558-48.1.56

[41] Ungemach FR, Müller-Bahrdt D, Abraham G. Guidelines for prudent use of antimicrobials and their implications on antibiotic usage in veterinary medicine. International Journal of Medical Microbiology. 2006;**296**(S2):33-38. DOI: 10.1016/j.ijmm.2006.01.059

[42] Waberski D, Luther A-M, Grünther B, Jäkel H, Henning H, Vogel C, et al. Sperm function in vitro and fertility after antibiotic-free, hypothermic storage of liquid preserved boar semen. Scientific Reports. 2019;**9**:14748. DOI: 10.1038/s41598-019-51319-1

[43] de Menezes TA, APG M, da Silva Oliveira G, Bernardi ML, Wentz I, da Ulguim RR, et al. Antibiotic-free extended boar semen preserved under low temperature maintains acceptable in-vitro sperm quality and reduces bacterial load. Theriogenology. 2020;**149**:131-138. DOI: 10.1016/j.theriogenology.2020.03.003

[44] Morrell JM. Antimicrobials in boar semen extenders – A risk/benefit analysis. Journal of Antimicrobial Agents. 2016;**2**:1-2. DOI: 10.4172/2472-1212.1000107

[45] Elmi A, Prosperi A, Zannoni A, Bertocchi M, Scorpio DG, Forni M, et al. Antimicrobial capabilities of non-spermicidal concentrations of tea tree (Melaleuca alternifolia) and rosemary (Rosmarinus officinalis) essential oils on the liquid phase of refrigerated swine seminal doses. Research in Veterinary Science. 2019;**127**:76-81. DOI: 10.1016/j.rvsc.2019.10.014

[46] Ros-Santaella JL, Kadlec M, Pintus E. Pharmacological activity of Honeybush (Cyclopia intermedia) in boar spermatozoa during semen storage and under oxidative stress. Animals Open Access Journal MDPI. 2020;**10**:772-796. DOI: 10.3390/ani10030463

[47] Schulze M, Junkes C, Müller P, Speck S, Rüdiger K, Dathe M, et al. Effects of cationic antimicrobial peptides on liquid-preserved boar spermatozoa. PLoS One. 2014;**9**:e100490. DOI: 10.1371/journal.pone.0100490

[48] Malo C, Gil L, Gonzalez N, Martínez F, Cano R, de Blas I, et al. Anti-oxidant supplementation improves boar sperm characteristics and fertility after cryopreservation: Comparison between cysteine and rosemary (Rosmarinus officinalis). Cryobiology. 2010;**61**:142-147. DOI: 10.1016/j.cryobiol.2010.06.009

[49] Elmi A, Ventrella D, Barone F, Filippini G, Benvenuti S, Pisi A, et al. Thymbra capitata (L.) cav. And Rosmarinus officinalis (L.) essential oils: In vitro effects and toxicity on swine spermatozoa. Molecules. 2017;**22**:2162-2176. DOI: 10.3390/molecules24061071

[50] Hagos L, Cojkic A, Hansson I, Johannisson A, Ntallaris T, Morrell JM. Effect of rosmarinic acid on bull sperm quality and bacteria in semen. Animal Reproduction Science. 2020;**220**:19

[51] Sokunbi OA, Ajani OS, Lawanson AA, Amao EA. Antibiotic potential of Moringa leaf (Moringa oleifera lam.) crude extract in bull semen extender European. Journal of Medicinal Plants. 2015;**9**:1-8. DOI: 10.9734/EJMP/2015/18546

[52] Carrera-Chávez JM, Jiménez-Aguilar EE, Acosta-Pérez TP, Núñez-GastélumJA, Quezada-CasasolaA, Escárcega-Ávila AM, et al. Effect of Moringa oleifera seed extract on antioxidant activity and sperm

characteristics in cryopreserved ram semen. Journal of Applied Animal Research. 2020;**48**:114-120. DOI: 10.1080/09712119.2020.1741374

[53] Nimrat S, Noppakun S, Sripuak K, Boonthai T, Vuthiphandchai V. Cryopreservation of banana shrimp (Fenneropenaeus merguiensis) spermatophores with supplementation of medicinal plant extracts: Development of a programmable controlled-rate method and a practical method. Aquaculture. 2020;**515**:734537. DOI: 10.1016/j.aquaculture.2019.734537

[54] Elmi A, Ventrella D, Barone F, Carnevali G, Filippini G, Pisi A, et al. In vitro effects of tea tree oil (Melaleuca alternifolia essential oil) and its principal component terpinen-4-ol on swine spermatozoa. Molecules. 2019;**24**:1071. DOI: 10.3390/molecules24061071

[55] Tvrdá E, Michalko J, Árvay J, Vukovic NL, Ivanišová E, Duracka M, et al. Characterization of the Omija (Schisandra chinensis) extract and its effects on the bovine sperm vitality and oxidative profile during in vitro storage. Evidence-Based Complementary and Alternative Medicine. 2020;**2020**:7123780. DOI: 10.1155/2020/7123780

[56] Hall-Stoodley L, Stoodley P, Kathju S, Hoiby N, Moser C, Costerton WJ, et al. Towards diagnostic guidelines for biofilm-associated infections. FEMS Immunology and Medical Microbiology. 2012;**65**:127-145. DOI: 10.1111/j.1574-695X.2012.00968

[57] Coelho ML, Duarte AFS, Bastos MCF. Bacterial labionin-containing peptides and sactibiotics: Unusual types of antimicrobial peptides with potential use in clinical settings (a review). Current Topics in Medicinal Chemistry. 2017;**17**:1-22. DOI: 10.2174/1568026616666160930144809

[58] Bourgeon F, Evrard B, Brillard-Bourdet M, Colleu D, Jegou B, Pineau C. Involvement of semenogelin-derived peptides in the antibacterial activity of human seminal plasma. Biology of Reproduction. 2004;**70**:768-774. DOI: 10.1095/biolreprod.103.022533

[59] Hirt H, Gorr S-U. Antimicrobial peptide GL13K is effective in reducing biofilms of *Pseudomonas aeruginosa*. Antimicrobial Agents and Chemotherapy. 2013;**57**:4903-4910. DOI: 10.1128/AAC.00311-13

[60] Luther A-M, Nguyen TQ, Verspohl J, Waberski D. Antimicrobially active semen extenders allow the reduction of antibiotic use in pig insemination. Antibiotics. 2021;**10**:1319. DOI: doi.org/10.3390/antibiotics10111319

[61] Hensel B, Jakop U, Scheinpflug K, Mühldorfer K, Schröter F, Schäfer J, et al. Low temperature preservation of porcine semen: Influence of short antimicrobial lipopeptides on sperm quality and bacterial load. Scientific Reports. 2020;**10**:13225. DOI: 10.1038/s41598-020-70180-1

[62] Vickram AS, Kuldeep D, Archana K, Parameswari R, Ramesh Pathy M, Iqbal HMN, et al. Antimicrobial peptides in semen extenders: A valuable replacement option for antibiotics in cryopreservation—A prospective review. J. Exp. biol. Agricultural Sciences. 2017;**5**:578-588. DOI: 10.18006/2017.5(4).578.588

[63] Guzman M, Dille J, Godet S. Synthesis and antibacterial activity of silver nanoparticles against gram-positive and gram-negative bacteria. Nanomedicine. 2012;**8**:37-45. DOI: 10.1016/j.nano.2011.05.007

[64] Tsakmakidis IA, Samaras T, Anastasiadou S, Basioura A, Ntemka A, Michos I, et al. Iron oxide nanoparticles as an alternative to antibiotics additive on extended boar semen. Nanomaterials. 2020;**10**:1568-1583. DOI: 10.3390/nano10081568

[65] Tsakmakidis IA, Samaras T, Anastasiadou S, Basioura A, Ntemka A, Michos I, et al. Toxic and microbiological effects of iron oxide and silver nanoparticles as additives on extended ram semen. Animals. 2021;**11**:1011-1026. DOI: 10.3390/ani11041011

[66] Khalil WA, El-Harairy MA, Zeidan AEB, Hassan MAE. Impact of selenium nano-particles in semen extender on bull sperm quality after cryopreservation. Theriogenology. 2019;**126**:121-127. DOI: 10.1016/j.theriogenology.2018.12.017

[67] Jahanbin R, Yazdanshenas P, Rahimi M, Hajarizadeh A, Tvrda E, Ataei Nazari S, et al. In vivo and I vitro evaluation of bull semen processed with zinc (Zn) nanoparticles. Biological Trace Element Research. 2021;**199**:126-135. DOI: 10.1007/s12011-020-02153-4

[68] Viudes-de-Castro MP, Marco-Jimenez F, Vicente JS, Marin C. Antibacterial activity of some molecules added to rabbit semen extender as alternative to antibiotics. Animals. 2021;**11**:1178. DOI: 10.3390/ani11041178

[69] Morrell JM, Wallgren M. Removal of bacteria from boar ejaculates by single layer centrifugation can reduce the use of antibiotics in semen extenders. Animal Reproduction Science. 2011;**123**:64-69. DOI: 10.1016/j.anireprosci.2010.11.005

[70] Al-Kass Z, Spergser J, Aurich C, Kuhl J, Schmidt K, Morrell JM. Effect of presence or absence of antibiotics and use of modified single layer centrifugation on bacteria in pony stallion semen. Reproduction in Domestic Animals. 2018;**54**:342-349. DOI: 10.1111/rda.13366

[71] Morrell JM, van Wienen M, Wallgren M. Single layer centrifugation can be scaled-up further to process up to 150mL semen. ISRN Veterinary Science. 2012;**2011**:6 pages. DOI: 10.5402/2011/183412

[72] Morrell JM, Nunes MM. Practical guide to single layer centrifugation of stallion semen. Equine Veterinary Education. 2018;**30**:392-398. DOI: 10.1111/eve.12658

[73] Varela E, Plaza E, Muñoz de Propios P, Ortiz-Rodríguez JM, Álvarez M, Anel-López L, et al. How does the microbial load affect the quality of equine cool-stored semen? Theriogenology. 2018;**2018**(114):212-220. DOI: 10.1016/j.theriogenology.2018.03.028

[74] Al-Kass Z, Spergser J, Aurich C, Kuhl J, Schmidt K, Morrell JM. Effect of presence or absence of antibiotics and use of modified single layer centrifugation on bacteria in pony stallion semen. Reproduction in Domestic Animals. 2019;**54**:342-349. DOI: 10.1111/rda.13366

[75] Morrell JM, Johannisson A, Dalin A-M, Hammar L, Sandebert T, Rodriguez-Martinez H. Sperm morphology and chromatin integrity in Swedish warmblood stallions and their relationship to pregnancy rates. Acta Veterinaria Scandinavica. 2008;**50**:2-8. DOI: 10.1186/1751-0147-50-2

[76] Morrell JM, Nunez-Gonzalez A, Crespo-Felez I, Martinez-Martinez S, MartinezAlborciaMJ,Fernandez-AlegreE, et al. Removal of bacteria from boar semen using a low-density colloid.

Theriogenology. 2019;**126**:272-278. DOI: 10.1016/j.theriogenology.2018.12.028

[77] Morrell JM, Martinez-Alborcia M-J, Martinez-Pastor F, Bolarin A. Centrifugation of boar spermatozoa through low density Porcicoll to separate them from bacteria does not affect fertility after insemination. JIOMICS. 2021;**11**:12 I

[78] Love CC, Thompson JA, Lowry VK, Varner DD. Effect of storage time and temperature on stallion sperm DNA and fertility. Theriogenology. 2002;**57**:1135e42. DOI: 10.1016/s0093-691x(01)00689-6

[79] World Organisation for Animal Health (OIE). 2019. Terrestrial Animal Health Code available at: https://www.oie.int/fileadmin/Home/eng/Health_standards/tahc/current/chapitre_general_hygiene_semen.pdf [Accessed: July 13, 2021]

Chapter 12

Quorum Sensing Inhibition Based Drugs to Conquer Antimicrobial Resistance

Kothandapani Sundar, Ramachandira Prabu and Gopal Jayalakshmi

Abstract

Quorum sensing is the cell to cell communication mechanism in microorganism through signalling molecules. Regulation of virulence factor, sporulation, proteolytic enzymes production, biofilm formation, auto-inducers, cell population density are key physiological process mediated through quorum-sensing (QS) signalling. Elevation of innate immune system and antibiotic tolerance of pathogens is highly increased with perspective of quorum-sensing (QS) activity. Development of novel drugs is highly attractive scenario against cell-cell communication of microbes. Design of synthetic drugs and natural compounds against QS signal molecules is vital combat system to attenuate microbial pathogenicity. Quorum sensing inhibitors (QSIs), quorum quenchers (QQs), efflux pump inhibitors (EPIs) act against multi-drug resistance strains (MDR) and other pathogenic microbes through regulation of auto-inducers and signal molecule with perceptive to growth arrest both in-vitro and in-vivo. QQs, QSIs and EPIs compounds has been validated with various animal models for high selection pressure on therapeutics arsenal against microbe's growth inhibition. Promising QSI are phytochemicals and secondary metabolites includes polyacetylenes, alkaloids, polyphenols, terpenoids, quinones.

Keywords: quorum sensing, quorum sensing inhibitors, antimicrobial resistance, QSI drugs, bioactive metabolites

1. Introduction

UK government has calculated approximately 10 million deaths every year and loss of $100 trillion to the global economy by 2050 due to AMR strains (anti-microbial resistance) infection given by commission's report [1]. Quorum sensing (QS) are responsible for persistent infection on humans causing urinary tract infection, otitis media, cystic fibrosis, endocarditis, periodontitis, and implantable device infections [2]. 60% of clinical infection are caused by biofilm formation reported by National Institutes of Health (NIH) [2]. In general scenario of QS system based on the synthesis of signal molecule in the bacteria act as ligand and docked with the bacterial receptor for signal transduction. Five types of QS signalling molecules belongs to *N*-acylhomoserine lactones (AHLs), oligopeptides, autoinducer (AI), MTAN (5′-methylthioadenosine/S-adenosylhomocysteine nucleosidase), PQS (Pseudomonas quinolone signal) for gene regulation and sense threshold level bacterial population within QS system (**Figure 1**) [3].

N-Acylhomoserine lactones (AHLs), an autoinducer system with distinct pathways and signalling mechanisms in Gram-negative, rather than oligopeptides-pheromone mediated quorum signalling molecule in Gram-positive bacteria using two component signal transduction system [4, 5]. In Gram-negative species, AHLs synthesised under LuxI/R circuitry system present in the cytoplasm as autoinducer receptor protein LuxR and autoinducer synthase LuxI. In cytoplasm, AHLs synthesised by LuxI catalyses amide bond formation between acylated acyl carrier protein (ACP) and S-adenosylmethionine (SAM). AHLs diffuse bacterial cell envelope freely and reached threshold level favours binding with LuxR-type receptor protein in cytosol and regulate gene expression [5]. In *Vibrio harveyi*, detection of AHL accumulation with LuxN, membrane bound sensor kinases followed by series of signalling cascade activation [5]. There are four main types of autoinducer are AI-1, AI-2, AI-3, AIP (autoinducing peptide). AI-1, AI-3, AIP used for intraspecies signalling and AI-2 for cross-species, interspecies signalling [5]. AI-2 is absent in many species of bacteria and product of LuxS enzymes [5]. LuxS is metalloenzyme synthase convert substrate S-adenosylhomocysteine to 4,5-dihydroxy-2,3-pentanedione. This 4,5-dihydroxy-2,3-pentanedione, an unstable compounds undergoes immediate rearrangement in the solution state into multiple interconvertible cyclic furanone called as AI-2 [5]. In *V. harveyi*, LuxLM synthesis AI-1 and form complex with LuxN, worked as hybrid system with AI-2 for luciferase production. In *V. harveyi*, BAI-2 (furanosyl borate diester) a different form of AI-2 was detected under LuxP/Q cascade. AI-2 binds with LuxP (autoinducer-specific binding protein) in the periplasm initiates signalling cascade by phosphorylation of LuxQ (sensor kinase) in the cytoplasm. Then phosphorylates LuxU (integrator protein) at histidine residue and bind with LuxO (regulator protein). Then phosphorylates LuxO transcribed *luxCDABE* operon for

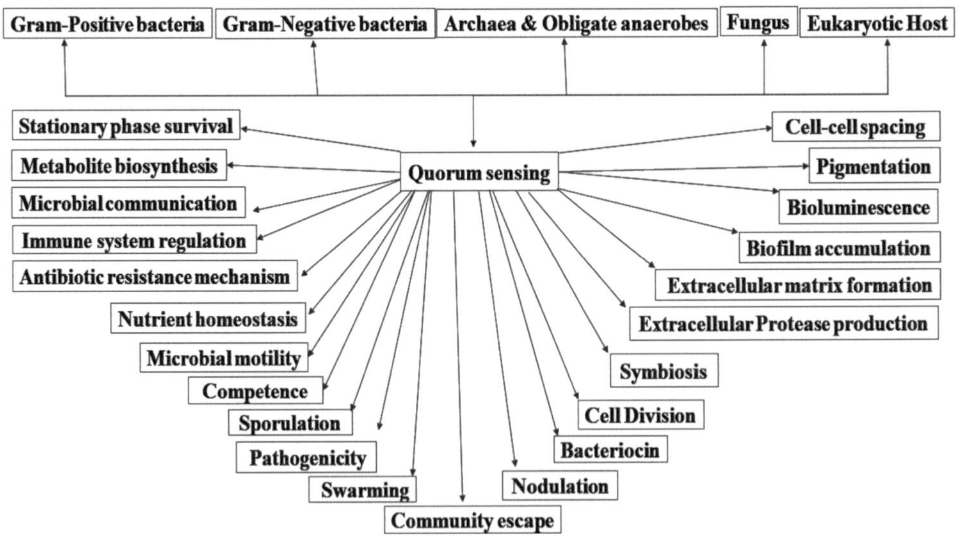

Figure 1.
Quorum sensing circuit involved in host-physiology interaction and host-microbiome interaction.

luciferase production and luminescence [5]. *ComABCDE* operon for quorum-sensing system in *Streptococcus mutans, S. gordonii, S. pneumoniae,* in which ComC encode for autoinducer peptides. ComA and ComB export ComC to the extracellular space. Then ComD, membrane-bound receptor kinase get phosphorylated through autoinducer peptide. ComD activates ComE, a regulator for genes transcription for biofilm formation and competence [6].

Biotic and abiotic factors which interfere the degradation of microbial cell signalling called quorum quenching [7]. QS-blocking approaches are meant to block bacterial virulence factor synthesis, instead of bactericidal activity and not like antibiotic medication, with perceptive to least selective pressures on AMR mutant development [8]. Four type of QS inhibition mechanism as follows (i) conventional antibiotics action on QS molecule, (ii) prevention of QS signal detection, (iii) abortion of QS signal biosynthesis and (iv) QS signals inactivation and degradation [3]. QS inhibitors act on broad range of Gram-positive and Gram-negative bacteria causes growth inhibition have been thoroughly studied for many years. Various studies have been discovered QS inhibitors of small molecule with high potent ability to defence against microbial growth phase were validated in animal model and other in-vivo & in-vitro experiments [1, 9]. Scenario of QS inhibitors meant for target screening of QS signal molecule and signalling pathways, such as bacterial biosensors. Screening of QS inhibitors in Gram negative bacteria, for example bioengineered strains of *Escherichia coli, Agrobacterium tumefaciens*, *Pseudomonas* spp., *Chromobacterium violaceum* [5]. Signalling molecules involved in biofilm production controlled by QS system, small RNA (sRNA), two-component systems (TCS), cyclic diguanylate (c-di-GMP) [10]. Core of QS system called TCS (Two-component system) involved in drug resistance, pathogenicity, nutrient metabolism, host recognition, virulence factor expression [11]. Regulatory protein (RR) present in the cytosol and histidine protein kinase (HPK) present in the inner cell membrane are two components of TCS. Signalling mechanism triggers HPK to phosphorylate RR at conserved aspartic acid residue by addition of phosphate group. RR bind to DNA promoter region and upregulated gene expression [11].

2. Multicellular interaction of horizontal and vertical mediated quorum-sensing in microbes

Bacteria communication to each other for the reasons of conjugation, sporulation, competence, symbiosis, motility, antibiotic production, virulence, biofilm formation. In the late 1960s, quorum sensing based research started on N-acyl-homoserine lactones (AHL). In bacterial growth media, inhibitors of luminescence was completely wiped out with *Vibrio fischeri,* a marine bioiluminescence bacteria clusters in high bacterial density [12]. Luminescence was reported in conditioned growth media by removal of inhibitors due to the increased accumulation of autoinducer [13]. First autoinducer was *N*-(3-oxohexanoyl)-homoserine lactone (3-oxo-C6-HSL) isolated from *V. fischeri* enable to sense their surrounding and cell density [14]. Carbapenem antibiotics synthesis was aborted in one of the *Erwinia carotovora* mutants and defective in antibiotic biosynthetic pathway. This paradigm shift was revealed at gene level comparative from other class mutants. Novel methodology of quorum sensing system was identified from cross-fed with new mutants, which triggered first mutant by sending signalling molecule stimulates the antibiotic production and reporter gene was expressed light emission of bioluminescent bacteria [15, 16].

Spontaneous gene mutations in horizontal gene transfer (HGT) can developed antibiotic resistance to bacteria [17]. Scenario of spatial biology differential complex communication signals in microbial quorum sensing system within spatial vicinity [18]. The long-range communication of signal molecule diffuse within bacterial local community using conjugative transfer for regulate traits. However in short range communication (least micron distance) stimulates horizontal gene transfer within close vicinity of extracellular fraction such as mobile genetic elements, conjugative transposon to the susceptible host cells [18]. Signal decay length scale also known as exponential decline in quorum sensing signal concentration, in which bacterial cluster of signal producers signals reduces to one order of magnitude within spatial and temporal distance [18]. Detection of quorum sensing signals degradation or attenuation into two types absorbing design and non-absorbing design [18]. Absorbing design called as irreversibly uptake of quorum sensing signal molecule immediately without sensing and preventing for spatial propagation, for example Gram-positive bacteria RNPP superfamily of peptides (PlcR, PrgX, NprR, Rap) for signal sequestering [18]. Non-absorbing design called as continuously quorum sensing molecule propagate after sensed from membrane-bound receptors and intracellularly and without action of signal sequestering, for example Gram-positive bacteria and *Vibrio* species [18]. Several non-absorbing systems in recent years has been revealed at both natural and synthetic means of quorum-sensing system, but still communication range and pathways remains unanswered to compared with signal-absorbing design [18].

Saliva mucins MUC5B and MUC7 from oral cavity, mucus barriers protect teeth and soft tissue of mouth to abolish infection of *S. mutans* from binding and agglutinin activity [19]. MUC5B involved in downregulation of *sigX*-inducing peptide and competence stimulating peptide with perceptive to quorum-sensing mediated gene transfer. MUC5B complex with *O*-linked glycans forms mucin *O*-glycans helps in preventing bacterial gene transfer mechanism and antimicrobial resistance acquisition through QS [19]. C4-HSL is one of the key virulence factor of QS involved in the regulation of haemolysin, rhamnolipid [20]. Various AHL-based autoinducers of QS were reported in distinct bacteria such as 3-oxo-C8-HSL detected in *Agrobacterium tumefaciens*; C8-HSL in *Burkholderia cepacia*; C6-HSL in *C. violaceum*; 3-hydroxy-7-*cis*-C14-HSL in *Rhizobium leguminosarum*; 7-*cis*-C14-HSL in *Rhodobacter sphaeroides*; 3-oxo-C10-HSL in *Vibrio anguillarum*; 3-hydroxy-C4-HSL in *Xenorhabdus nematophilus*; *N*-(3-oxohexanoyl)-homoserine lactone (HSL) & 3-OH-C_{10}-HSL in *Lysobacter brunescens* [21–28]. In *Pseudomonas aeruginosa*, discovered a novel AHL virulence factor called 3-oxo-C12-HSL involved in elastase production and regulation [29].

3. Immune system regulation on quorum-sensing

Quorum sensing enhances bacterial biofilm formation to increase tolerance against animal host immune system and antibiotics [2]. The molecular mechanism of multi-factorial microbial tolerance with perceptive of gene regulation, cell-population density fluctuation, antibiotics resistance gene expression, heterogeneous metabolic activity, restricted penetration [2]. *P. aeruginosa* autoinducer called *N*-(3-oxo-dodecanoyl) homoserine lactone expressed under LasI-LasR circuitry, which promotes lymphocytes cell death. *N*-(3-oxo-dodecanoyl) penetrates host cell membrane and lipid domains dissolution of binding tumour necrosis factor receptor 1 and blockage of caspase 3-caspase 8-mediated apoptosis [30]. Non-enzymatic method of QS signals sequestration using monoclonal antibody AP4-24 H11 degrade the (AIP)-4 produced from *Staphylococcus aureus* RN4850 [31]. Phage therapy resurgence were used to treat multidrug resistance bacterial infection, however quorum sensing increases CRISPR-*cas* immune system and virulence genes against phage infection [32]. Synthetic quorum sensing inhibitors speculate a new finding to prevent CRISPR immunity evolution during phage therapy (DMS3vir) in the population dynamics of *P. aeruginosa*.

In case of chemical inhibition downregulated Type IV pilus reduces phage adsorption leads to favours CRISPR immunity evolution and slow lysis of bacteria in the culture medium [32]. In mouse acute lung infection model, *P. aeruginosa* lasR mutant cause death, bacteremia and pneumonia [33]. lasR mutant responsible for interleukin-8 (IL-8) production in epithelial cells mouse infection model with perceptive to increased cells adherence of *P. aeruginosa* [33]. 2-Alkyl-4-quinolones (AQs) family of QS known as 2-heptyl-3-hydroxy-4(1H)-quinolone (Pseudomonas quinolone signal [PQS]) synthesis from *pqsABCDE* operon in *P. aeruginosa* [34]. PQS involved in host immune modulatory response (interleukin-12, dendritic cells, T cell proliferation), cytotoxicity, iron acquisition, biogenesis of outer-membrane vesicle. Signalling mechanism of PQS regulates LecA lectin, pyocyanin, and elastase production [34]. PQS downregulate interleukin-12 production in *E. coli*. Acute urinary tract *P. aeruginosa* infection mouse model validated with *pqsA* and *pqsH* mutants has revealed that decreased pathological markers, lesser tissue damage, bacterial count reduction [34].

4. Bioactive secondary metabolites controls the quorum-sensing signals

Various bioactive metabolites derived from microbes of endophytic origin with help of solvents affinity, polarity, non-polar groups and semi-polar groups. Solvents are methanol, ethyl acetate and chloroform deployed for metabolite extraction [5]. Extraction techniques of metabolites are ultrasound-assisted extraction, microwave-assisted extraction, supercritical fluid extraction, accelerated (or pressurised) solvent extraction [5]. Antimicrobial metabolites includes aliphatic compounds, proteins, peptides, phenols, flavonoids, terpenoids, steroids, quinones, alkaloids [35]. Bioactive secondary metabolites synthesis from endophytic microorganisms of plant origin includes azadirachtin, camptothecin, hypericin, podophyllotoxin, paclitaxel, deoxypodophyllotoxin for quorum quenching, antibacterial, antifungal, anticancer [36].

Ajoene (4,5,9-trithiadodeca-1,6,11-triene-9-oxide), sulphur-rich therapeutic secondary metabolite extracted from garlic act on *P. aeruginosa* to controls virulence factor and quorum sensing [4]. Ajoene structure used for broad-spectrum quorum sensing inhibitor act on Gram-positive and Gram-negative bacteria. AHL degrading enzymes are two major types called acylases and lactonases ability to produce homoserine lactone and acyl homoserines through enzymatic action of cleaving AHL amide bond and HSL ring of AHL can be results of QS signal degradation [37, 38]. Human lactonases called paraoxanase, family of enzymes which act on cleaving low density lipoproteins and organophosphate involved in the host-defence modulation [39]. In 1990s started treatment for QS mediated anti-microbial resistance infection reported that *Delisea pulchra,* a macro-algae synthesis brominated furanone. This furanone act as QS blocking agent in wide range of bacterial species to competitive against AHL-controlled phenotype [40]. Halogenated furanones were experimented in mice infection model to understand QS-blocking mediated bacteriostatic action against *P. aeruginosa* in the lungs [41]. Fragin compund, a diazeniumdiolate derivative show potential against anti-tumour activity and anti-microbial activity by regulating AHL dependent QS systems [42]. Erythromycin, ciprofloxacin, ceftazidime, azithromycin treatment target quorum sensing and signal molecule. *P. aeruginosa* virulence factor such as Phospholipase C, DNase, elastase, leucocidin, proteases, exotoxin A with perceptive to QS showing reduced expression and growth after antibiotics treatment [5].

In *P. aeruginosa*, AHL production level decreases after erythromycin treatment [5]. Zosteric acid, a phenolic derivatives of sub-lethal dose act against *Candida albicans.* Ursolic acid, derived from *Diospyros dendo* plant suppress biofilm formation of *V. harveyi, P. aeruginosa* and *E. coli* [5]. Endophytic bacteria, fungi, algae, actinomycetes, oomycetes are ubiquitous in nature in higher plants ability to synthesis bioactive metabolites to control biofilm formation [5]. QQs activity has been investigated in endophytic bacteria from *Cannabis sativa* L. plant such as *Brevibacillus borstelensis* strain B8, *Bacillus* sp. strain B3, *Bacillus* sp. strain B11, *Bacillus megaterium* strain B4 [36]. QS signals was distorted in *C. violaceum* (DSM 30191), Gram-negative bacteria by *Bacillus* sp. of endophytic origin [5]. *Bacillus amyloliquefaciens* bacteria exhibits aiiA gene encodes lactonase AiiA protein, which act as quorum quenching agent against *Botryosphaeria dothidea* fungus for canker disease [5]. *Microbacterium testaceum* BAC1065 and BAC1093 strains of endophytic bacterial origin belongs to *Phaseolus vulgaris* plant was determined for quorum quenching activity and fluctuation on AHL compound (bioluminescence and violacein) concentration in *E. coli* pSB403 and *C. violaceum* CV026 was detected [43]. Curcumin and gingerol act as QS inhibitors against LasR, PhzR, RhlR dependent pathways with perceptive for EPS production, biofilm formation, pyocyanin in *P. aeruginosa* [44].

5. Small RNAs coupled with quorum sensing system

Small regulatory RNAs (sRNA) employs for quorum sensing inhibition of *P. aeruginosa* and *S. aureus* often seen in polymicrobial chronic infection. Ajoene compound were lowered the expression of hemolysins and proteases virulence factors mediated through sRNAs [4]. In *P. aeruginosa, RsmZ* and *RsmY* are two sRNAs act on RsmA, global regulator protein of quorum sensing responsible for motility, polysaccharides and key virulence factor [45, 46]. In *Staphylococcus aureus*, RNAIII sRNAs necessitates the expression of protease, lipases, α-hemolysin usually peak in the onset of stationary phase cell density [47]. Ajoene repressed the expression of RsmZ, RsmY and RNAIII by indirectly act on transcript level of various regulator or either directly act on target sRNAs-mRNA interaction [48–50]. *RyhB*, a small non-coding RNA involved in iron-dependent gene modulation and quorum-sensing in

Vibrio vulnificus achieved by *LuxS* mRNA transcription and biosynthesis of autoinducer-2 (AI-2). Master regulator of QS as SmcR and Fur-iron complex bind to upstream region of *RyhB* and supress gene expression [51].

E1 Tor belongs to *Vibrio cholera* species controls small regulatory RNAs (sRNAs) of 21 nucleotides length are Qrr1, Qrr2, Qrr3, Qrr4 involved in QS [52]. Transcribed sRNAs complex with HapR mRNA (transcriptional regulator) complexed with Hfq (RNA chaperone) and hamper translation initiation. DNA uptake, protease synthesis, biofilm production and other virulence factor regulated by HapR transcriptional regulator. HapR-GFP translational fusion was constructed in *E. coli* and repressed by *Orr* sRNAs expression [52]. Early growth phase of bacteria and cell count was impaired based on CRISPRi mediated gene silencing at transcription level may leads to higher restriction of endogenous metabolic pathways [53]. Eventually repressive sRNA regulation systems of genetically engineered plasmid, not like CRISPRi mediated gene silencing. Its only target QS multiple genes and not disturbing endogenous metabolic pathways responsive genes and cell growth [53]. Combinatorial dynamic repression strategy elucidates P_{lux} promoter control of small RNA transcription complex with overexpression Hfq chaperone and $LuxR_{I58N}$ transcription factor in plasmid involved in quorum-sensing system with increased 3-oxohexanoyl-homoserine lactone (AHL) concentrations [53]. The optimising metabolic networks as well simultaneously control of multiple genes using sRNA from the massive genome in a cell-density-dependent manner without affecting cellular conditions [53].

6. Efflux pump-inhibition and anti-microbial activity with perceptive of quorum sensing system

Antibiotic resistance determinants encoded in core MDR bacterial genomes and associated mobile genetic elements (plasmids) constitutes mainly of efflux pumps category. Major function of efflux pumps extrudes cell signalling molecules, water metabolites, toxins, dyes, detergent and antibiotics. Efflux pumps belongs to five superfamilies categories are complex with MDR strains (i) RND (resistance-nodulation-division), ABC (ATP-binding cassette), MFS (major facilitator superfamily), SMR (small multidrug resistance) and MATE (multidrug and toxin extrusion). Proton/sodium motive force and ATP hydrolysis (ABC superfamily) are energy driven force utilised by efflux pumps mediated function within cells [54]. AcrD of *Salmonella enterica*, AdeFGH of *Acinetobacter baumannii*, MexAB-OprM of *P. aeruginosa*, AcrAB-TolC of *E. coli,* are example of efflux pumps involved in biofilm production [54]. Compound of efflux pump inhibitors (EPI's) and efflux pump gene mutagenesis, which involved in efflux pumps modulation and inhibition strategy.

Combinatorial use of antibiotics with anti-QS strategies for development of QS inhibitors for medical treatment effective against drug efflux pump and stimulate antimicrobial resistance [5]. In mice infection model, combination of tobramycin, ajoene, iberin derivatives of horseradish extracts, furanone C-30 was evaluated in *P. aeruginosa* QS inhibition [5]. Synergistic QS inhibitors activity in animal infectious model treated with combination of tobramycin, cinnamaldehyde, hamamelitannin, baicalin hydrate prevents bacterial growth of *P. aeruginosa, S. aureus*, *B. cepacia* and viable cell count in 2-log reduction [5]. In *S. aureus*, bactericidal activity reported from combinatorial treatments with vancomycin, hamamelitannin and AIPs analog [5]. Infant mouse infection model developed for cholera treatment from engineered probiotic *E. coli Nissle* strain were heterologously expressed CAI-1 synthase gene, *cqsA* by targeting *V. cholera* QS signal [55, 56]. Protected skin lesion from treatment of *S. aureus* lethal doses by disrupting agr quorum sensing system in mice model administered with antiAI-4 MAb AP4-24H11 [31]. In *P. aeruginosa*, MexCD-OprJ for multidrug efflux pump overexpression were downregulated QS response through extruction of kynurenine (an alkyl-quinolone signal precursor) and 4-hydroxy-2-heptylquinoline (*Pseudomonas* quinolone precursor) [57]. *P. aeruginosa* epidemiological studies reported that CCCP, a proton motive force (PMF) inhibitor decreases biofilm production. Synergistic function of biofilm inhibition reported on wild-type strains of uropathogenic *E. coli* strain 83,972, *Klebsiella pneumoniae* strain i222-86 and *E. coli* strain F18 with corresponding EPIs (Efflux pumps inhibitors) are PABN, thioridazine and 1-(1-napthylmethyl) piperazine (NMP) [54]. 2,2-Dipyridyl, acetohydroxamic acid are iron chelators act against *P. aeruginosa* biofilm production and particularly 2.5-fold downregulation in the biofilm biomass were treated with EDTA [58]. Synergistic action was reported in *Enterococcus faecalis* and *S. aureus* anti-biofilm activity from a potent EPIs called 4′,5′-*O*-dicaffeoylquinic acid were extracted from the *Artemisia absinthium* plant [58].

7. Quorum-sensing control, inhibition and quenching quorum sensing system

QS inhibitors concept was derived from attenuation of human and plant pathogen virulence in early discovery of QS mutant history for therapeutic development [59]. Quorum-sensing signal can be degrade by the chemical action of enzymes and pH secreted from the microbial surrounding [7]. Quorum quenching (QQ) activities against AHLs was evaluated in *P. segetis* strain P6, a phytopathogenic bacteria act as potential biocontrol agent [60]. Acylase was identified as potent QQ enzyme detected from evolutionary clade analysis and HPLC-MRM data [60]. Acylase enzyme act on broad range of AHLs belongs to *Pectobacterium carotovorum, P. atrosepticum, Dickeya solani* known for soft rot symptoms in carrot and potato [60]. AHL degradation activity was elucidated from *Klebsiella* (Se14), *Burkholderia* (GG4), *Acinetobacter* (GG2) genera of endophytic origin derived from *Pterocarpus santalinus* plant [5]. Quorum quenching activity of ginger plant rhizospheric extracted *Acinetobacter* and *Burkholderia* produce lactonase breaks 3-oxoAHLs autoinducer, lactone ring to 3-hydroxy compounds [61]. Principle of QS signal detection inhibition by downstream changes in signal transduction pathways by competitively blocking signal-receptor complexes [3]. Structurally modified AHL analogs, which is synthetic and non-natural exhibits a broad range of function in term of synergistic agonism, pure antagonism, pure agonism and even no activity [3]. From crystal structure of LasR, a transcriptional activator of *P. aeruginosa* involved in QS.

Developed a structure-based derivatization of antagonist probes were designed and termed as Itc-11 and Itc-12 (isothiocyanate), which covalent binds to LasR at ligand binding pocket of cysteine residue for inhibition of *P. aeruginosa* QS activity [62]. In *E. coli*, LsrK an autoinducer-2 kinase can phosphorylate AI-2 which inhibits QS activity of both *Salmonella typhimurium* (Interspecies) or *E. coli* (Intraspecies) [3]. A fungal metabolite called ambuic acid, which inhibits gelatinase production via. Cyclic peptide biosynthesis of QS in Gram-positive bacteria and *Enterococcus faecalis* [63]. AHL production suppression was detected in synthase enzyme inactivation, obstructing the acyl-ACP generation and hindering SAM biosynthesis [3]. FabI (NADH-dependent enoyl-ACP reductase), a member of short-chain alcohol dehydrogenase family involved in acyl-ACP biosynthesis with perceptive to AHL production. Catalysing of Acyl-ACP biosynthesis in the final step process was inhibited by triclosan and diazobroines, found to be inhibitor of N-butanoyl-l-homoserine lactone (C4-HSL) [3].

Two types of MTAN inhibitor are immucillin A (ImmA) and DADMe-ImmA, which mimic early and late dissociative transition state [64]. In *P. aeruginosa,* methyl anthranilate (anthranilate analogs) inhibited PQS production [65]. Quorum quenching compound were screened from crude extracts (20 mg/mL) belongs to 8 phyllosphere strains of bacteria (mainly JB 17B, JB 3B, JB 20B, JB 17B isolates) act against biofilm formation in *V. harveyi, C. violaceum*, *Streptococcus agalactiae*, *Aeromonas hydrophila* of fish pathogen origin [66]. In rat models, Dyer Ex Eichler extract (DSE) from *Dioon spinulosum* plant, inhibits biofilm formation (mainly exopolysaccharide production) against *P. aeruginosa* clinical isolates. DES inhibiting activity on *P. aeruginosa* biofilm development has performed in scanning electron microscopes [67]. *ndv*B gene responsible for biofilm and *las*I, *las*R, *rhl*I, *rhl*R gene responsible for quorum sensing were downregulated in the relative gene expression level and IC_{50} value shows cytotoxic activity 4.36 ± 0.52 μg/ml level against human shin fibroblast cell lines treated with DSE [67]. DSE extracts was reported in *C. violaceum* (ATCC 12,472) with it reduced violacein production [67] In *P. aeruginosa,* low levels of intracellular second messenger called c-di-GMP, a small nucleotide which causes reduction of biofilm formation. Moreover, higher intracellular concentration of c-di-GMP leads to EPS production and cell adhesion factors for bacterial adherence to biomaterial surface and biofilm formation [10].

8. Novel methodology, high-throughput screening and discovery of quorum-sensing inhibitors

Extraction and purification of desired metabolite of bacterial endophytic origin for screening quorum-sensing inhibitors. Choice of chromatographic techniques, semi-preparative and preparative HPLC such as low-pressure liquid chromatography (LPLC), medium-pressure liquid chromatography (MPLC), flash chromatography (FC), vacuum liquid chromatography (VLC) [5]. Identification of extracted metabolites using various spectroscopic methods such as Fourier transform infrared spectroscopy (FTIR), nuclear magnetic resonance (NMR), mass spectrometry (MS), near infrared spectroscopy (NIR). Chemical identification techniques at beginning stage using liquid chromatography-nuclear magnetic resonance (LC-NMR), liquid chromatography-mass spectrometry (LC-MS), gas chromatography-mass spectrometry (GC-MS), liquid chromatography-photodiode array (LC-PDA) [5]. Matrix-assisted laser desorption ionisation imaging were coupled with high-resolution mass spectrometry (MALDI-imaging-HRMS) for QS molecules quantification in *C. violaceum.* High-performance liquid chromatography were coupled with high-resolution mass spectrometry (HPLC-ESI-HRMS) has been used for QS compound quantitative analysis [36].

Methodology of screening QSIs as follow (i) QSIs identification, (ii) in-vitro characterisation of QSI compounds, (iii) QSIs compound validation in the animal model. *Streptococcus* ATP-binding cassette transporter called ComA, involved in QS signal transduction harbours peptidase domain (PEP) belongs to cysteine protease family, whereas other transmembrane and C-terminal nucleotide-binding domain are not involved. Inhibition of PEP activity detected from high-throughput screening of 164,514 fluorescence-labelled compound, from which six compound were identified as potent inhibitors belongs to quinuclidine derivatives and others. Enzymes kinetics revealed potent inhibitors act on allosteric binding site with hydrophobic core undergo PEP structural modification. Action of inhibitors form competence and attenuation against *S. mutans* growth and biofilm formation [6]. Juvenile hormone (JH) specifically synthesised in insects at larval stage, hence JH signalling mechanism is a major target of insecticides for development of JH signalling inhibitors (JHSIs) in agricultural field [68]. JH signalling activators (JHSAs) are fenoxycarb and pyriproxyfen, which impairs the metamorphosis such as aphids insects; whereas JHSIs provoke precocious metamorphosis and reduced feeding to consecutive generation [68].

High-throughput screening (HTS) system has developed for large-scale screening of novel JHSIs from *Bombyx mori* cell line (BmN_JF & AR cells) for targeting JH signalling pathway. JH ligand bind to methoprene-tolerant protein (Met) and form complex with steroid receptor coactivator protein (SRC), belongs to basic helix-loop-helix Per-ARNT-Sim (bHLH-PAS) transcription factors. JH/Met/SRC heterodimer complex targets Krüppel homologue 1 gene (Kr-h1), C_2H_2 zinc-finger type transcription factor responsible for JH-ligand receptor activation. Four-step HTS system have been developed for screening chemical libraries and 69 candidates compounds identified for JHSIs targeted Krüppel homologue 1 gene (Kr-h1) [68]. The Kr-h1 prevent *ecdysone-induced protein 93F* (*E93*) and *broad-complex* (*BR-C*) gene expression responsible for blocking precocious larva adults development [68]. From metagenomics analysis revealed that 3-oxoC12-homoserine lactone (3OC12HSL), an AHL is degraded by NADH-dependent oxidoreductase enzyme (BpiB09) [69]. Acyl-homoserine lactones production was disturbed with intensity of blue light modulates QS activity in *Acinetobacter baumannii*. In blue light condition, BlsA (photoreceptor) does not bind with AbaR (transcriptional regulator) and reducing *abaI* (AHL synthase) expression, but promotes the *aidA* lactonase expression mediated quorum quenching activity [70]. Probiotic *Lactobacillus brevis* strain 3M004 was investigated for QQs function using transcriptomics for screening AHL (OC12-HSL) inhibition [71]. *Lactobacillus brevis* strain inhibits pyocyanin production and biofilm formation in *P. aeruginosa* strain PA002. *L. brevis* cells/lysate treated with dosage level of 1 & 2 mg/mL with inhibition rate of biofilm formation (polysaccharides biosynthesis) were 16.92% & 33.0%. In *P. aeruginosa*, *LasA* and *LasB* gene was showing down-regulation responsible for QS system. *PhzAB* genes was down-regulated and inhibited the biosynthesis of pyocyanin to prevent irreversible action from chorismite to pyocyanin [71]. Advance development in computational biology to screen 9500 phytochemicals from 1700 medicinal plants of Indian origin that hits with QS-antagonist activity against *P. aeruginosa* registered in IMPPAT database (Indian Medicinal Plants, Phytochemistry and Therapeutics) [72]. The advanced computational principle of screening and validation of phytochemicals are (i) high throughput virtual screening (HVTS), (ii) ligand mapping by E-pharmacophore, (iii) extra precision (XP) docking, (iv) free energy calculation by MM-GBSA and (v) molecular dynamics (MD) simulations for computationally validated of top phytochemicals [72].

9. Conclusion for global challenges

Lack of suitable animal models on drug discovery with perceptive on mode of drug action, target delivery, level of host cytotoxicity, drug stability is a matter of great concern. Successful of human gut infection and other clinical illness are increasing rapidly in the western world, due to rare phenomenal development of QS control behaviours among diverse community of microbes at the natural ecosystem, remains a key challenge for the future. For controlling microbial biofilm formation through bioprospecting of quenching quorum sensing compounds. It's vital to understand complete chemical and kinetic mechanism of QQs would be major interface for immunological studies and drug targets. To understand various polymicrobial communities, signal calling distance, spatial conformation among microbes with perceptive of quorum quenching and therapy is still remains challenging. Scenario of understanding microbial ecological behaviour and pan-genome evolutionary sharing within species in a specified shape is important for quorum signalling. High-throughput screening of quorum quenching compounds to defence against pathogenesis of microbes and virulent anti-biofilm activity is still remain unclear. There are certain limitation in use of QQ metabolites coupled with large-scale commercial production in food packaging industry for effective against broad spectrum activity of Gram-positive and Gram-negative bacteria. Therefore, it is vital to understand endophytic origin of microbial metabolites for optimum production under In-vitro fermentation process. Biggest question of evolutionary biologist, how quorum sensing can be regulated in the natural selection within bacterial community under restrictive condition with perceptive to altruism of selfish local group and uncooperative individuals gained cost fitness. Bacterial QS activity with dynamics of adaptation to surrounding environment and complexity of microbial communities will be critical factor for understanding, how to interfere with clinical infection. Human microbiome studies continues to expand rapidly to future needs for fundamental development in QS system communication and sociality. Better to know about, how ability of eukaryotic quorum quenching enzymes involved in providing fitness cost, symbiosis, competition to host microbiome through modulation of defence function. This will be crucial way to discover signal molecule-dependent interaction between microbes-host cells. Current research face challenge with perceptive to the fundamental discovery of LuxI-type QS signal synthases and small molecule inhibitors against signal transduction of AHL signals degradation. Several researchers have face obstacles in designing structure-based derivatization of non-AHL pharmacophores against LuxR-type proteins and pathways. Computer-aided screens, system biology and high throughput omics platform can modifying inhibitors at structure level for broad range of target QS specificities and enhance inhibitory activity.

Acknowledgements

Authors are thankful to Sri Lakshmi Narayana Institute of Medical Sciences, Puducherry (affiliated to Bharath Institute of Higher Education and Research, Chennai) and PRIST University, Thanjavur, India.

Conflict of interest

The authors declare no conflict of interest.

Author details

Kothandapani Sundar[1*], Ramachandira Prabu[2] and Gopal Jayalakshmi[1]

1 Department of Microbiology, Sri Lakshmi Narayana Institute of Medical Sciences (Affiliated to Bharath Institute of Higher Education and Research (BIHER) Chennai), Puducherry, India

2 Department of Microbiology, PRIST University, Puducherry Campus, Puducherry, India

*Address all correspondence to: sundar.gk@gmail.com

References

[1] O'Neill J. Tackling Drug-Resistant Infections Globally: Final Report and Recommendations. The review on antimicrobial resistance. United Kingdom: Government of the United Kingdom; 2016. Available from: https://books.google.co.in/books/about/Tackling_drug_resistant_Infections_Globa.html?id=aa6IAQAACAAJ&redir_esc=y

[2] Costerton JW, Stewart PS, Greenberg EP. Bacterial biofilms: A common cause of persistent infections. Science. 1999;**284**:1318-1322. DOI: 10.1126/science.284.5418.1318

[3] LaSarre B, Federle MJ. Exploiting quorum sensing to confuse bacterial pathogens. Microbiology and Molecular Biology Reviews. 2013;**77**:73-111. DOI: 10.1128/mmbr.00046-12

[4] Jakobsen TH, Warming AN, Vejborg RM, Moscoso JA, Stegger M, Lorenzen F, et al. A broad range quorum sensing inhibitor working through sRNA inhibition. Scientific Reports. 2017;**7**:9857. DOI: 10.1038/s41598-017-09886-8

[5] Mookherjee A, Singh S, Maiti MK. Quorum sensing inhibitors: Can endophytes be prospective sources? Archives of Microbiology. 2018;**200**(2):355-369. DOI: 10.1007/s00203-017-1437-3

[6] Ishii S, Fukui K, Yokoshima S, Kumagai K, Beniyama Y, Kodama T, et al. High-throughput screening of small molecule inhibitors of the streptococcus quorum-sensing signal pathway. Scientific Reports. 2017;**7**(1):4029. DOI: 10.1038/s41598-017-03567-2

[7] Whiteley M, Diggle SP, Greenberg EP. Progress in and promise of bacterial quorum sensing research. Nature. 2017;**551**:313-320. DOI: 10.1038/nature24624

[8] Allen RC, Popat R, Diggle SP, Brown SP. Targeting virulence: Can we make evolution-proof drugs? Nature Reviews. Microbiology. 2014;**12**:300-308. DOI: 10.1038/nrmicro3232

[9] Christensen QH, Grove TL, Booker SJ, Greenberg EP. A high throughput screen for quorum-sensing inhibitors that target acyl-homoserine lactone synthases. Proceedings of the National Academy of Sciences of the United States of America. 2013;**110**:13815-13820. DOI: 10.1073/pnas.1313098110

[10] Yi L, Li J, Liu B, Wang Y. Advances in research on signal molecules regulating biofilms. World Journal of Microbiology and Biotechnology. 2019;**35**(8):130. DOI: 10.1007/s11274-019-2706-x

[11] Guo XP, Sun YC. New insights into the non-orthodox two component Rcs phosphorelay system. Frontiers in Microbiology. 2017;**8**:2014. DOI: 10.3389/fmicb.2017.02014

[12] Kempner ES, Hanson FE. Aspects of light production by *Photobacterium fischeri*. Journal of Bacteriology. 1968;**95**:975-979. DOI: 10.1128/jb.95.3.975-979.1968

[13] Nealson KH, Platt T, Hastings JW. Cellular control of the synthesis and activity of the bacterial luminescence system. Journal of Bacteriology. 1970;**104**:313-322. DOI: 10.1128/jb.104.1.313-322.1970

[14] Engebrecht J, Nealson K, Silverman M. Bacterial bioluminescence:

Isolation and genetic analysis of functions from *Vibrio fischeri*. Cell. 1983;**32**:773-781. DOI: 10.1016/0092-8674(83)90063-6

[15] Bainton NJ, Bycroft BW, Chhabra SR, Stead P, Gledhill L, Hill PJ, et al. A general role for the lux autoinducer in bacterial cell signalling: Control of antibiotic biosynthesis in Erwinia. Gene. 1992;**116**:87-91. DOI: 10.1016/0378-1119(92)90633-z

[16] Bainton NJ, Stead P, Chhabra SR, Bycroft BW, Salmond GPC, Stewart GSAB, et al. N-(3-oxohexanoyl)-L-homoserine lactone regulates carbapenem antibiotic production in *Erwinia carotovora*. The Biochemical Journal. 1992;**288**:997-1004. DOI: 10.1042/bj2880997

[17] Blair J, Webber M, Baylay A. Molecular mechanisms of antibiotic resistance. Nature Reviews. Microbiology. 2015;**13**:42-51. DOI: 10.1038/nrmicro3380

[18] van Gestel J, Bareia T, Tenennbaum B. Short-range quorum sensing controls horizontal gene transfer at micron scale in bacterial communities. Nature Communications. 2021;**12**:2324. DOI: 10.1038/s41467-021-22649-4

[19] Werlang CA, Chen WG, Aoki K. Mucin O-glycans suppress quorum-sensing pathways and genetic transformation in *Streptococcus mutans*. Nature Microbiology. 2021;**6**:574-583. DOI: 10.1038/s41564-021-00876-1

[20] Winson MK, Camara M, Latifi A, Foglino M, Chhabra SR, Daykin M, et al. Multiple N-acyl-L-homoserine lactone signal molecules regulate production of virulence determinants and secondary metabolites in *Pseudomonas aeruginosa*. Proceedings of the National Academy of Sciences of the United States of America. 1995;**92**:9427-9431. DOI: 10.1073/pnas.92.20.9427

[21] Piper KR, Beck von Bodma S, Hwang I, Farrand SK. Hierarchical gene regulatory systems arising from fortuitous gene associations: Controlling quorum sensing by the opine regulon in Agrobacterium. Molecular Microbiology. 1999;**32**:1077-1089. DOI: 10.1046/j.1365-2958.1999.01422.x

[22] Lewenza S, Conway B, Greenberg EP, Sokol PA. Quorum sensing in *Burkholderia cepacia*: Identification of the LuxRI homologs CepRI. Journal of Bacteriology. 1999;**181**:748-756. DOI: 10.1128/JB.181.3.748-756.1999

[23] Chernin LS, Winson MK, Thompson JM, Haran S, Bycroft BW, Chet I, et al. Chitinolytic activity in *Chromobacterium violaceum*: Substrate analysis and regulation by quorum sensing. Journal of Bacteriology. 1998;**180**:4435-4441. DOI: 10.1128/JB.180.17.4435-4441.1998

[24] Thorne SH, Williams HD. Cell density-dependent starvation survival of *Rhizobium leguminosarum* bv. phaseoli: Identification of the role of an N-acyl homoserine lactone in adaptation to stationary-phase survival. Journal of Bacteriology. 1999;**181**:981-990. DOI: 10.1128/JB.181.3.981-990.1999

[25] Puskas A, Greenberg EP, Kaplan S, Schaefer AL. A quorum sensing system in the free-living photosynthetic bacteria *Rhodobacter spaeroides*. Journal of Bacteriology. 1997;**179**:7530-7537. DOI: 10.1128/jb.179.23.7530-7537.1997

[26] Milton DL, Hardman A, Camara M, Chhabra SR, Bycroft BW, Stewart GSAB, et al. Quorum sensing in *Vibrio anguillarum*: Characterization of the vanI/R locus and identification of the autoinducer

N-(3-oxodecanoyl)-L-homoserine lactone. Journal of Bacteriology. 1997;**179**:3004-3012. DOI: 10.1128/jb.179.9.3004-3012.1997

[27] Dunphy G, Miyamoto C, Meighen E. A homoserine lactone autoinducer regulates virulence of an insect-pathogenic bacterium, *Xenorhabdus nematophilus* (Enterobacteriaceae). Journal of Bacteriology. 1997;**179**:5288-5291. DOI: 10.1128/jb.179.17.5288-5291.1997

[28] Ling J, Zhou L, Wu G, Zhao Y, Jiang T, Liu F. The AHL quorum-sensing system negatively regulates growth and autolysis in *Lysobacter brunescens*. Frontiers in Microbiology. 2019;**10**:2748. DOI: 10.3389/fmicb.2019.02748

[29] Pearson JP, Gray KM, Passador L, Tucker KD, Eberhard A, Iglewski BH, et al. Structure of the autoinducer required for expression of *Pseudomonas aeruginosa* virulence genes. Proceedings of the National Academy of Sciences of the United States of America. 1994;**91**(1):197-201. DOI: 10.1073/pnas.91.1.197

[30] Song D, Meng J, Cheng J. *Pseudomonas aeruginosa* quorum-sensing metabolite induces host immune cell death through cell surface lipid domain dissolution. Nature Microbiology. 2019;**4**:97-111. DOI: 10.1038/s41564-018-0290-8

[31] Park J. Infection control by antibody disruption of bacterial quorum sensing signalling. Chemistry & Biology. 2007;**14**:1119-1127. DOI: 10.1016/j.chembiol.2007.08.013

[32] Broniewski JM, Chisnall AM, Westra ER. The effect of quorum sensing inhibitors on the evolution of CRISPR-based phage immunity in *Pseudomonas aeruginosa*. The ISME Journal. 2021;**15**(8):1-9. DOI: 10.1038/s41396-021-00946-6

[33] Azimi S, Klementiev AD, Whiteley M, Diggle SP. Bacterial quorum sensing during infection. Annual Review of Microbiology. 2020;**74**:201-219. DOI: 10.1146/annurev-micro-032020-093845

[34] Lin J, Cheng J, Wang Y, Shen X. The pseudomonas quinolone signal (PQS): Not just for quorum sensing anymore. Frontiers in Cellular and Infection Microbiology. 2018;**8**:230. DOI: 10.3389/fcimb.2018.00230

[35] Yu H, Zhang L, Li L, Zheng C, Guo L, Li W, et al. Recent developments and future prospects of antimicrobial metabolites produced by endophytes. Microbiological Research. 2010;**165**:437-449. DOI: 10.1016/j.micres.2009.11.009

[36] Kusari P, Kusari S, Lamshoft M, Sezgin S, Spiteller M, Kayser O. Quorum quenching is an antivirulence strategy employed by endophytic bacteria. Applied Microbiology and Biotechnology. 2014;**98**:7173-7183. DOI: 10.1007/s00253-014-5807-3

[37] Dong YH. Quenching quorum-sensing-dependent bacterial infection by an N-acyl homoserine lactonase. Nature. 2001;**411**:813-817. DOI: 10.1038/35081101

[38] Lin YH. Acyl-homoserine lactone acylase from Ralstonia strain XJ12B represents a novel and potent class of quorum-quenching enzymes. Molecular Microbiology. 2003;**47**:849-860. DOI: 10.1046/j.1365-2958.2003.03351.x

[39] Li XC, Wang C, Mulchandani A, Ge X. Engineering soluble human paraoxonase 2 for quorum quenching. ACS Chemical Biology. 2016;**11**:3122-3131. DOI: 10.1021/acschembio.6b00527

[40] Givskov M. Eukaryotic interference with homoserine lactone-mediated prokaryotic signalling. Journal of Bacteriology. 1996;**178**(22):6618-6622. DOI: 10.1128/jb.178.22.6618-6622.1996

[41] Hentzer M. Attenuation of *Pseudomonas aeruginosa* virulence by quorum sensing inhibitors. The EMBO Journal. 2003;**22**:3803-3815. DOI: 10.1093/emboj/cdg366

[42] Jenul C, Sieber S, Daeppen C. Biosynthesis of fragin is controlled by a novel quorum sensing signal. Nature Communications. 2018;**9**:1297. DOI: 10.1038/s41467-018-03690-2

[43] Lopes RB, Costa LE, Vanetti MC, de Araujo EF, de Queiroz MV. Endophytic bacteria isolated from common bean (*Phaseolus vulgaris*) exhibiting high variability showed antimicrobial activity and quorum sensing inhibition. Current Microbiology. 2015, 2015;**71**:509-516. DOI: 10.1007/s00284-015-0879-6

[44] Shukla A, Shukla G, Parmar P, Patel B, Goswami D, Saraf M. Exemplifying the next generation of antibiotic susceptibility intensifiers of phytochemicals by LasR-mediated quorum sensing inhibition. Scientific Reports. 2021;**11**:22421. DOI: 10.1038/s41598-021-01845-8

[45] Irie Y. *Pseudomonas aeruginosa* biofilm matrix polysaccharide Psl is regulated transcriptionally by RpoS and post-transcriptionally by RsmA. Molecular Microbiology. 2010;**78**:158-172. DOI: 10.1111/j.1365-2958.2010.07320.x

[46] Burrowes E, Baysse C, Adams C, O'Gara F. Influence of the regulatory protein RsmA on cellular functions in *Pseudomonas aeruginosa* PAO1, as revealed by transcriptome analysis. Microbiology. 2006;**152**:405-418. DOI: 10.1099/mic.0.28324-0

[47] Novick RP. Autoinduction and signal transduction in the regulation of staphylococcal virulence. Molecular Microbiology. 2003;**48**:1429-1449. DOI: 10.1046/j.1365-2958.2003.03526.x

[48] Huntzinger E. *Staphylococcus aureus* RNAIII and the endoribonuclease III coordinately regulate spa gene expression. The EMBO Journal. 2005;**24**:824-835. DOI: 10.1038/sj.emboj.7600572

[49] Morfeldt E, Taylor D, von Gabain A, Arvidson S. Activation of alpha-toxin translation in *Staphylococcus aureus* by the trans-encoded antisense RNA, RNAIII. The EMBO Journal. 1995;**14**(18):4569-4577. DOI: 10.1002/j.1460-2075.1995.tb00136.x

[50] Boisset S, Geissmann T, Huntzinger E, Fechter P, Bendridi N, Possedko M. *Staphylococcus aureus* RNAIII co-ordinately represses the synthesis of virulence factors and the transcription regulator Rot by an antisense mechanism. Genes & Development. 2007;**21**(11):1353-1366. DOI: 10.1101/gad.423507

[51] Lee K, Wen Y, Park N, Kim K. Quorum sensing and iron-dependent coordinated control of autoinducer-2 production via small RNA RyhB in *Vibrio vulnificus*. Scientific Reports. 2022;**12**:831. DOI: 10.1038/s41598-021-04757-9

[52] Bardill JP, Zhao X, Hammer BK. The *Vibrio cholerae* quorum sensing response is mediated by Hfq-dependent sRNA/mRNA base pairing interactions. Molecular Microbiology. 2011;**80**(5):1381-1394. DOI: 10.1111/j.1365-2958.2011.07655.x

[53] Bao S, Jiang H, Zhu L, Yao G, Han P, Zhang D, et al. A dynamic and multilocus metabolic regulation strategy using quorum-sensing-controlled

bacterial small RNA. Cell Reports. 2021;**36**(3):109413. DOI: 10.1016/j.celrep.2021.109413

[54] Alav I, Sutton JM, Rahman KM. Role of bacterial efflux pumps in biofilm formation. Journal of Antimicrobial Chemotherapy. 2018;**73**(8):2003-2020. DOI: 10.1093/jac/dky042

[55] Duan F, March JC. Interrupting *Vibrio cholerae* infection of human epithelial cells with engineered commensal bacterial signalling. Biotechnology and Bioengineering. 2010;**101**:128-134. DOI: 10.1002/bit.21897

[56] Duan F, March JC. Engineered bacterial communication prevents *Vibrio cholerae* virulence in an infant mouse model. Proceedings of the National Academy of Sciences of the United States of America. 2010;**107**:11260-11264. DOI: 10.1073/ pnas.1001294107

[57] Alcalde-Rico M, Olivares-Pacheco J, Alvarez-Ortega C, Cámara M, Martínez JL. Role of the multidrug resistance efflux pump MexCD-OprJ in the *Pseudomonas aeruginosa* quorum sensing response. Frontiers in Microbiology. 2018;**9**:2752. DOI: 10.3389/fmicb.2018.02752

[58] Fiamegos YC, Kastritis PL, Exarchou V, Han H, Vervoort J, Lewis K, et al. Antimicrobial and efflux pump inhibitory activity of caffeoylquinic acids from *Artemisia absinthium* against Gram-positive pathogenic bacteria. PLoS One. 2011;**6**(4):e18127. DOI: 10.1371/journal.pone.0018127

[59] Pearson JP, Feldman M, Iglewski BH, Prince A. *Pseudomonas aeruginosa* cell-to-cell signaling is required for virulence in a model of acute pulmonary infection. Infection and Immunity. 2000;**68**:4331-4334. DOI: 10.1128/IAI.68.7.4331-4334.2000

[60] Rodríguez M, Torres M, Blanco L, Béjar V, Sampedro I, Llamas I. Plant growth-promoting activity and quorum quenching-mediated biocontrol of bacterial phytopathogens by *Pseudomonas segetis* strain P6. Scientific Reports. 2020;**10**(1):4121. DOI: 10.1038/s41598-020-61084-1

[61] Chan K, Atkinson S, Mathee K, Sam C, Chhabra SR, Cámara M, et al. Characterization of N-acylhomoserine lactone-degrading bacteria associated with the *Zingiber officinale* (ginger) rhizosphere: Co-existence of quorum quenching and quorum sensing in Acinetobacter and Burkholderia. BMC Microbiology. 2011;**11**:51. DOI: 10.1186/1471-2180-11-51

[62] Amara N, Mashiach R, Amar D, Krief P, Spieser SA, Bottomley MJ, et al. Covalent inhibition of bacterial quorum sensing. Journal of the American Chemical Society. 2009;**131**:10610-10619. DOI: 10.1021/ja903292v

[63] Nakayama J, Uemura Y, Nishiguchi K, Yoshimura N, Igarashi Y, Sonomoto K. Ambuic acid inhibits the biosynthesis of cyclic peptide quormones in Gram-positive bacteria. Antimicrobial Agents and Chemotherapy. 2009;**53**:580-586. DOI: 10.1128/ aac.00995-08

[64] Lee JE, Singh V, Evans GB, Tyler PC, Furneaux RH, Cornell KA, et al. Structural rationale for the affinity of pico- and femtomolar transition state analogues of *Escherichia coli* 5′-methylthioadenosine/S-adenosylhomocysteine nucleosidase. The Journal of Biological Chemistry. 2005;**280**:18274-18282. DOI: 10.1074/ jbc.M414471200

[65] Calfee MW, Coleman JP, Pesci EC. Interference with Pseudomonas quinolone signal synthesis inhibits

virulence factor expression by *Pseudomonas aeruginosa*. Proceedings of the National Academy of Sciences of the United States of America. 2001;**98**:11633-11637. DOI: 10.1073/pnas.201328498

[66] Nathalia O, Waturangi DE. Extract from phyllosphere bacteria with antibiofilm and quorum quenching activity to control several fish pathogenic bacteria. BMC Research Notes. 2021;**14**:202. DOI: 10.1186/s13104-021-05612-w

[67] Elekhnawy E, Negm WA, El-Aasr M. Histological assessment, anti-quorum sensing, and anti-biofilm activities of *Dioon spinulosum* extract: In vitro and in vivo approach. Scientific Reports. 2022;**12**:180. DOI: 10.1038/s41598-021-03953-x

[68] Kayukawa T, Furuta K, Nagamine K. Identification of a juvenile-hormone signalling inhibitor via high-throughput screening of a chemical library. Scientific Reports. 2020;**10**:18413. DOI: 10.1038/s41598-020-75386-x

[69] Bijtenhoorn P, Schipper C, Hornung C, Quitschau M, Grond S, Weiland N, et al. BpiB05, a novel metagenome-derived hydrolase acting on N-acylhomoserine lactones. Journal of Biotechnology. 2011;**155**:86-94. DOI: 10.1016/j.jbiotec.2010.12.016

[70] Tuttobene MR, Müller GL, Blasco L, Arana N, Hourcade M, Diacovich L, et al. Blue light directly modulates the quorum network in the human pathogen *Acinetobacter baumannii*. Scientific Reports. 2021;**11**:13375. DOI: 10.1038/s41598-021-92845-1

[71] Liang Y, Pan Y, Li Q, Wu B, Hu M. RNA-seq-based transcriptomic analysis of AHL-induced biofilm and pyocyanin inhibition in *Pseudomonas aeruginosa* by *Lactobacillus brevis*. International Microbiology. 2022. DOI: 10.1007/s10123-021-00228-3

[72] Mohanraj K, Karthikeyan BS, Vivek-Ananth RP, Chand RPB, Aparna SR, Mangalapandi P, et al. IMPPAT: A curated database of Indian medicinal plants, phytochemistry and therapeutics. Scientific Reports. 2018;**8**:4329. DOI: 10.1038/s41598-018-22631-z

Chapter 13

Acinetobacter baumannii: Emergence of a Superbug, Past, Present, and Future

Yashar Jalali, Monika Jalali and Juraj Payer

Abstract

By exhibiting resistance to most known antibiotics or quickly acquiring resistance to antimicrobials it was once susceptible to, *Acinetobacter baumannii* has attracted increasing attention worldwide since the last decades of the previous century. The resistance abilities of the bacterium were soon shown to be so advanced that it was even able to resist antibiotics that had only just been discovered and used for first time. Utilizing complex mechanisms of resistance, combining different modalities, and achieving new resistant traits established *A. baumannii* as one of the most clinically important and challenging pathogens of the new century, being categorized by the World Health Organization as a critical priority bacterium for which new antibiotics are urgently needed. After even last-resort, broad-spectrum antibiotics were rendered useless, the fight against this superbug began to be led by the reintroduction of once abandoned antimicrobials, new combination therapies and novel modalities of treatment. In this chapter, we will look at the history and background of Acinetobacter species and then specifically focus on *A. baumannii*, explaining its clinical importance in detail, reviewing the most recent findings regarding its mechanisms of resistance, latest modalities of treatment and newest areas of research towards opening new frontiers in the management of infections caused by multi-resistant strains of this bacterium.

Keywords: *Acinetobacter baumannii*, antimicrobial resistance, carbapenem resistance, ESKAPE bacteria, health-associated infections

1. Introduction

The story of the Acinetobacter genus dates back to the early 20th century when in 1911, Dutch microbiologist Martinus Willem Beijerinck named a newly discovered organism isolated from soil using minimal media enriched with calcium acetate: *Micrococcus calco-aceticus* [1]. Since he did not describe the microorganism but only named it, his report was greatly overlooked [2]. Hence, in the meantime, the same microorganism was described under various names over several decades, including: *Diplococcus mucosus*, *Micrococcus calcoaceticus*, *Alcaligenes haemolysans*, *Mima polymorpha*, *Bacterium anitratum*, *Herellea vaginicola, Moraxella lwoffi*, *Achromobacter anitratus*, *Moraxellalwoffi* var. *glucidolytica*, *Achromobacter mucosus,* and *Neisseria winogradskyi* [1]. Due to this, the original strain has been lost and the name proposed by Beijernik was used by few authors [2].

In 1954 Brisou and Prévot coined the name Acinetobacter from the Greek "ακινετοσ" [akinetos], meaning non-motile, to separate non-motile from motile microorganisms within the genus Achromobacter [3]. Years later in 1968 Paul Baumann, through a comprehensive survey, suggested that all of the above-listed species belonged to a single genus and could not be further sub-classified into different species based on the phenotypical characteristics, for which the name Acinetobacter was proposed [4]. In 1971, based on the results of Baumann's publication subcommittee on nomenclature of Moraxella and allied bacteria, the genus Acinetobacter was officially acknowledged [5]. In 1974 the genus Acinetobacter was listed in Bergey's manual of systematic bacteriology, described as a single species named *Acinetobacter calcoaceticus* [1]. Despite that, in the "Approved List of Bacterial Names" of 1980, based on the observation that some Acinetobacter species were able to acidify glucose, the species were listed as *A. calcoaceticus* (which was further subdivided into the two biovars *A. calcoaceticus bv. anitratus* and *A. calcoaceticus bv. lwoffii*) and *Acinetobacter lwoffii* [6]. These nomenclatures, however, were never approved by taxonomists [1].

Acinetobacter are defined as gram negative (however, since Acinetobacter species are often difficult to de-stain through the process of Gram staining, they are often incorrectly identified as gram positive), strictly aerobic, catalase-positive, oxidase-negative, non-motile, non-fermenting, non-fastidious bacteria with a DNA G + C content of 39–47% [7]. Acinetobacter has been classified in the family *Moraxellaceae*, within the order *Pseudomonadales*, and the class Gammaproteobacteria since 1991 [8].

By 2015 only 33 species of Acinetobacter had been identified. Rapid technological advances in recent years have caused faster and more precise identification of novel strains. Continued reports of such novel strains have doubled the number of described species in the last 6 years. Within the 4 years from 2017 to 2020 alone, 22 new Acinetobacter species were identified [9]. By May 2021 67 Acinetobacter species had already been validly named, and 20 additional species were under tentative species designation. Most identified species, however, are non-pathogenic, environmental microorganisms [9].

A. baumannii is a ubiquitous opportunistic coccobacillus, widely distributed in the environment, having reservoirs in nearly all environmental niches [8]. Despite isolation of *A. baumannii* from soil, water, crude oil, sewage, solid surfaces, human skin, raw meat, milk, milk products, vegetables, livestock, fish, shrimp, plants, and so on, natural habitats of this bacterium are still poorly understood as it is almost exclusively isolated from close contact communities and hospital environments [8].

Smith et al., performed and published the first whole genome sequencing of *A. baumannii* (strain ATCC 17978) in 2007 [10]. However, identification of *A. baumannii* in routine diagnostic laboratories has been found to be quite complicated due to bacterium's phylogenetical closeness to several other Acinetobacter species, collectively called *A. calcoaceticus-A. baumannii* (Acb) complex [8].

Traditionally four species of Acinetobacter *A. calcoaceticus* (formerly known as genomic species 1), *A. baumannii*, *Acinetobacter pittii* (formerly known as genomic species 3), and *Acinetobacter nosocomialis* (formerly known as genomic species 13TU), were members of Acb complex [11]. In recent years, other genomic species have been proposed for inclusion, including: *Acinetobacter seifertii* (formerly known as *Acinetobacter* genomic species "close to 13TU"), *A. lactucae* (formerly known as *Acinetobacter* NB14 and synonym of *A. dijkshoorniae*), and the *Acinetobacter* genomic species "between 1 and 3" [12]. As mentioned earlier, all species of the complex are phenotypically indistinguishable and genetically are closely related to the extent that molecular methods are required for their identification [12]. Acb complex members (except for *A. calcoaceticus*) are most frequently associated with human pathogenicity and health association infections (HAI). While the pathogenicity of *A. calcoaceticus* is still unknown, other members of the complex are frequently identified as major human pathogens, with *A. baumannii* being the most prominent one, having the highest associated HAI, followed by *A. pittii* and *A. nosocomialis* [12].

Although until 1975 *A. baumannii* was still treated by antibiotic monotherapy with relatively good susceptibility, the rate of antibiotic resistance (ABR) of the bacterium has been rising ever since [13]. In the early years of the 21st century, the Infectious Disease Society of America highlighted a group of bacteria including *Enterococcus faecium, Staphylococcus aureus, Klebsiella pneumoniae, A. baumannii, Pseudomonas aeruginosa, and Enterobacter spp.* as "the ESKAPE pathogens" for their ability to "escape" antibiotics and resist antibiotic treatment, mutually representing a new paradigm in pathogenesis, resistance, and transmission [14]. Multidrug resistance (MDR) species of *A. baumannii* with resistance to almost all known antibiotics (including last-resort broad-spectrum antibiotics like carbapenems) are being isolated from hospital environments worldwide, posing a great burden on the health care system [13]. Resistance to carbapenem (a broad-spectrum β-lactam antibiotic) is considered to be a marker of extensive antimicrobial resistance since it involves a large range of co-resistance to other unrelated antibiotic classes [8]. In 2017, the World Health Organization (WHO) listed carbapenem-resistant (CR) *A. baumannii* as a critical priority bacterium for which new antibiotics are urgently needed [8]. The rate of clinical outbreaks of CR *A. baumannii* in Europe ranges between 1% to over 30%, being most intensive in eastern and southeastern Europe [8].

The range of *A. baumannii* clinical isolates among all gram-negative aerobes differs from 0.7% in North America to 4.6% in the Middle East [8]. Historically the Middle East is one of the regions most known for MDR outbreaks of *A. baumannii*, giving it the memorable "Iraqibacter" title [15]. Several MDR outbreaks of the bacterium were reported among US military hospitals in Iraq, Afghanistan, and Kuwait during the Iraq war [15, 16], making *A. baumannii* a global concern. Several research studies have been conducted ever since to understand the antibiotic resistance mechanism of the bacterium. However, so far, our understanding of the pathology, epidemiology, and MDR mechanism of *A. baumannii* is limited. Worrisomely, restricted options exist for the treatment of MDR *A. baumannii* infections, where reportedly, Colistin is the only antibiotic still having therapeutic effects on such strains, making the bacterium extremely difficult to treat [8]. Isolation of colistin-resistant *A. baumannii* reports have been rising in recent years [17–19], emphasizing the emergent need for new antimicrobials, treatment strategies, stricter control, and better management of colonization of patients by a rational antibiotic stewardship programme.

In this chapter, we describe the latest findings regarding the pathogenicity and mechanisms of antibiotic resistance of *A. baumannii*, then review the available routine and novel treatment strategies for this new-age superbug.

2. Clinical relevance and mechanisms of the pathogenicity of *A. baumannii*

The survival of *A. baumannii* on solid surfaces for extended periods of time is linked with the ability of the bacterium to cause nosocomial outbreaks. A majority of isolates of the bacterium are collected from the hospital environment and are closely connected with HAI, specifically in patients hospitalized in intensive care units (ICU) or in immunocompromised patients [8]. The bacterium can colonize curtains, linen fomites, bed rails, sinks, tables, medical equipment, etc., and hence can be transmitted through the vicinity of infected or colonized patients [20]. Contamination of respiratory support devices or intravenous access devices such as central venous catheters (CVC) is a major source of infections in critically ill patients [20].

The most important and common manifestations of *A. baumannii* include ventilator-associated pneumonia (VAP); CVC-associated bloodstream infections (BSI) (both VAP and BSI having the highest mortality rate); urinary tract infections (UTI) (although not very frequent, it causes up to 1.6% of urinary tract-related infections in ICU patients associated with prolonged catheter-related complications); and central nervous system infections such as meningitis, skin, soft tissue, and wound infections [21]. Infections caused by *A. baumannii* are associated with extended hospitalization, older age, and male gender [21].

A. baumannii has also gained the ability to infect the general population (however, to a lesser extent in comparison to HAI), causing community-acquired pneumonia (some reports stating that community-acquired pneumonia of *A. baumannii* is more fulminant than nosocomial pneumonia caused by the same bacterium, causing death within 8 days on average from diagnosis [22]), bacteraemia, skin and soft tissue infections, endocarditis, secondary meningitis, and ocular infections connected to contact lens use [11, 23, 24]. These infections are associated more with male gender and are more common in patients with comorbidities such as diabetes melitus, chronic obstructive pulmonary disease (COPD), renal disfunctions, and unhealthy lifestyle, including alcoholism or heavy smoking [21]. Origin of infection in 20–70% of patients infected with *A. baumannii* remains unknown [1].

The mortality rate of *A. baumannii* largely varies depending on strain type, type of infection, degree of immunosuppression and comorbidities of infected patients. The mortality rate of the bacterium can vary from 26% in a general hospital setting up to 43% in ICUs [25]. *A. baumannii* is the principal agent of VAP, accounting for 15% of HAI, which accounts for approximately 50% of total use of antibiotics and the highest mortality rate in ICUs [13].

A. Baumannii also remains an important threat to neurosurgery patients, causing up to 4% of all meningitis and shunt-related infections, with a mortality rate as high as 70% [26].

The virulence of *A. baumannii*, as mentioned earlier, is strain dependant and involves complicated mechanisms including adhesion, cell invasion, cytotoxicity, and serum persistence [8].

Adhesion and invasion: Outer membrane proteins (OMPs) are by far the most well studied structures of the bacterium, contributing to cell adhesion, invasion, and cytotoxicity [27]. OmpA (also known as Omp38), a conserved, abundant porin with a molecular weight of 38 kDa, is one of the adhesins which plays an essential role in the bacterium's adhesion and cell invasion [27]. After attachment of *A. baumannii* to epithelial cells (especially with high affinity for epithelial cells of the respiratory tract), the bacterium secretes OmpA into the cell, leading to bacterial uptake through actin rearrangement and membrane reorganization of the attacked cell, a process known as the zipper-like mechanism [27]. After uptake, internalized bacteria are located in membrane-bound vacuoles [27]. Following that, internalized OmpA adhesins are translocated to the nucleus and mitochondria of the invaded cell, causing the release of a group of proapoptotic molecules, promoting cell apoptosis and hence acting as cytotoxins [28]. OmpA plays an important role in the enhancement of biofilms on plastic and is a crucial component in its formation [27]. Omp33–36 (also known as Omp34) is another adhesin which contributes to cell adhesion and cytotoxicity [8]. Studies show that mutation in the Omp33 gene results in significant reduction in the bacterium's ability in adhesion, invasion, and apoptosis [29]. Besides enabling the bacteria to attach to host cells, Omp33 induces cell apoptosis by activating caspase and modulating autophagy, leading to the accumulation of sequestosome 1 and LC3B-11 autophagosome [29]. Some of the other cell surface proteins contributing to the bacteria's cell adhesion and consequently to its virulence include: Biofilm-associated protein (Bap), by increasing the membrane hydrophobicity [30], *Acinetobacter* trimeric autotransporter (Ata), by mediating adherence through binding to type IV collagen in the extracellular matrix of attacked cells [31], and the FhaB/FhaC, CdiA/CdiB type Vb secretion system [32]. Ata has been shown to cause secretion of pro-inflammatory cytokines (such as interleukin (IL) 6, IL8), leading to caspase-dependent cellular apoptosis [33]. Hydrophobic abilities allow the bacterium to attach to foreign materials, the same characteristic that helps bacteria to attach to materials such as plastics used in intravascular access devices. It is has been shown that surface hydrophobicity is highly expressed in bacteria isolated from patients in comparison to the normal flora of the skin [7]. Bap is directly involved in the formation of a mature biofilm in medically relevant surfaces such as titanium, polystyrene, and polypropylene [7].

Recently, the response regulator BfmR and the sensor kinase BfmS, in the two-component regulatory system BfmRS, have been shown to play a role, as non-OMP adhesins, in cell attachment of the bacterium [34]. The two-component regulatory system BfmRS can sense extracellular signals important in the formation of biofilm, causing expressions of the chaperone-usher assembly system accountable for the formation of pili (the needed component for biofilm formation on polystyrene surfaces). It has been suggested that pili may be relevantly attributed to bacterial attachment in biotic and abiotic surfaces [8]. Some of those indicators include the ability of *A. baumannii* to produce different adhesins and pili in response to different environmental conditions (B1sA mediated light-regulated type 1 pilus assembly system PrpABCD [35]), and the association between twitching motility caused by type IV pili and cell adhesion [36]. An association between motility and adhesion of *A. baumannii* was shown as well, in bacteria with disrupted expression of a homolog of the histone-like nucleoid structuring protein (H-NS) known as a global transcriptional repressor [37].

Phospholipase enzymes play a significant role in the bacteria's invasion ability [8]. Both phospholipase C (PLC) and phospholipase D (PLD) inactivation lead to impaired cell invasion [38]. Phospholipases are essential enzymes necessary for the metabolism of phosphatidylcholine and the release of phosphorylated head groups and polar head groups from PLD trough PLC cleavage, imposing change in the constancy of the membrane of host cells [38] and so facilitating invasion. OMPs containing phosphorylcholine contribute to *A. baumannii* cell adhesion and invasion by binding to the platelet-activating factor receptor on human lung epithelial cells, causing an activation signaling cascade involving G protein, clathrin, β-arrestins, vacuolar movement proteins and intercellular calcium, leading to the invasion of cells [39].

To date, our understanding of host cell receptors' contribution in bacteria's adhesion and invasion is still very limited.

2.1 Cytotoxins

Other than the already-mentioned cytotoxic abilities of OmpA, Omp33 and Ata, the lipopolysaccharide (LPS) envelope of *A. Baumannii* is another main contributor to the bacterium's cytotoxicity and immunogenicity towards the host cell [40]. LPS, as the major component of the outer leaflet of the outer membrane of the bacterium, consisting of an O-antigen and a lipid A moiety (acting as a chemotactic agent), stimulates the secretion of tumor necrotic factors (TNFs) and IL-8 from macrophages [41]. Absence or disturbance in the production of any of the first three enzymes in the lipid A biosynthetic pathway (namely acyltransferase (LpxA), deacetylase (LpxC), and N-acyltransferase (LpxD)) reduces serum survival of the bacterium and lead to less lethal infections, highlighting the immunogenicity of LPS [42].

γ-glutamyl transferase enzyme (GGT), a secretion from the type II secretion system, is another protein that contributes to the bacteria's cytotoxicity via caspase activation leading to ATP depletion followed by cell-cycle arrest and apoptosis [43]. It has been shown that strains of *A. baumannii* with higher levels of extracellular GGT activity cause more severe tissue damage, through increased inflammation and oxidative activities with elevated phenoloxidase, lysozyme, lactate dehydrogenase, and lipid peroxidation [43]. Higher levels of GGT in serum was also associated with oxidative stress and exacerbation of COPD [44]. The type IV secretion system, based on its identification within pathogenicity islands after whole genome sequencing, as well as the type VI secretion system, by killing competing bacteria, have been suggested to contribute to the pathogenicity of the bacteria; however, the complete mechanism and molecular interactions of these systems are not yet fully understood [8, 10, 45].

2.2 Serum persistence

In addition to the mentioned mechanisms, serum persistence indirectly affects the pathogenicity of the bacterium [8]. Capsular exopolysaccharide (CPS) is an important component mediating in persistence and hence, better survival of many invasive bacteria under unfavorable and harsh conditions, protecting them from phagocytosis of humoral immune attacks and inhibiting activation of alternative complement pathway [46] (an ability reported from OmpA proteins as well [47]). Although many isolates of *A. baumannii* from patients express the K locus gene cluster, which is needed for biosynthesis and export of CPS [46], having CPS alone should not be considered as a pathogenic determinant as many non-pathogenic microorganisms possess this envelope component as well [8]. Protective characteristics of CPS have been suggested through a series of studies, showing that mutation in its associated genes significantly decreases or in some cases impairs the survival of the bacterium in tissue. For example, two capsule-associated genes (namely *ptk* and *epsA*), believed to encode a putative protein tyrosine kinase and putative polysaccharide export OMP mutation, caused a significant decrease in survival of the bacterium in soft tissue [48].

The ability of the bacterium to degrade H_2O_2, utilized by catalases KatE and KatG, which could decrease production of reactive oxygen species by the innate immune system, is another mechanism used by *A. baumannii* to increase its serum resistance [8]. Universal stress protein A also showed the capacity to protect the bacterium from low PH, H_2O_2, and 2,4-DNP [49].

Mutation of LpsB glycosyltransferase involved in LPS synthesis caused a reduction in the resistance of *A. baumannii* to human serum, suggesting the role of LPS in the bacteria's persistence [50].

Finally, metal acquisition systems, the BfmRS system through biofilm formation, phospholipases, and the novel plasminogen protein CipA all were shown to have an effect on the bacteria's ability to survive longer in vivo [49, 51]. CipA links to the active form of plasmin and plasminogen, leading to a breakdown of fibrinogen, facilitating the spread of bacteria [49]. *A. baumannii* utilizes a complex set of metal acquisition systems, including iron (BasD, BauA, NfuA), zinc (ZnuABC, ZigA), and manganese (MumC, MumT) acquisition systems [8]. As an example, an iron acquisition system consisting of three siderophores (namely baumannoferrin, fimsbactin, and acinetobactin as highly converged iron chelators) is used by the bacteria to thrive under iron-reduced circumstances [52].

3. Mechanisms of resistance

Currently there are reports of isolates of *A. baumannii* which are resistant to all known antibiotics [7]. Misuse and overuse of antibiotics, lack of a proper antibiotic stewardship programme and the innate capability of the bacterium to adapt to new environmental challenges and acquire new resistant capacities are all part of the bigger picture which has made *A. baumannii* into the superbug it is today, for which we barely have any proper treatment. The genetic setup of the bacterium is capable of rapid developments towards antimicrobial resistance, making *A. baumannii* a natural transformant [7]. This ability is most likely due to the bacteria's lack of mismatch repair system gene (called *mut*S), causing increased mutation rates. It is unknown if the bacterium is naturally capable of transformation or if this change in pathogenicity and antibiotic resistance is due to environmental conditions [53]. Until the 1970s, ampicillin, gentamycin, and nalidixic acid were still used in *A. baumannii* treatment, as single or combination therapy. By 1975, high rates of resistance to mentioned antibiotics had already been reported [7]. To date, antibiotics such as penicillins, cephalosporins, tetracyclines, chloramphenicol, and most aminoglycosides have lost their efficacy against the bacterium [54].

A. baumannii uses a complex set of mechanisms for its antibiotic resistance [54]. Enzymatic hydrolysis of antibiotics, genetic modifications, and efflux pumps are adopted mechanisms used by the bacterium to escape the effect of antimicrobials.

3.1 Resistance to beta-lactams

In *A. baumannii,* resistance to beta-lactam antibiotics is achieved by the utilization of beta-lactamases, causing degradation of the antibiotic, modification in penicillin-binding proteins, and decreased permeability to antibiotics through change in outer membrane porins or expulsion of the antibiotic from the cell with the help of efflux pumps [55].

3.1.1 Beta-lactamases

Beta-lactamases are generally classified by two main schemes; the first one is the widely accepted Ambler Molecular classification scheme, which classifies enzymes based on their amino acid sequence, and the second one is the Bush-Jacoby classification scheme, which is based on biochemical characteristics of the enzymes [56, 57]. Ambler classification classifies beta-lactams into four major sub-classes A, B, C, and D based on their distinguishing amino acid motifs. Class A, C, and D are serin carbapenemases, achieving their effect through hydrolysing their substrate by the active site, forming acyl groups for which they need serin as a cofactor. Class B are collectively called bacterial metallo beta-lactamases (MBLs), using at least one zinc ion in their active site to hydrolyse beta-lactam antibiotics. MBLs are further subdivided into the three subclasses of B1, B2, and B3 [57]. Bush-Jacoby classification classifies these enzymes into the three subgroups of 1 (cephalosporinases, equivalent of class C), group 2 (serin beta-lactamases, equivalent of class A and D), and group 3 (MBLs, equivalent of class B) [56].

AmpC cephalosporinase, from the molecular class C beta lactamase, is the most prevalent enzyme utilized by *A. baumannii* against beta-lactam antibiotics [53]. This enzyme is encoded by the bla gene, making the bacteria achieve resistance against penicillins as well as narrow- and extended-spectrum cephalosporins. Moreover, ampC enzymes can also cause resistance to a combination of these antibiotics along with beta-lactam inhibitors [53]. Other than ampC, *A. baumannii* strains may contain a wide range of different classes of beta lactamases, namely PER-1, VEb-1, CTX-M, TEM, and SHV from class A, MBLs; IMP, SIM, VIM from class B and OXA from class D [58]. Having OXA-type carbapenemases is the mainstay for resistance against these last-resort antibiotics [58]. Class D carbapenemases are capable of hydrolysing isoxazolyl penicillin drugs such as oxacillin. Although more than 574 variants of the enzyme have been identified so far, few of them have carbapenemase activity. Therefore, class D beta lactamases are subdivided into 12 main Oxacillinase groups, taking OXA abbreviation. Among them, OXA23, OXA24, and OXA58 are plasmids encoding carbapenemases and are the main variants used by the bacterium for carbapenem degradation. OXA25, OXA26 and OXA40 are other enzymes utilized by *A. baumannii* from class D, which are chromosome mediated, unlike the first three mentioned enzymes. Coexistence of OXA23 and a class B MBL NDM-1 has been reported in a pan-resistant isolate from India [59].

3.1.2 Target alteration of the antibiotics and modification in membrane permeability

Reduced expression of OMPs and alteration in penicillin-binding proteins (PBPs) are two mechanisms used by *A. baumannii* to decrease uptake and permeability to beta-lactam antibiotics. Since hydrophilic antibiotics can cross the bacterial outer cell membrane only through OMPs, a decrease in their expression can decrease the permeability of antibiotics. OprD a 43 kDa protein and CarO a 29 kDa protein are among the most studied OMPs the bacterium uses to downregulate permeability [60]. Loss of CarO, secondary to the porin's gene disruption by a distinct insertion element, was shown to be associated with imipenem and meropenem resistance [61]. Since no specific imipenem binding site was found in CarO, it is believed that this porin forms nonspecific channels [62]. Heat-modifiable protein HMP-AB (homologous to OmpA of Enterobacteriaceae and OmpF of P. aeruginosa) and OmpW are some of the other notable OMPs identified to have an effect on the beta-lactam resistance [63, 64]. Proteomic studies showed difference in CarO expression and structural changes in isoforms of OmpW in multidrug-resistant strains of *A. baumannii* in comparison to control strains; interestingly, there was no difference in expression of OprD porins among the two strains [65].

PBPs play a crucial role in the synthesis of peptidoglycan used in the bacterial cell wall. They also catalyze transglycosylation and the cross-linking by transpeptidation of peptidoglycan. *A. baumannii* can develop beta-lactam resistance by decreasing the affinity of PBPs to beta-lactams, overproducing a critical PBP, or by producing new or altered PBPs [66].

3.1.3 Increased efflux pumps

So far, six families of efflux pumps have been identified: the resistance nodulation cell division family, the small multidrug resistance superfamily, the ATP-binding cassette (ABC) family, the major facilitator superfamily, the multidrug toxic compound extrusion family, and the recently identified proteobacterial antimicrobial compound efflux family [67]. The presence of efflux pumps provides resistance to multiple classes of antibiotics. The AdeABC efflux pump, a member of the resistance nodulation cell division family, is one of most studied and well-defined pumps in *A. baumannii,* conferring resistance to a group of multiclass antibiotics including aminoglycosides, fluoroquinolones, chloramphenicol, erythromycin, cefotaxime, tetracyclines, trimethoprim and ethidium bromide [68]. Once this efflux pump is overexpressed, it can confer resistance to carbapenems as well [68]. AdeABC contains a three-component structure: AdeB forms the transmembrane component, AdeA forms the inner membrane fusion protein, and AdeC forms the OMP [1, 68]. This efflux pump is encoded chromosomally and is regulated by a two-component system: a sensor kinase (AdeS) and its associated response regulator (AdeR) [69]. Point mutation within this regulatory system is associated with pump overexpression and hence, carbapenem resistance. However, mutation is not the only mechanism necessary for its overexpression [69]. Several other efflux pumps identified in *A. baumannii* so far include: AbeS, from a small multidrug resistance efflux pump group, AdeABC, AdeIJK, and AdeFGH of the ABC group, as well as CraA, AmvA/AedF, and Tet(B) [7].

3.2 Resistance to aminoglycosides

Most frequently, *A. baumannii* modifies amino or hydroxyl group by means of aminoglycoside-modifying enzymes, namely: adenylases, acetylases, methyltransferases, and phosphotransferases, to confer resistance to aminoglycosides [53]. Presence of genes coding for aminoglycoside-modifying enzymes within class 1 integrins is commonly seen in antibiotic-resistant strains of *A. baumannii* [70].

The bacterium uses alteration of the target ribosomal protein as the other mechanism for resistance against this class of antibiotics [71]. 16S rRNA methylation has been described for *A. baumannii* (armA) strains impairing aminoglycoside binding to its target site, causing high levels of resistance to all clinically important aminoglycosides, including amikacin, gentamycin, and tobramycin [72, 73]. armA is a plasmid born gene within a transposon (Tn1548) which has very similar characteristics to several genes across gram-negative organisms [73]. Besides the mentioned mechanisms, *A. baumannii* utilizes the AdeABC efflux pump and the AbeM pump (member of the multidrug and toxic compound extrusion (MATE) family) to increase the efflux of aminoglycosides from the cell [74]. AdeABC pumps, however, are somewhat less effective against amikacin and kanamycin due to their more hydrophilic characteristics [74].

3.3 Resistance to quinolones

Mutations in the gyrA and parC genes, resulting in phenotypic changes in DNA gyrase and topoisomerase IV causing reduced drug affinity by interference with target site binding, are the main mechanisms used by *A. baumannii* to resist quinolones [75].

Decreased permeability by reduced expression of OMPs and increased efflux through efflux pumps such as AdeABC and MATE pump AdeM (as mentioned for other classes of antibiotics) is used here as well, for resistance against quinolones [76]. Inhibition of binding of DNA gyrase and/or topoisomerase by protection of DNA through plasmid-encoded quinolone-resistant determinants gnrA, gnrB, and gnrS is another notable mechanism utilized by the bacterium to resist these antibiotics [77].

3.4 Resistance to colistin

Chromosomal DNA genes of the bacterium encode resistance to colistin via two mechanisms, target modification and remodeling of the outer membrane. The first mechanism is due to mutation in lipid A encoding genes (namely lpxA, lpxC, and lpxD) resulting in loss of LPS, which is an initial target for colistin [78]. The two-component system pmrAB (a response regulator and sensor kinase), in response to environmental conditions such as change in pH, Mg^{2+}, and Fe^{3+}, regulates the expression of genes causing synthesis of lipid A. Point mutation (as the second mechanism) in pmrA and pmrB causes upregulation of their gene expression and hence, remodeling of the outer membrane [79].

3.5 Resistance to Tetracyclines

Efflux and ribosomal protection are two mechanisms used for resistance against tetracyclines by *A. baumannii*. Tetracycline-specific efflux pumps encoded by the tet(A) and tet(B) determinants have been identified in the bacterium [80]. Tet(A) genes are located within a transposon similar to Tn1721, in association with an IS element [80]. Apart from these class-specific efflux pumps, tetracyclines are susceptible to the effect of multi-efflux systems such as AdeABC as well [81]. The effect on microsomal protection is conferred by tet(O) and tet(M) determinants, the later being described rarely for *A. baumannii* [82].

3.6 Resistance to trimethoprim-sulfamethoxazole and chloramphenicol

Resistant gene-containing integrons are very commonly seen among multidrug-resistant strains of the bacterium. The 3′-conserved region of an integron containing a qac gene fused to a sul gene has been shown to provide resistance to sulphonamides [83]. Similarly, the trimethoprim-resistant gene (dhfr) and chloramphenicol resistant gene (cat) are located on integron structures [70]. Efflux pumps also contribute to conferring resistance to these antibiotics [53].

4. Antibiotic treatment options for multi-resistant strains of *A. baumannii* and new non-antibiotic modalities under development

4.1 Antibiotic treatment

4.1.1 Carbapenems

For years carbapenems were a mainstay of antimicrobial therapy against *A. baumannii*. Since 1990 these antibiotics have been used as last-resort, broad-spectrum antimicrobials for treatment of complicated life-threatening infections caused by this bacterium. Interestingly, Acinetobacter's first resistance to carbapenems was reported in 1985, the year imipenem was discovered, indicating that the resistance mechanism had existed even before the antibiotic's first use [13]. Rate of resistance to carbapenems is highly geographical and is connected to regional distribution of strain types and percentage of antibiotic prescription and misuse. Currently on average, the general percentage of resistance of *A. baumannii* to carbapenems is 74–92% globally [84]. The percentage of resistance rises from 15% in Europe and North America to 40% in Latin America and up to 85–100% in south and east Asia [85–87]. This high resistance level renders use of carbapenems and consequently other beta-lactams ineffective in multi-resistant *A. baumannii* strains. Therefore, choosing treatment with carbapenems is rational in areas with lower resistance to this antibiotic.

4.1.2 Polymyxins Colistin (polymyxin E)

Polymyxins are bactericidal antibiotics that function through cell membrane disruption. Polymyxins' positively charged cationic region bind to the negatively charged hydrophilic part of LPS, causing loss of integrity of the cell membrane [88]. The emergence of pan-resistant strains of *A. baumannii* led to the reintroduction of these historically discarded antibiotics (due to colistin's cell toxicity, especially nephrotoxicity ranging between 11% up to 76% in different reports, use of colistin was discontinued shortly after its introduction in the 1950s). Colistin nephrotoxicity is suggested to be due to accumulation of the drug in proximal renal tubules causing oxidative damage [89]. Other than nephrotoxicity, colistin may induce neurotoxicity (although infrequently) when administered via the respiratory tract, causing bronchoconstriction and cough, and may result in chemical meningitis [90]. Several other medication side effects, such as ataxia, apnoea, delirium, visual disturbances, seizures, etc., have been reported in connection to use of this antibiotic [91]. Monotherapy of colistin raises two distinct complications resulting in resistance against it: on the one hand, due to administration of colistin as a pro-drug in the form of colistin methane sulfonate, achievement of therapeutic serum concentration is mostly complicated and suboptimal, causing emergence of regrowth. On the other hand, selective resistance of bacterial subpopulation, called heteroresistance, renders the antibiotic ineffective [92]. Thus, combination therapy with other active agents to achieve synergy has been proposed. Although there are still debates regarding the advantages of monotherapy vs. combination therapy, extensive review of literature has suggested combination therapy as a superior option both in terms of microbiological clearance and clinical cure [93, 94]. The combination of colistin with carbapenem or colistin with rifampicin are the most studied and well-established choices [95]. Several other combinations including colistin and aminoglycosides, daptomycin, co-trimoxazole, fosfomycin, sulbactam, etc. exist in literature [96]. Despite all of this, resistance to this last-resort antibiotic has been reported (the first report of *A. baumannii* being resistant to colistin was published in Czechia in 1999) and is rising worldwide [97]. Resistance to colistin in the US and Europe is documented at around 11%, with the highest resistance so far reported from India with 53%, Iran 48% and Spain 40% [13, 98].

4.1.3 Tetracyclines minocycline: tigecycline

Tetracyclines function by entering the outer membrane of gram-negative bacteria via protein channels, causing conformational changes in the RNA of the bacteria by binding to its 30S ribosomal unit [99], blocking entry of aminoacyl transferase RNA into the site A of the ribosome and consequently stopping protein production [99]. Since minocycline is a more lipophilic type (along with doxycycline) among tetracyclines, this antibiotic has better tissue penetration and antimicrobial activity [99]. As mentioned in the mechanisms of resistance section of this chapter, the main mechanism of resistance against tetracyclines is efflux of the drug by *A. baumannii's* efflux pumps, namely tet(A) and tet(B). It has been shown (in vitro) that tet(A) pumps are not effective against minocycline and strains lacking tet(B) pumps are susceptible to this antibiotic [100]. Although tigecycline was synthetically designed to overcome the effect of efflux pumps (commonly seen in other tetracyclines), development of other resistance mechanisms (such as overexpression of various efflux pumps, including AdeABC), has made this once promising antibiotic far less effective [101]. Interestingly, so far, the effect of overexpression of RND pumps and increase of resistance has not been reported on minocycline and therefore, tigecycline-resistant *A. baumannii* can still display susceptibility to this antibiotic [99]. Due to the mentioned properties, minocycline has attracted a lot of attention during the last decade or so and has become more clinically used. Still, superiority in use of this antibiotic as monotherapy or combination therapy is not clear and more large clinical studies are needed to provide a better understanding of the efficacy of this drug. To date, published case reports (with small cohorts of patients) have shown good relative efficacy and tolerance of this antibiotic in clinical practice [102–104].

4.2 Bacteriophages

In 2010 the first report of isolation and characterization of phage against *A. baumannii* was published, showing lytic behavior of AB1 and AB2 phages against the bacterium [105]. Since then, many lytic phages have been identified; however, the first use of phage in a human in Western countries was reported no sooner than 2017. Being a successful attempt, this experiment paved the way for the initiation of a phage therapy programme just a year later at the University of California. They recently published a case report of the first 10 patients being treated by this modality [106]. Currently, only five clinical case reports exist on phage therapy against *A. baumannii* [107]. All these reports share a similar trend, using phage cocktails combined with antibiotics in more than one dose, administered intravenously or in the case of respiratory infection, nebulized. Although general positive impact was seen in all of these reports, phage therapy was not always correlated with complete recovery in all cases [107].

4.3 Monoclonal antibodies

Neutralizing virulence factors of the bacterium by the binding of monoclonal antibodies is the mainstay mechanism of action of these molecules against *A. baumannii*. Many studies in recent years have been dedicated to the development and safety of monoclonal antibodies in laboratory animals. So far, MAb-based treatment (a monoclonal antibody) is one modality showing promising effectivity in the management of infection in animals without driving resistance to antibiotics [108]. Historically, the clinical efficacy of MAbs and other passive immunization therapies was based on enhancing microbial clearance induced by opsonophagocytosis and activating complement system [108]. However, it has been suggested that MAbs may confer their effect not only through microbial clearance but also by altering the inflammatory settings of the host regardless of microbial count [108]. This hypothesis has been further supported by a recent publication which demonstrated that despite a decrease of bacterial burden through the effect of antibodies, the antibacterial effect of MAbs was insufficient to mediate clinical improvement, as measured by survival of the host [108]. It was suggested that efficacy of MAb therapy depends more on "normalizing the inflammatory response to infection than on reducing bacterial burden" [108]. Due to their complicated mechanism of action, safety concerns, and very expensive price in terms of mass production, so far, monoclonal antibodies are still only in animal research stages, but are providing a promising modality for the future of antimicrobial treatment.

4.4 Antimicrobial peptides

Antimicrobial peptides (AMPs) are products of eukaryotic and prokaryotic organisms as part of their innate host immune response [13]. Since they have bactericidal capacities and are broad spectrum, with low immunogenicity and low resistance, interest in them has been increasingly rising in recent years as alternatives to antibiotic treatment. As an example, efficacy of a hybrid of cecropin A and melittin has been reported in management of peritoneal sepsis caused by pan-resistant *A. baumannii* in an animal model [109]. Several reports have already been published showing the antimicrobial effect of Brevinin 2, alyteserin 2, catonic α-helical peptide, A3-APO, and D-RR4 against multidrug-resistant strains of the bacterium in animals [109, 110]. However, cytotoxicity, enzymatic degradation, and high production costs still need to be evaluated before we may expect to see them in clinical use for humans.

4.5 Vaccination

Since *A. baumannii* has an intracellular lifestyle, vaccine development strategies against it should include the induction of acquired cellular immunity with long-term memory, and be targeted for a population cohort of patients prone to be infected with the bacterium [13]. The first attempts to produce a vaccine against *A. baumannii* were initiated during the last decade, testing almost all modalities of vaccines, including whole-cell vaccines using attenuated or killed bacteria in mice, subunit vaccines using OmpA, Ata, and Omp 33–36, conjugate vaccines, multicomponent vaccines, and nucleic acid vaccines [13]. However, none of them have entered clinical trials so far, indicating our lack of knowledge in understanding the complex bacterium-host relationship. To date, most of the vaccine candidates developed against *A. baumannii* have been focusing on protein-based technologies (including the mentioned opsonophagocytic antibody-mediated killing and/or antibody-mediated toxin inhibition) [13]. Since it has been shown that polarized (T helper) Th2 will not be sufficient for long-term protection, new efforts have been shifted to achieve a mixed Th1/Th2 or Th1/Th17 response. Thanks to the technology acceleration associated with the COVID vaccine, the horizon has been expanded on nucleic acid vaccines against *A. baumannii* and other pathogens, and interesting developments in this area in the future are expected.

4.6 Other modalities

Some of the other modalities that have been explored in the clamor caused by the urgent need for a new treatment against multi-resistant *A. baumannii* include the use of clustered, regularly intercepted short palindromic repeat (Cas) systems to eliminate the resistant gene, producing susceptible species; utilization of metal chelators, essential in the expression of bacterial virulence factors, as targets for designing new antimicrobials; and toxin-carrying liposomes [111, 112].

5. Conclusion

A. baumannii is one of the most complicated and challenging to treat pathogens of the modern era of medicine. The bacterium's complex mechanisms of resistance, its genetic set-up, and its ability to survive on different surfaces and tolerate harsh environmental conditions make *A. baumannii* an advanced nosocomial pathogen causing clinical outbreaks all around the world. Since most antibiotics are ineffective against it, there is an urgent need for new antibiotics or other modalities of treatment (phage therapy, vaccination, etc.) in the near future. Meanwhile, available options for the management of *A. baumannii* include a rational antimicrobial stewardship programme to minimize selective pressure in support of growth of multi-resistant strains, decreased misuse of antibiotics, early diagnosis and proper isolation of infected and colonized patients, and proper choice in antibiotic therapy (carbapenems, minocycline or tigecycline in case of susceptibility to these antimicrobials or combination therapy with colistin + another antibiotic agent (i.e. carbapenems) in case of resistance) once infection with such strains has been established.

6. Conflict of interest

The authors declare no conflict of interest.

Author details

Yashar Jalali*, Monika Jalali and Juraj Payer
Faculty of Medicine, Comenius University in Bratislava, 5th Department of Internal Medicine, University Hospital Bratislava, Ružinov, Slovakia

*Address all correspondence to: jalali_yashar@yahoo.com

References

[1] Peleg AY, Seifert H, Paterson DL. *Acinetobacter baumannii*: Emergence of a successful pathogen. Clinical Microbiology Reviews. 2008;**21**(3): 538-582

[2] Henriksen SD. Moraxella, Acinetobacter, and the Mimeae. Bacteriological Reviews. 1973;**37**(4): 522-561

[3] Brisou J, Prevot AR. Studies on bacterial taxonomy. X. The revision of species under Acromobacter group. Annales de l'Institut Pasteur. 1954;**86**(6):722-728

[4] Baumann P, Doudoroff M, Stanier RY. A study of the Moraxella group. II. Oxidative-negative species (genus Acinetobacter). Journal of Bacteriology. 1968;**95**(5):1520-1541

[5] Lessel E. International Committee on Nomenclature of Bacteria Subcommittee on the Taxonomy of Moraxella and Allied Bacteria: Minutes of the meeting, 11 august 1970. Room constitution C, Maria-Isabel Hotel, Mexico City, Mexico. International Journal of Systematic and Evolutionary Microbiology. 1971;**21**(2):213-214

[6] Skerman VBD, McGowan V, Sneath PHA. Approved lists of bacterial names. International Journal of Systematic Bacteriology. 1980;**30**(1): 225-230

[7] Howard A et al. *Acinetobacter baumannii*: An emerging opportunistic pathogen. Virulence. 2012;**3**(3):243-250

[8] Ma C, McClean S. Mapping global prevalence of *Acinetobacter baumannii* and recent vaccine development to tackle it. Vaccine. 2021;**9**(6):570

[9] Qin J et al. Precise species identification for Acinetobacter: A genome-based study with description of two novel Acinetobacter species. mSystems. 2021;**6**(3):e00237-e00221

[10] Smith MG et al. New insights into *Acinetobacter baumannii* pathogenesis revealed by high-density pyrosequencing and transposon mutagenesis. Genes & Development. 2007;**21**(5):601-614

[11] Dexter C et al. Community-acquired *Acinetobacter baumannii*: Clinical characteristics, epidemiology and pathogenesis. Expert Review of Anti-Infective Therapy. 2015;**13**(5):567-573

[12] Villalón P et al. Dynamics of a sporadic nosocomial *Acinetobacter calcoaceticus - Acinetobacter baumannii* Complex population. Frontiers in Microbiology. 2019;**10**:593

[13] Asif M, Alvi IA, Rehman SU. Insight into *Acinetobacter baumannii*: Pathogenesis, global resistance, mechanisms of resistance, treatment options, and alternative modalities. Infection and Drug Resistance. 2018;**11**:1249-1260

[14] Pendleton JN, Gorman SP, Gilmore BF. Clinical relevance of the ESKAPE pathogens. Expert Review of Anti-Infective Therapy. 2013;**11**(3):297-308

[15] Scott P et al. An outbreak of multidrug-resistant *Acinetobacter baumannii-calcoaceticus* complex infection in the US military health care system associated with military operations in Iraq. Clinical Infectious Diseases. 2007;**44**(12):1577-1584

[16] Centers for Disease Control and Prevention (CDC). Acinetobacter

baumannii infections among patients at military medical facilities treating injured U.S. service members, 2002-2004. MMWR. Morbidity and Mortality Weekly Report. 2004;**53**(45):1063-1066

[17] Elham B, Fawzia A. Colistin resistance in *Acinetobacter baumannii* isolated from critically ill patients: Clinical characteristics, antimicrobial susceptibility and outcome. African Health Sciences. 2019;**19**(3):2400-2406

[18] Garcia Casallas JC et al. Bacteremia by colistin-resistant *Acinetobacter baumannii* isolate: A case report. Journal of Medical Case Reports. 2019;**13**(1):141

[19] Papathanakos G et al. Colistin-resistant *Acinetobacter baumannii* bacteremia: A serious threat for critically ill patients. Microorganisms. 2020;**8**(2):287

[20] Jung J, Park W. Acinetobacter species as model microorganisms in environmental microbiology: Current state and perspectives. Applied Microbiology and Biotechnology. 2015;**99**(6):2533-2548

[21] Antunes LC, Visca P, Towner KJ. *Acinetobacter baumannii*: Evolution of a global pathogen. Pathogens and Disease. 2014;**71**(3):292-301

[22] Falagas ME, Rafailidis PI. Attributable mortality of *Acinetobacter baumannii*: No longer a controversial issue. Critical Care. 2007;**11**(3):134

[23] Falagas ME et al. Community-acquired Acinetobacter infections. European Journal of Clinical Microbiology & Infectious Diseases. 2007;**26**(12):857-868

[24] Chang WN et al. Community-acquired Acinetobacter meningitis in adults. Infection. 2000;**28**(6):395-397

[25] Greene C et al. The influence of biofilm formation and multidrug resistance on environmental survival of clinical and environmental isolates of *Acinetobacter baumannii*. American Journal of Infection Control. 2016;**44**(5):e65-e71

[26] Basri R et al. Burden of bacterial meningitis: A retrospective review on laboratory parameters and factors associated with death in meningitis, Kelantan Malaysia. Nagoya Journal of Medical Science. 2015;**77**(1-2):59-68

[27] Choi CH et al. *Acinetobacter baumannii* invades epithelial cells and outer membrane protein a mediates interactions with epithelial cells. BMC Microbiology. 2008;**8**(1):1-11

[28] Choi CH et al. *Acinetobacter baumannii* outer membrane protein a targets the nucleus and induces cytotoxicity. Cellular Microbiology. 2008;**10**(2):309-319

[29] Smani Y, Dominguez-Herrera J, Pachón J. Association of the outer membrane protein Omp33 with fitness and virulence of *Acinetobacter baumannii*. The Journal of Infectious Diseases. 2013;**208**(10):1561-1570

[30] Brossard KA, Campagnari AA. The *Acinetobacter baumannii* biofilm-associated protein plays a role in adherence to human epithelial cells. Infection and Immunity. 2012;**80**(1):228-233

[31] Bentancor LV et al. Identification of Ata, a multifunctional trimeric autotransporter of *Acinetobacter baumannii*. Journal of Bacteriology. 2012;**194**(15):3950-3960

[32] Pérez A et al. The FhaB/FhaC two-partner secretion system is involved in adhesion of *Acinetobacter*

baumannii AbH12O-A2 strain. Virulence. 2017;**8**(6):959-974

[33] Weidensdorfer M et al. The Acinetobacter trimeric autotransporter adhesin Ata controls key virulence traits of *Acinetobacter baumannii*. Virulence. 2019;**10**(1):68-81

[34] Liou M-L et al. The sensor kinase BfmS mediates virulence in *Acinetobacter baumannii*. Journal of Microbiology, Immunology and Infection. 2014;**47**(4):275-281

[35] Wood CR et al. A light-regulated type I pilus contributes to *Acinetobacter baumannii* biofilm, motility, and virulence functions. Infection and Immunity. 2018;**86**(9):e00442-e00418

[36] Eijkelkamp BA et al. Adherence and motility characteristics of clinical *Acinetobacter baumannii* isolates. FEMS Microbiology Letters. 2011;**323**(1):44-51

[37] Eijkelkamp BA et al. H-NS plays a role in expression of *Acinetobacter baumannii* virulence features. Infection and Immunity. 2013;**81**(7):2574-2583

[38] Jacobs AC et al. Inactivation of phospholipase D diminishes *Acinetobacter baumannii* pathogenesis. Infection and Immunity. 2010;**78**(5):1952-1962

[39] Smani Y et al. Platelet-activating factor receptor initiates contact of *Acinetobacter baumannii* expressing phosphorylcholine with host cells. Journal of Biological Chemistry. 2012;**287**(32):26901-26910

[40] Lee C-R et al. Biology of Acinetobacter baumannii: Pathogenesis, antibiotic resistance mechanisms, and prospective treatment options. Frontiers in Cellular and Infection Microbiology. March 2017;7. Article ID: 55

[41] Erridge C et al. *Acinetobacter baumannii* lipopolysaccharides are potent stimulators of human monocyte activation via toll-like receptor 4 signalling. Journal of Medical Microbiology. 2007;**56**(2):165-171

[42] McConnell MJ, Actis L, Pachón J. *Acinetobacter baumannii*: Human infections, factors contributing to pathogenesis and animal models. FEMS Microbiology Reviews. 2013;**37**(2):130-155

[43] Elhosseiny NM et al. γ-Glutamyltransferase as a novel virulence factor of *Acinetobacter baumannii* inducing alveolar wall destruction and renal damage in systemic disease. The Journal of Infectious Diseases. 2020;**222**(5):871-879

[44] Ermis H et al. Relationship between serum γ-glutamyltransferase levels and acute exacerbation of chronic obstructive pulmonary disease. Polskie Archiwum Medycyny Wewnętrznej. 2013;**123**(3):85-90

[45] Carruthers MD et al. *Acinetobacter baumannii* utilizes a type VI secretion system for bacterial competition. PLoS One. 2013;**8**(3):e59388

[46] Bannister BA, Begg NT, Gillespie SH. Infectious Disease. Oxford: Blackwell Science; 2000

[47] Kim SW et al. Outer membrane protein a plays a role in pathogenesis of *Acinetobacter nosocomialis*. Virulence. 2016;7(4):413-426

[48] Russo TA et al. The K1 capsular polysaccharide of *Acinetobacter baumannii* strain 307-0294 is a major virulence factor. Infection and Immunity. 2010;**78**(9):3993-4000

[49] Koenigs A et al. CipA of *Acinetobacter baumannii* is a novel plasminogen

binding and complement inhibitory protein. The Journal of Infectious Diseases. 2016;**213**(9):1388-1399

[50] Luke NR et al. Identification and characterization of a glycosyltransferase involved in *Acinetobacter baumannii* lipopolysaccharide core biosynthesis. Infection and Immunity. 2010;**78**(5):2017-2023

[51] Fiester SE et al. Iron-regulated phospholipase C activity contributes to the cytolytic activity and virulence of *Acinetobacter baumannii*. PLoS One. 2016;**11**(11):e0167068

[52] Zimbler DL et al. Stress response and virulence functions of the *Acinetobacter baumannii* NfuA Fe-S scaffold protein. Journal of Bacteriology. 2012;**194**(11):2884-2893

[53] Maragakis LL, Perl TM. Acinetobacter baumannii: Epidemiology, antimicrobial resistance, and treatment options. Clinical Infectious Diseases. 2008;**46**(8):1254-1263

[54] Valencia R et al. Nosocomial outbreak of infection with pan-drug-resistant *Acinetobacter baumannii* in a tertiary care university hospital. Infection Control and Hospital Epidemiology. 2009;**30**(3):257-263

[55] Davies J, Davies D. Origins and evolution of antibiotic resistance. Microbiology and Molecular Biology Reviews. 2010;**74**(3):417-433

[56] Bush K, Jacoby GA. Updated functional classification of beta-lactamases. Antimicrobial Agents and Chemotherapy. 2010;**54**(3):969-976

[57] Hall BG, Barlow M. Revised ambler classification of β-lactamases. Journal of Antimicrobial Chemotherapy. 2005;**55**(6):1050-1051

[58] Perez F et al. Global challenge of multidrug-resistant *Acinetobacter baumannii*. Antimicrobial Agents and Chemotherapy. 2007;**51**(10):3471-3484

[59] Karthikeyan K, Thirunarayan MA, Krishnan P. Coexistence of blaOXA-23 with blaNDM-1 and armA in clinical isolates of *Acinetobacter baumannii* from India. The Journal of Antimicrobial Chemotherapy. 2010;**65**(10):2253-2254

[60] Morán-Barrio J et al. The *Acinetobacter* outer membrane contains multiple specific channels for carbapenem β-lactams as revealed by kinetic characterization analyses of Imipenem permeation into *Acinetobacter baylyi* cells. Antimicrobial Agents and Chemotherapy. 2017;**61**(3):7

[61] Mussi MA, Limansky AS, Viale AM. Acquisition of resistance to carbapenems in multidrug-resistant clinical strains of *Acinetobacter baumannii*: Natural insertional inactivation of a gene encoding a member of a novel family of beta-barrel outer membrane proteins. Antimicrobial Agents and Chemotherapy. 2005;**49**(4):1432-1440

[62] Siroy A et al. Channel formation by CarO, the carbapenem resistance-associated outer membrane protein of *Acinetobacter baumannii*. Antimicrobial Agents and Chemotherapy. 2005;**49**(12):4876-4883

[63] Gribun A et al. Molecular and structural characterization of the HMP-AB gene encoding a pore-forming protein from a clinical isolate of *Acinetobacter baumannii*. Current Microbiology. 2003;**47**(5):434-443

[64] Vila J, Martí S, Sánchez-Céspedes J. Porins, efflux pumps and multidrug resistance in *Acinetobacter baumannii*. The Journal of Antimicrobial Chemotherapy. 2007;**59**(6):1210-1215

[65] Siroy A et al. Global comparison of the membrane subproteomes between a multidrug-resistant *Acinetobacter baumannii* strain and a reference strain. Journal of Proteome Research. 2006;**5**(12):3385-3398

[66] Vashist J et al. Analysis of penicillin-binding proteins (PBPs) in carbapenem resistant *Acinetobacter baumannii*. The Indian Journal of Medical Research. 2011;**133**(3):332-338

[67] Buckner MM et al. Beyond antimicrobial resistance: Evidence for a distinct role of the AcrD efflux pump in *Salmonella* biology. MBio. 2016;**7**(6):1-2

[68] Yoon EJ et al. Contribution of the Ade resistance-nodulation-cell division-type efflux pumps to fitness and pathogenesis of *Acinetobacter baumannii*. MBio. 2016;**7**(3):1-3

[69] Marchand I et al. Expression of the RND-type efflux pump AdeABC in *Acinetobacter baumannii* is regulated by the AdeRS two-component system. Antimicrobial Agents and Chemotherapy. 2004;**48**(9):3298-3304

[70] Lee K et al. Novel acquired metallo-beta-lactamase gene, bla(SIM-1), in a class 1 integron from *Acinetobacter baumannii* clinical isolates from Korea. Antimicrobial Agents and Chemotherapy. 2005;**49**(11):4485-4491

[71] Shrestha S et al. Emergence of aminoglycoside resistance due to armA methylase in multi-drug resistant Acinetobacter Baumannii isolates in a University Hospital in Nepal. Journal of Nepal Health Research Council. 2016;**14**(33):72-76

[72] Lee H et al. Dissemination of 16S rRNA methylase-mediated highly amikacin-resistant isolates of *Klebsiella pneumoniae* and Acinetobacter baumannii in Korea. Diagnostic Microbiology and Infectious Disease. 2006;**56**(3):305-312

[73] Doi Y et al. Identification of 16S rRNA methylase-producing *Acinetobacter baumannii* clinical strains in North America. Antimicrobial Agents and Chemotherapy. 2007;**51**(11):4209-4210

[74] Su XZ et al. AbeM, an H+-coupled *Acinetobacter baumannii* multidrug efflux pump belonging to the MATE family of transporters. Antimicrobial Agents and Chemotherapy. 2005;**49**(10):4362-4364

[75] Ugolotti E et al. The whole genome sequencing of *Acinetobacter-calcoaceticus-baumannii* complex strains involved in suspected outbreak in an intensive care unit of a pediatric hospital. Journal of Hospital Administration. 2016;**5**(6):81

[76] Charrier C et al. In vitro biological evaluation of novel broad-spectrum isothiazolone inhibitors of bacterial type II topoisomerases. The Journal of Antimicrobial Chemotherapy. 2016;**71**(10):2831-2839

[77] Yang H et al. Detection of the plasmid-mediated quinolone resistance determinants in clinical isolates of *Acinetobacter baumannii* in China. Journal of Chemotherapy. 2016;**28**(5):443-445

[78] Bojkovic J et al. Characterization of an *Acinetobacter baumannii* lptD deletion strain: Permeability defects and response to inhibition of lipopolysaccharide and fatty acid biosynthesis. Journal of Bacteriology. 2015;**198**(4):731-741

[79] Choi HJ et al. Characterisation of successive *Acinetobacter baumannii* isolates from a deceased haemophagocytic lymphohistiocytosis patient. International Journal of

Antimicrobial Agents. 2017;**49**(1): 102-106

[80] Ribera A et al. Partial characterization of a transposon containing the tet(a) determinant in a clinical isolate of *Acinetobacter baumannii*. The Journal of Antimicrobial Chemotherapy. 2003;**52**(3):477-480

[81] Magnet S, Courvalin P, Lambert T. Resistance-nodulation-cell division-type efflux pump involved in aminoglycoside resistance in *Acinetobacter baumannii* strain BM4454. Antimicrobial Agents and Chemotherapy. 2001;**45**(12): 3375-3380

[82] Ribera A, Ruiz J, Vila J. Presence of the Tet M determinant in a clinical isolate of *Acinetobacter baumannii*. Antimicrobial Agents and Chemotherapy. 2003;**47**(7):2310-2312

[83] Walsh TR et al. Metallo-beta-lactamases: The quiet before the storm? Clinical Microbiology Reviews. 2005;**18**(2):306-325

[84] Lob SH et al. Regional differences and trends in antimicrobial susceptibility of *Acinetobacter baumannii*. International Journal of Antimicrobial Agents. 2016;**47**(4):317-323

[85] Karageorgopoulos DE, Falagas ME. Current control and treatment of multidrug-resistant *Acinetobacter baumannii* infections. The Lancet Infectious Diseases. 2008;**8**(12):751-762

[86] Tal-Jasper R et al. Clinical and epidemiological significance of carbapenem resistance in *Acinetobacter baumannii* infections. Antimicrobial Agents and Chemotherapy. 2016; **60**(5):3127-3131

[87] Jalali Y et al. Isolated carbapenem resistant bacteria, their multidrug resistant profile, percentage of healthcare associated infection and associated mortality, in hospitalized patients in a University Hospital in Bratislava. Bratislavské Lekárske Listy. 2021;**122**(6):379-385

[88] Thi Khanh Nhu N et al. The induction and identification of novel colistin resistance mutations in *Acinetobacter baumannii* and their implications. Scientific Reports. 2016;**6**:28291

[89] Ozkan G et al. How does colistin-induced nephropathy develop and can it be treated? Antimicrobial Agents and Chemotherapy. 2013;**57**(8):3463-3469

[90] Abdellatif S et al. Efficacy and toxicity of aerosolised colistin in ventilator-associated pneumonia: A prospective, randomised trial. Annals of Intensive Care. 2016;**6**(1):26

[91] Myint T et al. Respiratory muscle paralysis associated with colistin, polymyxin B, and muscle relaxants drugs: A case report. Journal of Investigative Medicine High Impact Case Reports. 2016;**4**(1):2324709616638362

[92] Silva A et al. Heteroresistance to colistin in *Klebsiella pneumoniae* is triggered by small colony variants sub-populations within biofilms. Pathogens and Disease. 2016;**74**(5):1-2

[93] Zusman O et al. Polymyxin monotherapy or in combination against carbapenem-resistant bacteria: Systematic review and meta-analysis. The Journal of Antimicrobial Chemotherapy. 2017;**72**(1):29-39

[94] Cai Y et al. Colistin resistance of *Acinetobacter baumannii*: Clinical reports, mechanisms and antimicrobial strategies. The Journal of Antimicrobial Chemotherapy. 2012;**67**(7):1607-1615

[95] Hong DJ et al. In vitro antimicrobial synergy of colistin with rifampicin and carbapenems against colistin-resistant *Acinetobacter baumannii* clinical isolates. Diagnostic Microbiology and Infectious Disease. 2016;**86**(2):184-189

[96] Nepka M et al. In vitro bactericidal activity of trimethoprim-sulfamethoxazole alone and in combination with colistin against carbapenem-resistant *Acinetobacter baumannii* clinical isolates. Antimicrobial Agents and Chemotherapy. 2016;**60**(11):6903-6906

[97] Hejnar P, Kolár M, Hájek V. Characteristics of *Acinetobacter* strains (phenotype classification, antibiotic susceptibility and production of beta-lactamases) isolated from haemocultures from patients at the teaching hospital in Olomouc. Acta Universitatis Palackianae Olomucensis Facultatis Medicae. 1999;**142**:73-77

[98] Oikonomou O et al. Rapid dissemination of colistin and carbapenem resistant *Acinetobacter baumannii* in Central Greece: Mechanisms of resistance, molecular identification and epidemiological data. BMC Infectious Diseases. 2015;**15**:559

[99] Lashinsky JN et al. Minocycline for the treatment of multidrug and extensively drug-resistant *A. baumannii*: A review. Infectious Disease and Therapy. 2017;**6**(2):199-211

[100] Lomovskaya O et al. Absence of TetB identifies minocycline-susceptible isolates of *Acinetobacter baumannii*. International Journal of Antimicrobial Agents. 2018;**52**(3):404-406

[101] Hornsey M et al. AdeABC-mediated efflux and tigecycline MICs for epidemic clones of *Acinetobacter baumannii*. The Journal of Antimicrobial Chemotherapy. 2010;**65**(8):1589-1593

[102] Ritchie DJ, Garavaglia-Wilson A. A review of intravenous minocycline for treatment of multidrug-resistant Acinetobacter infections. Clinical Infectious Diseases. 2014;**59**(Suppl. 6): S374-S380

[103] Chan JD, Graves JA, Dellit TH. Antimicrobial treatment and clinical outcomes of carbapenem-resistant *Acinetobacter baumannii* ventilator-associated pneumonia. Journal of Intensive Care Medicine. 2010;**25**(6):343-348

[104] Goff DA, Bauer KA, Mangino JE. Bad bugs need old drugs: A stewardship program's evaluation of minocycline for multidrug-resistant *Acinetobacter baumannii* infections. Clinical Infectious Diseases. 2014;**59**(Suppl. 6):S381-S387

[105] Lin NT et al. Isolation and characterization of phi AB2: A novel bacteriophage of *Acinetobacter baumannii*. Research in Microbiology. 2010;**161**(4):308-314

[106] Aslam S et al. Lessons learned from the first 10 consecutive cases of intravenous bacteriophage therapy to treat multidrug-resistant bacterial infections at a single center in the United States. Open forum. Infectious Diseases. 2020;**7**(9):ofaa389

[107] Broncano-Lavado A et al. Advances in bacteriophage therapy against relevant multiDrug-resistant pathogens. Antibiotics. 2021;**10**(6):2-6

[108] Nielsen TB et al. Monoclonal antibody requires immunomodulation for efficacy against *Acinetobacter baumannii* infection. The Journal of Infectious Diseases. 2021;**224**(12):2133-2147

[109] García-Quintanilla M et al. Emerging therapies for multidrug resistant *Acinetobacter baumannii*. Trends in Microbiology. 2013;**21**(3):157-163

[110] Mohamed MF et al. A short D-enantiomeric antimicrobial peptide with potent immunomodulatory and antibiofilm activity against multidrug-resistant Pseudomonas aeruginosa and *Acinetobacter baumannii*. Scientific Reports. 2017;**7**(1):6953

[111] Luo G et al. Combination therapy with iron chelation and vancomycin in treating murine staphylococcemia. European Journal of Clinical Microbiology & Infectious Diseases. 2014;**33**(5):845-851

[112] Czaplewski L et al. Alternatives to antibiotics-a pipeline portfolio review. The Lancet Infectious Diseases. 2016;**16**(2):239-251

Chapter 14

Efflux Pumps among Urinary *E. coli* and *K. pneumoniae* Local Isolates in Hilla City, Iraq

Hussein Al-Dahmoshi, Sahar A. Ali and Noor Al-Khafaji

Abstract

Urinary tract infections (UTI) are the most common bacterial infections affecting humans. *Escherichia coli* and *Klebsiella pneumoniae* were common enterobacteria engaged with community-acquired UTIs. Efflux pumps were vital resistance mechanisms for antibiotics, especially among enterobacteria. Overexpression of an efflux system, which results in a decrease in antibiotic accumulation, is an effective mechanism for drug resistance. The ATP-binding cassette (ABC) transporters, small multidrug resistance (SMR), and multidrug and toxic compound extrusion (MATE) families, the major facilitator superfamily (MFS), and the resistance-nodulation- cell division (RND) family are the five superfamilies of efflux systems linked to drug resistance. This chapter highlights the results of studying the prevalence of efflux pump genes among local isolates of *E. coli* and *K. pneumoniae* in Hilla City, Iraq. class RND AcrAB-TolC, AcrAD-TolC, and AcrFE-TolC genes detected by conventional PCR of *E. coli* and *K. pneumoniae* respectively. The result revealed approximately all studied efflux transporter were found in both *E. coli* and *K. pneumoniae* in different percentages. Biofilm formation were observed in 50(100%) of *K. pneumoniae* and 49(98%) of *E. coli* isolates were biofilm former and follow: 30(60%), 20(40%) were weak, 12(24%), 22(44%) were moderate and 7(14%) and 8(16%) were Strong biofilm former for *E. coli* and *K. pneumoniae*, respectively.

Keywords: UTIs, AcrAB-TolC, AcrAD-TolC, AcrFE-TolC, EmrAB-TolC, EmrD, MdfA, EmrE, YnfA, MacAB-TolC, MdlAB-TolCTehA

1. Introduction

Urinary tract infections (UTI) are the most common bacterial infections affecting humans (Zhanel et al. [1]). They may be simple or complicated urinary tract infections (cUTIs), with the latter occurring in patients with urinary tract anatomic or functional abnormalities or major comorbidities [2]. UTIs may be categorized as either population- or hospital-acquired. In community-acquired UTIs, *Escherichia coli, Klebsiella pneumoniae,* and *Staphylococcus saprophyticus* are the most common bacteria, while in hospital-acquired UTIs, bacteria like *Staphylococcus aureus, Enterococcus* spp., *Proteus* spp., *Pseudomonas aeruginosa, Acinetobacter* spp., and *Candida* spp. are more common [1, 3], UTIs are disproportionately prevalent in women, approximately half of them experiencing a UTI during their lifetimes. Although UTIs are prevalent in all age groups, they tend to be more common in postmenopausal and elderly women, because it is more susceptible to infection under this situation [4]. Three-quarters of all UTIs in outpatients are caused by *E. coli* [5]. However, some *E. coli* lineages are more likely to induce UTI than others [6]. Antibiotic resistance is more common in uropathogenic *E. coli* than in commensal *E. coli*, and one uropathogenic lineage [7] sequence type131 (ST131), especially sublineage ST131:O25b:H30, is linked to multidrug resistance [8].

In addition to being a clinically significant pathogen, *K. pneumoniae* is a common cause of both hospital-acquired (HA) and community-acquired (CA) urinary tract infections (UTI). Limited therapeutic options have been created due to the enlarged resistance of this pathogen [9]. Enterobacteriaceae that produce extended-spectrum beta-lactamase (ESBL) and carbapenemase are frequently multidrug resistance, posing significant therapeutic challenges [10]. *K. pneumoniae* strains that produce ESBL/carbapenemase are widely recorded around the world, and their spread is critical [11]. Quinolone resistance among *K. pneumoniae* clinical isolates has become a serious problem in both developing and developed countries since quinolones are commonly prescribed as broad spectrum antimicrobial agents for the treatment of UTI induced by ESBL-producing *K. pneumonia*. Efflux pumps are present in all major types of bacterial membrane transporters, and increased efflux levels are thought to cause MDR. The most important types for the maintenance of *E. coli* in the human gut are AcrAB-TolC, EmrAB-TolC, and MdtM for extruding bile salts, mammalian steroids, and different antibiotics [12]. The influx of antimicrobial agents is reduced when the outer membrane's permeability is reduced. As a result, resistance develops in a number of essential clinical microorganisms. The first discovered efflux pump system was the tetracycline efflux pump by Stuart Levy et al. in *E. coli* [13].

Several studies indicate that efflux pumps can play at least four roles in biofilm formation efflux of EPSs and/or QS and quorum quenching (QQ) molecules to promote biofilm environment creation and regulate QS, correspondingly; indirect regulation of genes engagement in biofilm formation; abolition of harmful particles such as antibiotics and metabolic intermediates; and efflux of harmful molecules such as antibiotics and metabolic intermediates and by encouraging or preventing adhesion to surfaces and other cells, aggregation can be influenced [14]. Overexpression of an efflux system, which results in a decrease in antibiotic accumulation, is an effective mechanism for drug resistance [15, 16].

1.1 Efflux pumps

Efflux pumps are membrane proteins that are involved in the export of noxious substances from within the bacterial cell into the external environment. Efflux proteins are found in both Gram-negative and Gram-positive bacteria as well as in eukaryotic organisms [17]. The ATP-binding cassette (ABC) superfamily [18], the resistance-nodulation-division (RND) family [19], the small multidrug resistance (SMR) family [20], the major facilitator superfamily (MFS) [21], and the multidrug and toxic compound extrusion (MATE) [22] family. The ABC family uses ATP hydrolysis to power substrate export, while the other families depend on the proton motive force for energy. The MFS, ABC, SMR, and MATE families are found in Gram-positive and Gram-negative bacteria, respectively, while the RND superfamily is only found in Gram-negative bacteria. Members of the RND family are often found as part of a three-part complex that spans Gram-negative bacteria's two membranes [23].

1.1.1 Resistance-nodulation-division (RND) efflux pump

This type of pump is occupied as a tripartite complex constituting the RND protein (the inner membrane component), membrane fusion protein (the periplasmic compartment), and the outer membrane protein. These three proteins form a constitute channel crossways the Gram-negative cell envelope guaranteeing that the molecule, taken from the outer leaflet of the inner membrane bilayer, is replied directly transversely to the periplasm and the outer membrane into the exterior medium with the assistance of the proton-gradient as an energy source. RND family shows a significant role in the intrinsic resistance of Gram-negative bacteria [24, 25]. The most frequent member of RND family in enterobacteria include AcrAB-TolC, AcrAD-TolC, and AcrFE-TolC.

AcrAB-TolC, one of the efflux systems, constitutively expressed in *E. coli*, is composed of the outer membrane protein TolC, the inner membrane transporter AcrB, and the periplasmic adaptor protein AcrA [26]. Overexpression of the AcrABTolC efflux pump is an intrinsic mechanism of multidrug resistance in Gram-negative bacteria [27]. AcrA, a highly elongated protein, is thought to bring the outer and inner membranes closer. It composed a trimer that interacts with a monomeric AcrB, which was shown by in vitro reconstitution to be a proton antiporter [28]. Efflux pumps, such as AcrAB-TolC and MexAB-OprM, are essential for bacteria survival and virulence/colonization, especially during the course of infection when the pathogen is attacked by toxic substances or adherence to the host [29]. The resistance mechanism by efflux pumps is the most important antibiotic resistance type because the efflux pump is able to remove more than one antibiotic such as B-lactam, erythromycin fluoroquinolones, and chloramphenicol [30]. Ruzin et al. [31] showed up that the resistance to quinolones related to the overexpression of the AcrAB efflux pump of *K. pneumoniae* was associated with resistance to other antibiotics, including erythromycin, chloramphenicol, tetracycline, and also tigecycline, a recently commercialized molecule.

AcrD is a component of an efflux pump that mediates the export of aminoglycosides and a few amphiphilic compounds such as sodium dodecyl sulfate (SDS), deoxycholate, and novobiocin, AcrA is a periplasmic fusion protein that also exports aminoglycosides in association with the cytoplasmic membrane protein AcrD [30]. Despite increased acrA expression, the mutant strain showed no increase in resistance, suggesting that the deletion of acrD contributes to adaptive resistance directly, rather than indirectly via mechanisms like control of other efflux components. As a result, the previously reported adaptive cross-resistance to non-aminoglycoside antibiotics cannot be due to the development of the AcrAB-TolC complex as shown by an acrD mutant that showed significantly reduced biofilm formation and expression of key biofilm proteins encoded by csgBD, the AcrD efflux pump has an effect on biofilm formation, and appears to play a special biological function, according to Buckner et al. [32] findings. The transcriptome showed major changes, which supported this theory. The transcriptomes of the acrD mutant were compared to the transcriptome of the acrB mutant, which had previously been released AcrD is not a "backup" efflux pump, but serves a physiological function in the cell, as shown by the fact that the effect was quite distinct. This comparison found 232 major gene expression changes that were only caused by the inactivation of acrD and not by the disruption of acrB. Both the acrB and acrD mutant transcriptomes had 169 genes that were differentially expressed as compared to the acrB mutant transcriptome. Experiments have shown that the AcrB and AcrD efflux pumps have different substrate profiles when it refers to aminoglycoside antibiotics [33]. AcrEF-TolC pump is known to exhibit higher expression levels in quinolone-resistant *E. coli*. AcrEF shares high homology (65–77%) with AcrAB and so AcrEF-TolC expression complement the activity of AcAB-TolC. Overexpression of acrEF restores resistance to acriflavine, erythromycin, novobiocin, and crystal violet. Additionally, AcrEF is responsible for increased resistance for compounds such as dyes, detergents, and antibiotic substrates like that of AcrAB especially levofloxacin [34].

1.1.2 ATP-binding cassette (ABC) efflux pump

There is a variety of transport systems in *E. coli*, including ABC-type transporters as well as substrate-binding proteins, outer membrane receptors, and a number of transporters with various functions. We have gained a better understanding of the molecular basis of transport through recent structures of ATPases, substrate-binding proteins, and full-length transporters [35]. Specialized ABC transporter types transport a diverse range of substrates, ranging from small molecules such as ions, sugars, or amino acids to larger compounds such as antibiotics, drugs, lipids, and oligopeptides [36]. The most important ABC family efflux pumps in *E. coli* include MAcAB-TolC and MdlAB-TolC.

The MacA-MacB-TolC assembly of *E. coli* is a transmembrane machine that spans the cell envelope and actively extrudes substrates, including macrolide antibiotics and polypeptide virulence factors. These transport processes are energized by the ATPase MacB, a member of the ATP-binding cassette (ABC) superfamily. A hexamer of the periplasmic protein MacA bridges between a TolC trimer in the outer membrane and a MacB dimer in the inner membrane, generating a quaternary structure with a central channel for substrate translocation. A gating ring found in MacA is proposed to act as a one-way valve in substrate transport. The MacB structure features an atypical transmembrane domain with a closely packed dimer interface and a periplasmic opening that is the likely portal for substrate entry from the periplasm, with subsequent displacement through an allosteric transport mechanism [37].

In a macrolide-susceptible AcrAB deficient *E. coli* strain, only overexpression of MacAB may increase resistance to macrolide antibiotics. MacAB, on the other hand, has recently been linked to the secretion of an *E. coli* heat-stable enterotoxin [38]. MacA is the MFP in the MacA–MacB–TolC pump, and due to high sequence similarity, it is predicted to share structural similarities with AcrA (44%) [39]. The C-terminal periplasmic membrane-proximal domain of MacA is required for these MacA–MacB interactions [40]. By modifying the conformation of MacA's membrane-proximal domain and disrupting the proper assembly of the MacA–MacB complex, a single G353A substitution in this domain impaired MacAB–TolC function [41]. In the *E. coli* genome, five putative open reading frame (ORF) clusters, mdlAB, ybjYZ, yddA, yojHI, and yhiH, have been assumed to be possible genes for ABC drug efflux transporters. MdlAb-TolC is multidrug efflux transporter with few studies concerned it is a function [42].

1.1.3 Small multidrug resistance (SMR) efflux pump

Small multidrug resistance transporters (SMR) systems provide to study the minimal requirements for active transport [43]. They are also small multidrug transporters, with four transmembrane helices and no significant extra membrane domain, although they function as dimers the minimum functional unit is a bundle of eight α-helices [44]. SMR transporter exports a broad class of polyaromatic cation substrates, thus conferring resistance to drug compounds matching this chemical description. Genes encoding SMR proteins (variously annotated emrE, ynfA and tehA) are frequently found in mobile drug resistance gene arrays, and provide a broad selective advantage by conferring resistance to ubiquitous environmental pollutants with low-grade toxicity to microbes [45]. The SMR family consists of small hydrophobic proteins of about 100 amino acid residues with four transmembrane α-helical spanners [46].

SMR family includes more than 40 proteins in eubacteria, a few of which have been studied in detail. One of them, EmrE, is an *E. coli* multidrug transporter (MDT), that utilizes proton gradients as an energy source to drive substrate translocation and confers resistance to a wide variety of toxicants by actively exchanging them with hydrogen ions [47]. EmrE is the smallest ion-coupled transporter known; it functions as an oligomer and each monomer comprises four transmembrane segments [48]. EmrE is a tetramer comprised of two conformational heterodimers related by a pseudo-two-fold symmetry axis perpendicular to the cell membrane. Based on the structure and biochemical evidence, we propose a mechanism by which EmrE accomplishes multidrug efflux by coupling conformational changes between two heterodimers with proton gradient [49]. The overexpression of EmrE causes bacteria to become resistant to a wide variety of toxic cationic hydrophobic compounds such as ethidium bromide, methyl viologen, tetracycline, and tetraphenylphosphonium, as well as other antiseptics and intercalating dyes [50].

The gene, ynfA of *E. coli* is the newest member of the small multidrug resistance (SMR) gene family, identified in both Gram-negative and Gram-positive bacterial species. It might be involved alone or with tolC or any other way by complex regulation in which the initial susceptible bacteria become resistant. The level of ynfA gene expression was observed between 2 and 6 folds equivalent to tolC gene [51].

1.1.4 Major facilitator superfamily (MFS) efflux pump

The major facilitator superfamily (MFS) is the largest known superfamily of secondary active transporters. MFS transporters are responsible for transporting a broad spectrum of substrates, either down their concentration gradient or uphill using the energy stored in the electrochemical gradients. The major facilitator transporters form a superfamily that composed a number of subfamilies; of these subfamilies, transporters of sugars and drugs are by far the most numerous [52].

These MFS transporters are typically composed of approximately 400 amino acids that are putatively arranged in 12 membrane-spanning helices, with a large cytoplasmic loop among helices six and seven [43]. The MFS family of drug transporters is made up of two domains that are centered around a central pore and two domains that transfer conformations from the cytoplasmic to the periplasmic side of the membrane in response to a Na + or H+ ion gradient [53]. The MFS drug transporters are classified into subfamilies —12-helix and 14-helix transporters (e.g.,

TetA(B) and TetA(K), class two and class K tetracycline transporters from *E. coli* and S. aureus respectively [54]. *E. coli* have many MFS transporter-like EmrAB-TolC, EmrD, MdfA.

EmrAB–TolC from *E. coli* is such a tripartite system, comprised of EmrB an MFS transporter, EmrA, a membrane fusion protein, and TolC, an outer membrane channel. The whole complex is predicted to form a continuous channel allowing direct export from the cytoplasm to the exterior of the cell [55]. The components of EmrAB-TolC were identified in *E. coli* for the first time more than a decade ago resistance to hydrophobic toxins like carbonyl cyanide m-chlorophenyl-hydrazone (CCCP). Its overexpression causes increased resistance to nalidixic acid, thiolactomycin, nitroxoline, hydrophobic proton uncouplers [56].

EmrD is a multidrug transporter from the Major Facilitator Superfamily that expels amphipathic compounds across the inner membrane of *E. coli*. It can transport detergents such as benzalkonium and sodium dodecylsulfate [57]. EmrD may have vital role in biofilm formation via the efflux of arabinose, which promotes cell aggregation and biofilm formation [58]. MdfA is drug/proton antiporter consisting of 410 amino acid long membrane protein responsible for resistance to a diverse group of cationic or zwitterionic lipophilic compounds such as ethidium bromide, tetraphenylphosphonium, rhodamine, daunomycin, benzalkonium, rifampin, tetracycline, and puromycin. Surprisingly, however, MdfA also confers resistance to chemically unrelated, clinically important antibiotics such as chloramphenicol, erythromycin, and certain aminoglycosides and fluoroquinolones. Synergistic overexpression of *mdfA* along with *acrAB* leads to increases in quinolone resistance [59].

1.1.5 Multidrug and toxic compound extrusion (MATE) efflux pump

Export of substrates and toxins by the cell is a fundamental life process and members of the MATE family represent the last class of multidrug resistance (MDR) transporters to be structurally characterized. MATE transporters involved a variety of important biological functions across all kingdoms of life [60]. MATE transporters are very similar in size to the MFS transporters and are typically composed of approx. 450 amino acids which are putatively arranged into 12 helices however, they do not have any sequence similarity to members of the MFS transporters [61]. MdtK is one of the important MATE inner membrane transporter in *E. coli* conferring resistance to quinolone and fluoroquinolone when overexpressed [62].

1.2 Antibiotic resistance and efflux pumps

The study include antibiotic susceptibility profile (for 20 antibiotics) according to CLSI-2021 [63] and efflux pumps gene profile for (19 genes) for 50 isolates of *E. coli* and 50 isolates of *K. pneumoniae* isolated from patients with UTIs. The results revealed high level of resistance to β-lactams and cephalosporin and low level of resistance to piperacillin, aminoglycosides and carbapenem (**Table 1**). Multidrug resistance for more than 3 antibiotics (at least one for each class) were studied and the results revealed that 68% of *E. coli* and 90% of *K. pneumoniae* were MDR (**Table 2**). Results of biofilm formation shown that, approximately all isolates were biofilm former (**Table 3**). Concern presence of efflux pump genes, the results of polymerase chain reaction revealed that: acrA 50 (100%), 48 (96%)–acrB

Antibiotic	Resistance %	
	E. coli	***K. pneumoniae***
Amoxicillin	100%	92%
Piperacillin	14%	16%
Ceftriaxone	54%	92%
Ceftazidime	100%	98%
Cefepime	58%	80%
Cefixime	52%	84%
Cefotaxime	100%	92%
Cefoxitin	42%	46%
Nitrfuraniton	38%	58%
streptomycin	36%	78%
Gentamycin	4%	40%
Kanamycin	42%	50%
Tobramycin	20%	44%
Amikacin	14%	10%
Netlimicin	2%	4%
Imipenem	0%	12%
Meropenem	6%	8%
Aztreonam	20%	58%
Azithromycin	10%	24%
Nalidixic acid	26%	16%

Table 1.
Antibiotic resistance among E. coli *and* Klebsiella pneumoniae.

Classes of MDR	*E. coli*	*K. pneumoniae*
MDR-8 classes	0%	4%
MDR-7 classes	4%	8%
MDR-6 classes	12%	20%
MDR-5 classes	2%	22%
MDR-4 classes	24%	28%
MDR-3 classes	20%	8%
non-MDR	38%	10%
Total	100%	100%

Table 2.
Classes of MDR among E. coli *and* Klebsiella pneumoniae.

43 (86%), 44 (88%)–acrD 48 (96%), 46(92%)–acrF 33 (66%), 32 (64%)–acrE 50 (100%), 46 (92%)–and tolC 50 (100%), 50 (100%), while class MFS pumps (EmrAB-TolC, EmrD and MdfA) were investigated for *E. coli* and *K. pneumonia* the

Biofilm Formation Pattern	E. coli	K. pneumoniae
non-biofilm former	2%	0%
weak biofilm former	60%	40%
moderate biofilm former	24%	44%
strong biofilm former	14%	16%
Total	100%	100%

Table 3.
Biofilm formation patterns among E. coli *and* Klebsiella pneumoniae.

Efflux pumps gene	Presence %	
	E. coli	*K. pneumoniae*
acrA	100	96
acrB	86	88
acrD	96	100
tolC	100	92
acrF	66	64
acrE	100	92
mdfA	98	100
emrD	100	100
emrA	100	96
emrB	100	98
emrE	96	70
ynfA	100	66
tehA	98	76
macA	100	76
macB	98	96
mdlA	100	76
mdlB	98	100
mdtK	100	100
dinF	100	86

Table 4.
Efflux pump genes among E. coli *and* Klebsiella pneumoniae.

results emrA 50 (100%), 48 (96%)–emrB 50 (100%), 49 (98%)–emrD 50 (100%), 50 (100%)–and mdfA 49 (98%), 50 (100%), class SMR pumps (EmrE, YnfA and TehA) genes were distributed as follow:: emrE 48 (96%), 35 (70%)–ynfA 50 (100%), 33 (66%)–tehA 49 (98%), 38 (76%), Two class ABC pumps (MacAB-TolC and MdlAB-TolC) The result revealed that the: macA 50 (100%), 38 (76%), macB 49 (98%), 48 (96%)–mdlA 50 (100%), 38 (765%)–and mdlB 49 (98%), 50 (100%), Two MATE pumps (MdtK and DinF) genes were studies and the results revealed that: mdtK and dinF genes were present in all *E. coli* isolates while *K. pneumoniae* revealed 50 (100%), 43 (86%) of MdtK and DinF respectively (**Table 4**).

1.3 Coexisted genotypes of efflux pumps

Concern results of coexisted pumps in the same *E. coli* or *K. pneumoniae* isolate the results were shown in **Tables 5** and **6**.

Genotype	No.	%
AcrAB-TolC/AcrAD-TolC/AcrFE-TolC/MdfA/EmrD/EmrAB-TolC/EmrE/YnfA/ TehA/ MacAB-TolC/MdlAB-TolC/Mdtk/DinF	32	64
AcrAB-TolC/ AcrAD-TolC/ AcrFE-TolC/ MdfA/ EmrD/ EmrAB-TolC/ YnfA/ TehA/ MacAB-TolC/ MdlAB-TolC/ Mdtk/ DinF	1	2
AcrAB-TolC/ AcrAD-TolC/ MdfA/ EmrD/ EmrAB-TolC/ EmrE/ YnfA/ TehA/ MacAB-TolC/ MdlAB-TolC/ Mdtk/ DinF	16	32
AcrAB-TolC/ AcrAD-TolC/ MdfA/ EmrD/ EmrAB-TolC/ EmrE/ YnfA/ MacAB-TolC	1	2

Table 5.
Co-existed efflux pump genes among E. coli.

Genotypes	No.	%
AcrAB-TolC/AcrAD-TolC/AcrFE-TolC/MdfA/EmrD/EmrAB-TolC/EmrE/YnfA/ TehA/MacAB-TolC/MdlAB-TolC/Mdtk/DinF	8	16
AcrAB-TolC/AcrAD-TolC/AcrFE-TolC/MdfA/EmrD/EmrAB-TolC/YnfA/TehA/ MacAB-TolC/MdlAB-TolC/Mdtk/DinF	3	6
AcrAB-TolC/AcrAD-TolC/AcrFE-TolC/MdfA/EmrD/EmrAB-TolC/EmrE/TehA/ MacAB-TolC/MdlAB-TolC/Mdtk/DinF	2	4
AcrAB-TolC/AcrAD-TolC/AcrFE-TolC/MdfA/EmrD/EmrAB-TolC/EmrE/YnfA/ TehA/MacAB-TolC/MdlAB-TolC/Mdtk	2	4
AcrAB-TolC/AcrAD-TolC/MdfA/EmrD/EmrAB-TolC/EmrE/ YnfA/TehA/ MacAB-TolC/MdlAB-TolC/Mdtk/DinF	8	16
AcrAB-TolC/AcrAD-TolC/AcrFE-TolC/MdfA/EmrD/EmrAB-TolC/EmrE/YnfA/ MacAB-TolC/Mdtk/DinF	3	6
AcrAB-TolC/AcrAD-TolC/AcrFE-TolC/MdfA/EmrD/EmrAB-TolC/EmrE/YnfA/ TehA/MdlAB-TolC/Mdtk	2	4
AcrAB-TolC/AcrAD-TolC/AcrFE-TolC/MdfA/EmrD/EmrAB-TolC/YnfA/TehA/ MacAB-TolC/MdlAB-TolC/Mdtk/DinF	1	2
AcrAB-TolC/AcrAD-TolC/AcrFE-TolC/MdfA/EmrD/EmrAB-TolC/YnfA/TehA/ MacAB-TolC/MdlAB-TolC/Mdtk/DinF	1	2
AcrAB-TolC/AcrAD-TolC/AcrFE-TolC/MdfA/EmrD/EmrAB-TolC/EmrE/TehA/ MacAB-TolC/MdlAB-TolC/Mdtk/DinF	1	2
AcrAB-TolC/AcrAD-TolC/AcrFE-TolC/MdfA/EmrD/EmrAB-TolC/EmrE/YnfA/ MacAB-TolC/MdlAB-TolC/Mdtk/DinF	1	2
AcrAB-TolC/AcrAD-TolC/AcrFE-TolC/MdfA/EmrD/EmrAB-TolC/EmrE/ MacAB-TolC/MdlAB-TolC/Mdtk/DinF	1	2

Genotypes	No.	%
AcrAB-TolC/AcrAD-TolC/AcrFE-TolC/MdfA/EmrD/EmrAB-TolC/EmrE/YnfA/ TehA/MdlAB-TolC/Mdtk/DinF	1	2
AcrAB-TolC/AcrAD-TolC/AcrFE-TolC/MdfA/EmrD/EmrAB-TolC/EmrE/TehA/ MacAB-TolC/Mdtk/DinF	1	2
AcrAB-TolC/AcrAD-TolC/MdfA/EmrD/EmrAB-TolC/EmrE/ TehA/MacAB-TolC/ MdlAB-TolC/Mdtk/DinF	1	2
AcrAB-TolC/AcrAD-TolC/AcrFE-TolC/MdfA/EmrD/EmrAB-TolC/EmrE/TehA/ MdlAB-TolC/Mdtk/DinF	1	2
AcrAB-TolC/AcrAD-TolC/AcrFE-TolC/MdfA/EmrD/EmrAB-TolC/MacAB-TolC/ MdlAB-TolC/Mdtk/DinF	1	2
AcrAB-TolC/AcrAD-TolC/AcrFE-TolC/MdfA/EmrD/EmrAB-TolC/YnfA/TehA/ MdlAB-TolC/Mdtk/DinF	1	2
AcrAB-TolC/AcrAD-TolC/MdfA/EmrD/EmrAB-TolC/TehA/ MacAB- TolC/ MdlAB-TolC/Mdtk/DinF	1	2
AcrAB-TolC/AcrAD-TolC/MdfA/EmrD/EmrAB-TolC/YnfA/TehA/ MacAB-TolC/ MdlAB-TolC/Mdtk	1	2
AcrAB-TolC/AcrAD-TolC/MdfA/EmrD/EmrAB-TolC/TehA/ MacAB- TolC/ MdlAB-TolC/Mdtk/DinF	1	2
AcrAB-TolC/AcrAD-TolC/MdfA/EmrD/EmrAB-TolC/EmrE/ MacAB-TolC/ MdlAB-TolC/Mdtk/DinF	1	2
AcrAB-TolC/AcrAD-TolC/AcrFE-TolC/MdfA/EmrD/EmrAB-TolC/MacAB-TolC/ Mdtk/DinF	1	2
AcrAB-TolC/AcrAD-TolC/MdfA/EmrD/EmrAB-TolC/EmrE/TehA/MacAB-TolC/ Mdtk/DinF	1	2
AcrAB-TolC/AcrAD-TolC/AcrFE-TolC/MdfA/EmrD/YnfA/TehA/ MdlAB-TolC/ Mdtk/ DinF	1	2
AcrAB-TolC/AcrAD-TolC/AcrFE-TolC/MdfA/EmrD/EmrE/ YnfA/MdlAB-TolC/ Mdtk/DinF	1	2
AcrAB-TolC/AcrAD-TolC/MdfA/EmrD/EmrAB-TolC/MacAB-TolC/Mdtk/DinF	1	2
AcrAB-TolC/MdfA/EmrD/EmrAB-TolC/EmrE/TehA/MacAB-TolC/Mdtk	1	2
AcrFE-TolC/MdfA/EmrD/EmrAB-TolC/YnfA/MdlAB-TolC/Mdtk/ DinF	1	2

Table 6.
Co-existed efflux pump genes among Klebsiella pneumoniae.

2. Conclusion

There is a strong correlation between antibiotic resistance, especially to β-lactams, and the presence of efflux pump genes, which may be reflected in biofilm formation in both *E. coli* and *K. pneumoniae.*

Conflict of interest

There is no conflict of interest for this work.

DOI: http://dx.doi.org/10.5772/intechopen.104408

Author details

Hussein Al-Dahmoshi[1*], Sahar A. Ali[2] and Noor Al-Khafaji[1]

1 Biology Department, College of Science, University of Babylon, Hilla, Iraq

2 General Directorate for the Education of Babylon, Hilla, Iraq

*Address all correspondence to: dr.dahmoshi83@gmail.com

References

[1] Zhanel GG, Walkty AJ, Karlowsky JA. Fosfomycin: A first-line oral therapy for acute uncomplicated cystitis. Canadian Journal of Infectious Diseases and Medical Microbiology. 2016;**2016**:2082693. DOI: 10.1155/2016/2082693

[2] Bader MS, Hawboldt J, Brooks A. Management of complicated urinary tract infections in the era of antimicrobial resistance. Postgraduate Medicine. 2010;**122**(6):7-15. DOI: 10.3810/pgm.2010.11.2217

[3] Liu HY, Lin HC, Lin YC, Yu SH, Wu WH, Lee YJ. Antimicrobial susceptibilities of urinary extended-spectrum beta-lactamase-producing *Escherichia coli* and *Klebsiella pneumoniae* to fosfomycin and nitrofurantoin in a teaching hospital in Taiwan. Journal of Microbiology, Immunology, and Infection. 2011;**44**(5):364-368. DOI: 10.1016/j.jmii.2010.08.012

[4] Neugent ML, Hulyalkar NV, Nguyen VH, Zimmern PE, De Nisco NJ. Advances in understanding the human urinary microbiome and its potential role in urinary tract infection. MBio. 2020;**11**(2):e00218-e00220. DOI: 10.1128/mBio.00218-20

[5] Foxman B. Urinary tract infection syndromes: Occurrence, recurrence, bacteriology, risk factors, and disease burden. Infectious Disease Clinics of North America. 2014;**28**(1):1-13. DOI: 10.1016/j.idc.2013.09.003

[6] Lo Y, Zhang L, Foxman B, Zöllner S. Whole-genome sequencing of uropathogenic *Escherichia coli* reveals long evolutionary history of diversity and virulence. Infection, Genetics and Evolution. 2015;**34**:244-250. DOI: 10.1016/j.meegid.2015.06.023

[7] Dang TND, Zhang L, Zöllner S, Srinivasan U, Abbas K, Marrs CF, et al. Uropathogenic *Escherichia coli* are less likely than paired fecal *E. coli* to have CRISPR loci. Infection, Genetics and Evolution. 2013;**19**:212-218. DOI: 10.1016/j.meegid.2013.07.017

[8] Graham SE, Zhang L, Ali I, Cho YK, Ismail MD, Carlson HA, et al. Prevalence of CTX-M extended-spectrum beta-lactamases and sequence type 131 in Korean blood, urine, and rectal *Escherichia coli* isolates. Infection, Genetics and Evolution. 2016;**41**:292-295. DOI: 10.1016/j.meegid.2016.04.020

[9] Caneiras C, Lito L, Melo-Cristino J, Duarte A. Community- and hospital-acquired *Klebsiella pneumoniae* urinary tract infections in Portugal: Virulence and antibiotic resistance. Microorganisms. 2019;**7**(5):138. DOI: 10.3390/microorganisms7050138

[10] Marques C, Menezes J, Belas A, Aboim C, Cavaco-Silva P, Trigueiro G, et al. Klebsiella pneumoniae causing urinary tract infections in companion animals and humans: Population structure, antimicrobial resistance and virulence genes. The Journal of Antimicrobial Chemotherapy. 2019;**74**(3):594-602. DOI: 10.1093/jac/dky499

[11] Ewers C, Stamm I, Pfeifer Y, Wieler LH, Kopp PA, Schønning K, et al. Clonal spread of highly successful ST15-CTX-M-15 Klebsiella pneumoniae in companion animals and horses. The Journal of Antimicrobial Chemotherapy. 2014;**69**(10):2676-2680. DOI: 10.1093/jac/dku217

[12] Chetri S, Dolley A, Bhowmik D, Chanda DD, Chakravarty A,

Bhattacharjee A. Transcriptional response of AcrEF-TolC against fluoroquinolone and carbapenem in Escherichia coli of clinical origin. Indian Journal of Medical Microbiology. 2018;**36**(4):537-540. DOI: 10.4103/ijmm.IJMM_18_308

[13] Kumar S, Varela MF. Molecular mechanisms of bacterial resistance to antimicrobial agents. Chemotherapy. 2013;**14**:522-534

[14] Alav I, Sutton JM, Rahman KM. Role of bacterial efflux pumps in biofilm formation. The Journal of Antimicrobial Chemotherapy. 2018;**73**(8):2003-2020. DOI: 10.1093/jac/dky042

[15] Nikaido H, Takatsuka Y. Mechanisms of RND multidrug efflux pumps. Biochimica et Biophysica Acta. 2009;**1794**(5):769-781. DOI: 10.1016/j.bbapap.2008.10.004

[16] Kumar S, Varela MF. Biochemistry of bacterial multidrug efflux pumps. International Journal of Molecular Sciences. 2012;**13**(4):4484-4495. DOI: 10.3390/ijms13044484

[17] Saier MH. A functional-phylogenetic classification system for transmembrane solute transporters. Microbiology and Molecular Biology Reviews. 2000;**64**(2):354-411. DOI: 10.1128/MMBR.64.2.354-411.2000

[18] Lubelski J, Konings WN, Driessen AJ. Distribution and physiology of ABC-type transporters contributing to multidrug resistance in bacteria. Microbiology and Molecular Biology Reviews. 2007;**71**(3):463-476. DOI: 10.1128/MMBR.00001-07

[19] Anes J, McCusker MP, Fanning S, Martins M. The ins and outs of RND efflux pumps in Escherichia coli. Frontiers in Microbiology. 2015;**6**:587. DOI: 10.3389/fmicb.2015.00587

[20] Chung YJ, Saier MH Jr. SMR-type multidrug resistance pumps. Current Opinion in Drug Discovery & Development. 2001;**4**(2):237-245

[21] Law CJ, Maloney PC, Wang DN. Ins and outs of major facilitator superfamily antiporters. Annual Review of Microbiology. 2008;**62**:289-305. DOI: 10.1146/annurev.micro.61.080706.093329

[22] Kuroda T, Tsuchiya T. Multidrug efflux transporters in the MATE family. Biochimica et Biophysica Acta. 2009;**1794**(5):763-768. DOI: 10.1016/j.bbapap.2008.11.012

[23] Nikaido H. Structure and mechanism of RND-type multidrug efflux pumps. Advances in Enzymology and Related Areas of Molecular Biology. 2011;**77**:1-60. DOI: 10.1002/9780470920541.ch1

[24] Nikaido H, Pagès JM. Broad-specificity efflux pumps and their role in multidrug resistance of gram-negative bacteria. FEMS Microbiology Reviews. 2012;**36**(2):340-363. DOI: 10.1111/j.1574-6976.2011.00290.x

[25] Zgurskaya HI. Multicomponent drug efflux complexes: Architecture and mechanism of assembly. Future Microbiology. 2009;**4**(7):919-932. DOI: 10.2217/fmb.09.62

[26] Shi X, Chen M, Yu Z, Bell JM, Wang H, Forrester I, et al. In situ structure and assembly of the multidrug efflux pump AcrAB-TolC. Nature Communications. 2019;**10**(1):2635. DOI: 10.1038/s41467-019-10512-6

[27] Blair JM, Webber MA, Baylay AJ, Ogbolu DO, Piddock LJ. Molecular mechanisms of antibiotic resistance. Nature Reviews. Microbiology. 2015;**13**(1):42-51. DOI: 10.1038/nrmicro3380

[28] Nikaido H, Zgurskaya HI. AcrAB and related multidrug efflux pumps of Escherichia coli. Journal of Molecular Microbiology and Biotechnology. 2001;**3**(2):215-218

[29] Zgurskaya HI, Nikaido H. Multidrug resistance mechanisms: Drug efflux across two membranes. Molecular Microbiology. 2000;**37**(2):219-225. DOI: 10.1046/j.1365-2958.2000.01926.x

[30] Elkins CA, Nikaido H. Substrate specificity of the RND-type multidrug efflux pumps AcrB and AcrD of Escherichia coli is determined predominantly by two large periplasmic loops. Journal of Bacteriology. 2002;**184**(23):6490-6498. DOI: 10.1128/JB.184.23.6490-6499.2002

[31] Ruzin A, Keeney D, Bradford PA. AcrAB efflux pump plays a role in decreased susceptibility to tigecycline in Morganella morganii. Antimicrobial Agents and Chemotherapy. 2005;**49**(2):791-793. DOI: 10.1128/AAC.49.2.791-793.2005

[32] Buckner MM, Blair JM, La Ragione RM, Newcombe J, Dwyer DJ, Ivens A, et al. Beyond antimicrobial resistance: Evidence for a distinct role of the AcrD efflux pump in Salmonella biology. mBio. Nov 22, 2016;**7**(6):1-10. DOI: 10.1128/mBio.01916-16. PMID: 27879336

[33] Kobayashi N, Tamura N, van Veen HW, Yamaguchi A, Murakami S. β-Lactam selectivity of multidrug transporters AcrB and AcrD resides in the proximal binding pocket. The Journal of Biological Chemistry. 2014;**289**(15):10680-10690. DOI: 10.1074/jbc.M114.547794

[34] Kobayashi K, Tsukagoshi N, Aono R. Suppression of hypersensitivity of Escherichia coli acrB mutant to organic solvents by integrational activation of the acrEF operon with the IS1 or IS2 element. Journal of Bacteriology. 2001;**183**(8):2646-2653. DOI: 10.1128/JB.183.8.2646-2653.2001

[35] Moussatova A, Kandt C, O'Mara ML, Tieleman DP. ATP-binding cassette transporters in Escherichia coli. Biochimica et Biophysica Acta. 2008;**1778**(9):1757-1771. DOI: 10.1016/j.bbamem.2008.06.009

[36] Fuellen G, Spitzer M, Cullen P, Lorkowski S. Correspondence of function and phylogeny of ABC proteins based on an automated analysis of 20 model protein data sets. Proteins. 2005;**61**(4):888-899. DOI: 10.1002/prot.20616

[37] Fitzpatrick AWP, Llabrés S, Neuberger A, Blaza JN, Bai XC, Okada U, et al. Structure of the MacAB–TolC ABC-type tripartite multidrug efflux pump. Nature Microbiology. 2017;**2**(7):17070. DOI: 10.1038/nmicrobiol.2017.70

[38] Yamanaka H, Kobayashi H, Takahashi E, Okamoto K. MacAB is involved in the secretion of Escherichia coli heat-stable enterotoxin II. Journal of Bacteriology. 2008;**190**(23):7693-7698. DOI: 10.1128/JB.00853-08

[39] Zgurskaya HI, Yamada Y, Tikhonova EB, Ge Q, Krishnamoorthy G. Structural and functional diversity of bacterial membrane fusion proteins. Biochimica et Biophysica Acta. 2009;**1794**(5):794-807. DOI: 10.1016/j.bbapap.2008.10.010

[40] Modali SD, Zgurskaya HI. The periplasmic membrane proximal domain of MacA acts as a switch in stimulation of ATP hydrolysis by MacB transporter. Molecular Microbiology. 2011;**81**(4):937-951. DOI: 10.1111/j.1365-2958.2011.07744.x

[41] Yum S, Xu Y, Piao S, Sim SH, Kim HM, Jo WS, et al. Crystal structure of the periplasmic component of a tripartite macrolide-specific efflux pump. Journal of Molecular Biology. 2009;**387**(5):1286-1297. DOI: 10.1016/j.jmb.2009.02.048

[42] Foo JL, Jensen HM, Dahl RH, George K, Keasling JD, Lee TS, et al. Improving microbial biogasoline production in Escherichia coli using tolerance engineering. MBio. 2014;**5**(6):e01932. DOI: 10.1128/mBio.01932-14

[43] Kroeger JK, Hassan K, Vörös A, Simm R, Saidijam M, Bettaney KE, et al. Bacillus cereus efflux protein BC3310 - a multidrug transporter of the unknown major facilitator family, UMF-2. Frontiers in Microbiology. 2015;**6**:1063. DOI: 10.3389/fmicb.2015.01063

[44] Higgins CF. Multiple molecular mechanisms for multidrug resistance transporters. Nature. 2007;**446**(7137):749-757. DOI: 10.1038/nature05630

[45] Kermani AA, Macdonald CB, Burata OE, Ben Koff BB, Koide A, Denbaum E, et al. The structural basis of promiscuity in small multidrug resistance transporters. Nature Communications. 2020;**11**(1):6064. DOI: 10.1038/s41467-020-19820-8

[46] Schuldiner S, Granot D, Steiner SS, Ninio S, Rotem D, Soskin M, et al. Precious things come in little packages. Journal of Molecular Microbiology and Biotechnology. 2001;**3**(2):155-162

[47] Tal N, Schuldiner S. A coordinated network of transporters with overlapping specificities provides a robust survival strategy. Proceedings of the National Academy of Sciences of the United States of America. 2009;**106**(22):9051-9056. DOI: 10.1073/pnas.0902400106

[48] Gottschalk KE, Soskine M, Schuldiner S, Kessler H. A structural model of EmrE, a multi-drug transporter from Escherichia coli. Biophysical Journal. 2004;**86**(6):3335-3348. DOI: 10.1529/biophysj.103.034546

[49] Ma C, Chang G. Structure of the multidrug resistance efflux transporter EmrE from Escherichia coli. Proceedings of the National Academy of Sciences of the United States of America. 2004;**101**(9):2852-2857. DOI: 10.1073/pnas.0400137101

[50] Yerushalmi H, Schuldiner S. A model for coupling of H(+) and substrate fluxes based on "time-sharing" of a common binding site. Biochemistry. 2000;**39**(48):14711-14719. DOI: 10.1021/bi001892i

[51] Sarkar SK, Bhattacharyya A, Mandal SM. YnfA, a SMR family efflux pump is abundant in Escherichia coli isolates from urinary infection. Indian Journal of Medical Microbiology. 2015;**33**(1):139-142. DOI: 10.4103/0255-0857.148415

[52] Drew D, North RA, Nagarathinam K, Tanabe M. Structures and general transport mechanisms by the major facilitator superfamily (MFS). Chemical Reviews. 2021;**121**(9):5289-5335. DOI: 10.1021/acs.chemrev.0c00983

[53] Reddy VS, Shlykov MA, Castillo R, Sun EI, Saier MH Jr. The major facilitator superfamily (MFS) revisited. FEBS Journal. 2012;**279**(11):2022-2035. DOI: 10.1111/j.1742-4658.2012.08588.x

[54] Lynch AS. Efflux systems in bacterial pathogens: An opportunity for therapeutic intervention? An industry view. Biochemical Pharmacology. 2006;**71**(7):949-956. DOI: 10.1016/j.bcp.2005.10.021

[55] Tanabe M, Szakonyi G, Brown KA, Henderson PJ, Nield J, Byrne B. The multidrug resistance efflux complex, EmrAB from Escherichia coli forms a dimer in vitro. Biochemical and Biophysical Research Communications. 2009;**380**(2):338-342. DOI: 10.1016/j.bbrc.2009.01.081

[56] Puértolas-Balint F, Warsi O, Linkevicius M, Tang PC, Andersson DI. Mutations that increase expression of the EmrAB-TolC efflux pump confer increased resistance to nitroxoline in Escherichia coli. The Journal of Antimicrobial Chemotherapy. 2020;**75**(2):300-308. DOI: 10.1093/jac/dkz434

[57] Yin Y, He X, Szewczyk P, Nguyen T, Chang G. Structure of the multidrug transporter EmrD from Escherichia coli. Science. 2006;**312**(5774):741-744. DOI: 10.1126/science/1125629

[58] Koita K, Rao CV. Identification and analysis of the putative pentose sugar efflux transporters in Escherichia coli. PLoS One. 2012;**7**(8):e43700. DOI: 10.1371/journal.pone.0043700

[59] Lewinson O, Adler J, Poelarends GJ, Mazurkiewicz P, Driessen AJ, Bibi E. The Escherichia coli multidrug transporter MdfA catalyzes both electrogenic and electroneutral transport reactions. Proceedings of the National Academy of Sciences of the United States of America. 2003;**100**(4):1667-1672. DOI: 10.1073/pnas.0435544100

[60] He X, Szewczyk P, Karyakin A, Evin M, Hong WX, Zhang Q, et al. Structure of a cation-bound multidrug and toxic compound extrusion transporter. Nature. 2010;**467**(7318):991-994. DOI: 10.1038/nature09408

[61] Jack DL, Yang NM, Saier MH Jr. The drug/metabolite transporter superfamily. European Journal of Biochemistry. 2001;**268**(13):3620-3639. DOI: 10.1046/j.1432-1327.2001.02265.x

[62] Eshaghi E, Shahvaroughi Farahani A, Mousaei S, Khalifeh-Gholi M, Adnani Sadati J, Fateh R. Detecting efflux pumps genes in the fluoroquinolones resistant and sensitive strains of Escherichia coli in patients with urinary tract infection in Qom, Iran. Qom University of Medical Sciences Journal. 2021;**15**(3):210-221. DOI: 10.52547/qums.15.3.210

[63] CLSI. Performance Standards for Antimicrobial Susceptibility Testing. 31st. CLSI Supplement M100-S31. Wayne, PA: Clinical and Laboratory Standards Institute; 2021